Physiology of the Digestive Tract

PHYSIOLOGY TEXTBOOK SERIES

These volumes may be purchased individually in a hard binding or paper covered.

PHYSIOLOGY of RESPIRATION

JULIUS H. COMROE, JR., M.D., *Professor of Physiology, University of California, San Francisco*

PHYSIOLOGY of the DIGESTIVE TRACT

HORACE W. DAVENPORT, PH.D., D.SC.(OXON.), *William Beaumont Professor of Physiology, University of Michigan*

PHYSIOLOGY of the NERVOUS SYSTEM

CARLOS EYZAGUIRRE, M.D., *Professor and Chairman, Department of Physiology, and* SALVATORE J. FIDONE, PH.D., *Assistant Professor of Physiology, University of Utah College of Medicine*

PHYSIOLOGY of the KIDNEY and BODY FLUIDS

ROBERT F. PITTS, PH.D., M.D., *Professor of Physiology, Maxwell M. Upson Department of Physiology and Biophysics, Cornell University Medical College*

METABOLIC and ENDOCRINE PHYSIOLOGY

JAY TEPPERMAN, M.D., *Professor of Experimental Medicine, Department of Pharmacology, State University of New York Upstate Medical Center, Syracuse*

Physiology of the Digestive Tract

An Introductory Text

HORACE W. DAVENPORT, Ph.D., D.Sc. (Oxon.)
William Beaumont Professor of Physiology
University of Michigan
Ann Arbor, Michigan

FIFTH EDITION

YEAR BOOK MEDICAL PUBLISHERS, INC.
CHICAGO • LONDON

Library of Congress Cataloging in Publication Data
Davenport, Horace Willard, 1912–
   Physiology of the digestive tract.
   (Physiology textbook series)
   Includes bibliographies and index.
   1. Digestive organs.   2. Digestion.   I. Title.
II. Series.   [DNLM: 1. Gastrointestinal system—
Physiology.   WI 102 D247p]
QP145.D33   1982   612′.3   81-16506
ISBN 0-8151-2330-2   AACR2
ISBN 0-8151-2329-9 (pbk.)

Contents

Preface

I HOPE THAT this fifth edition is what the title page of the first edition proclaimed it to be: an introductory text. From the beginning, the purpose of this book has been to make the basic facts of gastroenterological physiology readily available to the student or house officer beginning serious study of the digestive tract.

Since the first edition of this book was published in 1961, there has been an enormous increase in our knowledge of human gastroenterology; and man, well or ill, has become the chief subject of observation and experiment. As a result, this edition contains much information on the disordered physiology of the diseased digestive tract. Nevertheless, the experimental animal has continued to provide much important information. In many instances I have implied that knowledge derived from a dog, a rat, or an opossum applies to man; but I have not always warned the reader of my inference.

The plan of this book is to describe first the means by which the gastrointestinal tract is controlled: its neuromuscular apparatus and its chemical mediators. Then the reader is conducted on three journeys through the alimentary canal, the first two showing him how propulsion and secretion are accomplished and regulated. Finally, we repeat the trip, accompanying aliments in the process of being digested and absorbed, observing to the best of our ability how and to what extent the mechanisms of motility, secretion and their control are actually used. The usually relentlessly downward gradient of the digestive tract imposes tedious repetitiveness on this method of exposition, but I find the task of adding variety by beginning with an enema and ending with an eructation beyond my powers.

I cannot document with references the thousands of assertions of fact contained in this book. A fragment of the original literature has filtered through my understanding, and the casual student must trust me. The serious scholar can begin with the massive *Handbook of Physiology: The Alimentary Canal,* edited by Charles F. Code and published by the American Physiological Society. This contains references to most of the literature up to 1964–66, and it will remain the starting point for a search of the literature. The references I have added to each chapter include recent reviews whose text and references are useful. The papers I cite are usually the latest ones in an important series. The student can easily attain a view of a large field by working backward through the paper's references.

This is my last edition. I have lived through a glorious period of gastroenterological research; and in the successive editions of this book I have tried to summarize clearly, concisely, and correctly the rapidly expanding knowledge of the physiology and pathophysiology of the digestive tract. How well I have discharged that responsibility is for others to judge, but I must record that keeping up with the literature has been a pleasure as well as a burden.

I must thank the hundreds of persons I know only through their published work and whose accomplishments I admire. I thank the many friends who have given me advice, help, and access to unpublished work. In particular, I thank three preeminent physiologists, Charles F. Code, Morton I. Grossman, and Alan F. Hofmann, whose contributions to the physiology of the digestive tract and to this book overshadow all others.

HORACE W. DAVENPORT

PART I

Mechanisms of Control

1

Mechanisms of Control

1

The Neuromuscular Apparatus of the Digestive Tract

EXCEPT AT ITS pharyngeal and anal ends, the motor function of the digestive tract is performed by smooth muscle. In order to understand the propulsion of food through the gut, it is necessary to know the structure, innervation, and physiological properties of intestinal muscle. Gastrointestinal smooth muscle structures are extremely diverse in form and function, and only the general properties of the neuromuscular apparatus will be considered here. Later, coordinated activity as expressed in peristalsis, spasm, and other movements will be discussed, together with special nervous and humoral factors influencing them.

Structure of Smooth Muscle Cells

At its greatest extension a living smooth muscle cell is less than 250 nm long and 6 nm thick. This is an order of magnitude smaller than a striated muscle cell, and, because it is so small, a smooth muscle cell has a very large surface-to-volume ratio. After death, intestinal smooth muscle cells greatly elongate, with the result that the length of the dead intestine is far greater than that of the living one.

An irregular array of collagen fibrils surrounds the cells and is attached to the membrane. The collagen fibrils form the tendon through which the muscle cells exert their force. The collagen is synthesized by fibroblasts, but intestinal smooth muscle cells may themselves be able to synthesize collagen in response to injury.

The cell membrane is approximately 0.015 nm thick, and it contains a large number of flask-like invaginations, 168,000 in a single cell of taenia coli. These increase the cell surface by 30%. They are probably sites of calcium accumulation and are not pinocytotic vesicles. Mitochondria are less numerous than in striated muscle cells, and they are grouped near the periphery of the cell or in the perinuclear region. Thick myosin filaments lie parallel to the long axis of the cell, and they are surrounded by thin filaments of actin and tropomyosin. There is no troponin-like component. There are 15 actin filaments surrounding one myosin filament, and the bundles are arranged in an approximately 60–80 nm lattice. Longitudinally, there is no regular transverse alignment or cross-banding. An important physiological consequence is that the filaments may slide to an unlimited extent with respect to one another; this accounts for the great range of length over which tension can be exerted. Many, but not all, of the actin filaments are attached to the cell wall at dense bodies that are smooth muscle's equivalent of the Z disc. Contraction causes lateral indentations on the cell border, so that the cell seems to fold on itself like an accordion.

The sarcoplasmic reticulum is a closed tubular system that does not communicate directly with the extracellular space. It is the intracellular site of calcium accumulation and release. The cell's large surface-to-volume ratio allows calcium from the extracellular fluid to activate the contractile mechanism simply by diffusing into the cell at the time of the action potential, so there is no need for T tubules.

When smooth muscle is relaxed, intracellular concentration of ionized calcium is about 10^{-7} M. In that situation actin and myosin do not significantly interact with adenosine triphosphate (ATP). Activation occurs when intracellular calcium concentration rises, the additional calcium coming from extracellular fluid during the action potential and from intracellular stores. At rest, smooth muscle has a low content of ATP and phosphocreatine, but ATP is simultaneously synthesized by mitochondria and hydrolyzed when actin and myosin are permitted to interact by calcium. Use of ATP by smooth muscle is very economical, and the rate of oxygen consumption is directly proportional to the level of maintained tension.

Connections of Smooth Muscle Cells

Groups of muscle cells are organized as bundles. In the circular muscle of the cat intestine, these bundles are 500 nm thick and contain about 7,000 cells in cross section. Cells may be connected with each other through gap junctions or nexuses, and gap junctions are particularly abundant in the circular layer. In the guinea pig ileal circular muscle there are approximately 244 gap junctions connecting one cell with its neighbors. Electrical resistance between cells through gap junctions is less than resistance of the cell membrane, and therefore gap junctions permit electrical coupling between cells in a bundle. However, gap junctions are absent in the longitudinal layer of the dog intestine and infrequent in the longitudinal muscle of the dog stomach. Nevertheless, cells in these muscle layers are electrically coupled.

Bundles are attached to adjacent bundles by connective tissue. Larger units, made up of many bundles, form the circular layer on the mucosal side and the surrounding longitudinal layer on the serosal side. In some tissues, thin muscle strands connect the circular and longitudinal layers and are responsible for interaction between the layers.

The details of organization of bundles into layers are disputed. Some anatomists believe the circular layer is wound in a helix, whose direction is counterclockwise (viewed from the oral end), and that the longitudinal layer consists of bundles twisting in a more open, elongated helix. Others think that the circular layer is formed by closed rings and the longitudinal layer by bands, whose axes are parallel with that of the gut. The muscularis mucosae, lying beneath the glandular mucosa, consists of both circular and longitudinal layers of muscle fibers. The layers vary greatly in thickness in different species and parts of the tract, and they are especially thick in the pig's esophagus and in the human stomach. Smooth muscle

cells pass from the circular layer of the muscularis mucosae into the mucosa.

The submucosa is composed of interlacing collagenous fibers, which, when the intestine is relaxed, form helical coils at an approximately 45-degree angle with the axis, half clockwise and half counterclockwise. In the dog, this braided sheet is 18–24 layers thick.

Sympathetic Innervation

The autonomic innervation of intestinal smooth muscle is summarized in Table 1–1 and Figure 1–1. Efferent sympathetic fibers go to the stomach from the celiac plexus, to the small intestine from the celiac and superior mesenteric plexuses, and to the cecum, appendix, ascending colon, and transverse colon from the superior mesenteric plexus. The remainder of the colon receives sympathetic fibers from the superior and inferior hypogastric plexuses. Most sympathetic fibers to the intestine are postganglionic, and their cell bodies are in the ganglia named. Some sympathetic fibers entering the gut may be preganglionic ones, and their postganglionic fibers would be those of the cell bodies of the intramural plexuses upon which the preganglionic fibers terminate.

Sympathetic fibers innervating the intestine end in one of four places: (1) Some sympathetic fibers enter glandular tissue where they appear to innervate some secretory cells. (2) Some sympathetic fibers innervate smooth muscle cells of blood vessels, causing vasoconstriction, and smooth muscle cells of the muscularis mucosae, causing contraction. (3) Most sympathetic fibers terminate in contact with neuronal cell bodies of the intramural plexuses and with presynaptic fibers surrounding the cell bodies. They inhibit ganglionic activity, possibly by presynaptic inhibition. (4) Postganglionic adrenergic fibers directly innervate smooth muscle cells of the longitudinal and circular layers, but the extent to which this occurs is disputed. Such innervation is said to be rare in the longitudinal muscle layer of the guinea pig ileum but to be by no means rare in the circular layer. The muscle of the aganglionic segment of the colon of a patient with megacolon is rich in adrenergic innervation.

Sphincters of the gastrointestinal tract are adrenergically innervated, and the density of innervation is often greater than in adjacent muscle. Although there

TABLE 1–1. FUNCTIONAL CLASSIFICATION OF NERVES AFFECTING THE LOWER DIGESTIVE TRACT

Systems Mediating Extrinsic Sympathetic Innervation (see Fig 1–1,A):
Preganglionic cholinergic fibers pass
  from cell bodies in lateral columns of thoracolumbar region of cord, via white rami communicantes, to (chiefly) the
    prevertebral ganglia, where they synapse with
Postganglionic adrenergic (and cholinergic?) fibers, which pass
  from cell bodies in prevertebral ganglia, via thoracic splanchnic nerves, to endings on the following effectors:
    smooth muscle cells of blood vessels, constrictor (and dilator?)
    smooth muscle cells of muscularis mucosae, excitatory (and inhibitory?)
  cell bodies of neurons within the myenteric plexuses whose activity they modulate, inhibitory
    gland cells of salivary glands, excitatory
    gland cells of pancreas, small intestine and colon (inhibitory and excitatory?)
    smooth muscle cells of some circular layers, inhibitory
    smooth muscle cells of some sphincters, excitatory
Systems Mediating Extrinsic Parasympathetic Innervation (see Fig 1–1,B):
Preganglionic cholinergic, purinergic, and peptidergic fibers pass
  from cell bodies in the cranial division of neuraxis, via vagus nerves and from sacral region of the cord, via pelvic and
    splanchnic nerves to the digestive tract, where they synapse with
Postganglionic cholinergic, purinergic, and peptidergic fibers, which pass
  from cell bodies in ganglia of nerve plexuses to synapse with
  other ganglion cells in the same ganglion or in the same plexus or other plexuses and to the following effectors:
  smooth muscle cells in all muscle layers, chiefly excitatory, gland cells in the gastric antrum (and elsewhere?) that
    liberate
  hormones of the digestive tract, excitatory and inhibitory
  gland cells that liberate external secretions (but not directly to blood vessels)
Postganglionic fibers in the lower esophagus and stomach (and other sphincters?), inhibitory, nonadrenergic,
    noncholinergic
Systems Mediating Reflexes of Peripheral Location (see Fig 1–1,C):
Afferent limb, consisting of
  nerve terminals from which impulses originate
    in the mucosal epithelium
    in the muscle layers and plexuses
  cell bodies in the submucous (and myenteric?) plexus
    potentially effective stimuli of which are:
      stretch or distention
      pH of contents of viscus
      specific chemical constituents, such as amino acids, peptides, fats
  axons synapsing with
Efferent limb, consisting of
  cell bodies in myenteric and submucous plexuses
  second-order effector neurons and
  other (identical?) effector neurons of myenteric and submucous plexuses, which pass to
  endings that innervate the following effectors:
    smooth muscle cells of the digestive tract
    gland cells of internal secretion in gastric antrum (and elsewhere?) that liberate hormones of the digestive tract
    gland cells that liberate external secretions
    (smooth muscle cells of blood vessels?)
Systems Mediating Reflexes through Celiac Plexus (see Fig 1–1,D):
Afferent limb, consisting of
  nerve terminals from which impulses originate in epithelium of duodenal and other mucosa
  cell bodies in intrinsic plexuses
    potentially effective stimuli of which are:
      chemical composition of gastrointestinal contents, pH, osmotic pressure, digestion products of protein and fat
  axons synapsing in celiac plexus with
Efferent limb, consisting of
  cell bodies in celiac plexus
  postganglionic sympathetic fibers, which pass to endings on neuronal cell bodies within the intrinsic plexuses of
    stomach, inhibitory

TABLE 1–1.—*Continued*

Systems Mediating Reflexes Through Central Neuraxis (see Fig 1–1,*E*):
Afferent limb, consisting of
  nerve terminals from which impulses originate
    in epithelium
    in plexuses
    in smooth muscle, potentially effective stimuli of which are distention or chemical composition of contents of viscus
  fibers that pass centrally,
    via thoracic, lumbar, or pelvic splanchnic nerves and dorsal roots, to cord, and having cell bodies in dorsal root
      ganglia or
    via vagus nerves to brain stem,
  mediating the following:
    vasodilatation via axon reflexes
    pain and other visceral sensations
    vomiting reflex
    defecation reflex
    gallbladder reflex
    response to obstruction, and
  initiating specific action on
    digestive tract reflexes, facilitating and inhibiting
    somatic reflexes
    cardiovascular reflexes and
    centripetal fibers from cell bodies within intrinsic plexuses

---

is considerable species difference, in most species examined the effect of sympathetic innervation is to excite the lower esophageal sphincter, the choledochoduodenal sphincter, and the internal anal sphincter. Excitation is mediated by the action of norepinephrine upon alpha receptors of the smooth muscle cells.

Adrenergic nerve terminals avidly take up norepinephrine, thereby reducing the effectiveness of circulating norepinephrine.

Parasympathetic Innervation

Efferent parasympathetic innervation to the stomach, small intestine, cecum, appendix, ascending colon, and transverse colon is by way of the vagus, whose fibers follow blood vessels to end in the myenteric plexus. The rest of the colon receives parasympathetic innervation from the pelvic nerves via the hypogastric plexuses, and these, too, end in the myenteric plexus. The fibers are all preganglionic, and most are cholinergic and excitatory.

Current research shows that chemical mediation in the parasympathetic nervous system is not so straightforward as is implied in the old dicta that all pre- and postganglionic fibers are cholinergic and that the action of acetylcholine upon glands and muscles of the gut is excitatory. The vagus nerves contain fibers to the lower esophageal sphincter and to the stomach which are inhibitory, and those fibers are noncholinergic and nonadrenergic. Dopamine, 5-hydroxytryptamine, and purines are thought to be transmitters in such fibers. Many nerves have been shown to contain, and probably to release, peptide transmitters. For example, the human vagus nerve has been shown by immunochemical methods to contain substance P, vasoactive intestinal peptide (VIP), and enkaphalin-like compounds. VIP released from nerve endings mediates relaxation of the lower esophageal, pyloric, and internal anal sphincters. A peptide similar to cholecystokinin is synthesized in nerve cell bodies in the nodose ganglion and is transported centrally in the axoplasm of vagal afferent fibers. On the other hand, gastrin is carried peripherally in the axonal stream. Research in this area is developing so rapidly that few dogmatic statements can be made.

Plexuses

Five nerve plexuses are distinguished: subserous, myenteric, deep muscular, internal, and submucous. The deep muscular and the internal plexuses are probably formed by terminal fibers of the myenteric and submucous plexuses. Postganglionic sympathetic and preganglionic parasympathetic fibers make up a large fraction of the plexuses and disappear after extrinsic denervation. The myenteric plexus in the interspaces between longitudinal and circular muscle layers consists of a mesh of fibers, with ganglion cells at the nodal points, all ensheathed by the cytoplasm and membranes of a connective tissue syncytium comparable with the neurilemma of myelinated fibers. The plexuses themselves are composed of individually distinct cells, not of a syncytial network. Nerve fibers extend within the plexus for several centimeters, and from the plexus arises the reticulum of fine fibers passing to muscle cells of the two layers.

Branches of the preganglionic parasympathetic fibers synapse with the ganglion cells, some with the first cell encountered after they enter the plexus and others with more remote cells. Still other preganglionic fibers pass through the myenteric plexus to the submucous plexus. In turn, fibers go from the reticulum of ganglion cells and from fibers forming the submucous plexus to the muscularis mucosae, to the gland cells of the mucosa, to unidentified endocrine cells and to muscular tissue within the mucosa. In addition, axonal processes from one ganglion cell synapse with cells in the same node or in neighboring ganglia. Nerve impulses initiated in one ganglion cell spread through multiple pathways, chiefly in the longitudinal direction, and experience synaptic transmission at conduction distances no longer than 4 mm. The result is that when extrinsic fibers have degenerated after denervation, ganglion cells are still functionally connected and are capable of mediating local reflexes. Afferent limbs of the reflex arcs are provided by cells in the submucous plexus whose dendrites are receptor organs in the mucosa and in the muscle layers and whose axons project to synapses formed with cells of either the submucous or the myenteric plexus.

Function of Sympathetic Innervation

Sympathetic fibers to mucosal glands control to some extent the secretion of mucus, and those to the muscularis mucosae cause it to contract. Sympathetic fibers to the lower esophageal sphincter, to the pyloric sphincter, and to the internal anal sphincter are apparently excitatory, but their role in controlling the function of those sphincters is poorly understood. Otherwise, the chief function of sympathetic innervation is to cause vasoconstriction by stimulating contraction of vascular smooth muscle in the gut and to inhibit contraction of intestinal smooth muscle. Inhibition of motility is effected by three minor and one major means.

Circulating epinephrine and norepinephrine from the adrenal medulla inhibit intestinal motility, but the rate of uptake of the hormones from interstitial fluid by sympathetic nerve endings is so rapid that only the great rise in concentration occurring during massive sympathetic discharge has any significant effect upon gastrointestinal motility. Vasoconstriction itself has little or no effect upon motility, but diffusion of norepinephrine from sympathetic endings on blood vessels to nearby intestinal smooth muscle may be inhibitory. Again, on account of rapid uptake of the mediator, this is probably unimportant. Those smooth muscle cells, other than the ones in the sphincters, which are directly innervated by sympathetic fibers, are inhibited by norepinephrine released from the nerve fibers.

Most sympathetic fibers end in a dense network surrounding ganglion cells of the intramural plexuses. Those ganglion cells are excited by acetylcholine and by 5-hydroxytryptamine (serotonin) released from presynaptic terminals on or near the ganglion cells. The effect of norepinephrine from the sympathetic endings is to block the release of acetylcholine or 5-hydroxytryptamine from the excitatory presynaptic terminals, not to block the action of the two excitatory transmitters upon the ganglion cells. (In turn, acetylcholine released from excitatory terminals blocks release of norepinephrine from inhibitory terminals.) The effect of sympathetic stimulation is seen when there is ongoing activity within the plexuses. If, for example, gastric motility is enhanced by stimulation of vagal excitatory nerves to the stomach, stimulation of the hypothalamic defense area or of the surrounding pressor area produces prompt and often complete inhibition of gastric motility. When these hypothalamic areas are stimulated in the absence of simultaneous vagal excitation of the intramural plexuses, stimulation rarely has any effect on gastric motility. Reflex or direct stimulation of adrenergic fibers to the gut does not inhibit the effects of acetylcholine or its congeners upon gastric or intestinal motility, for the reason that the cholinergic mediator acts directly upon intestinal smooth muscle.

Afferent Fibers

Nerves of the digestive tract contain many visceral afferent fibers, which can be divided into two classes:

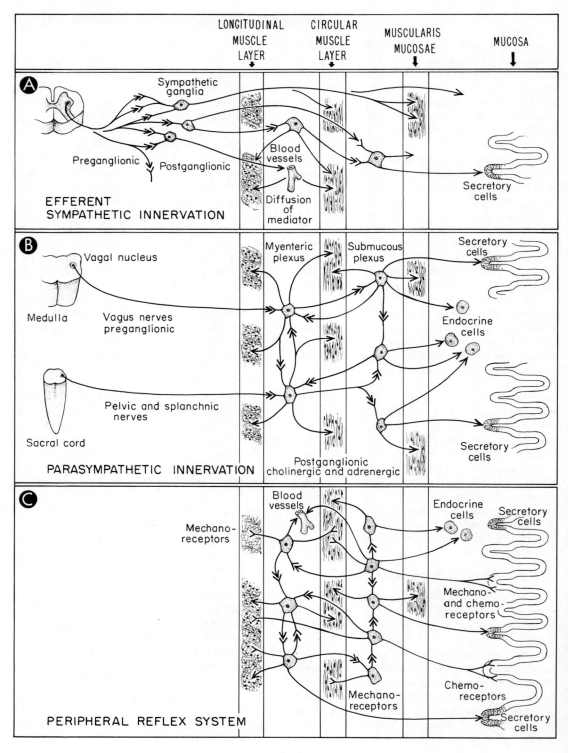

| LONGITUDINAL MUSCLE LAYER ↓ | CIRCULAR MUSCLE LAYER ↓ | MUSCULARIS MUCOSAE ↓ | MUCOSA ↓ |

A

Sympathetic ganglia

Preganglionic   Postganglionic

Blood vessels

Diffusion of mediator

Secretory cells

EFFERENT SYMPATHETIC INNERVATION

B

Vagal nucleus

Medulla   Vagus nerves preganglionic

Myenteric plexus   Submucous plexus

Secretory cells

Endocrine cells

Pelvic and splanchnic nerves

Sacral cord

Secretory cells

PARASYMPATHETIC INNERVATION

Postganglionic cholinergic and adrenergic

C

Blood vessels

Mechano-receptors

Endocrine cells   Secretory cells

Mechano- and chemo-receptors

Mechano-receptors

Chemo-receptors

Secretory cells

PERIPHERAL REFLEX SYSTEM

8

NEUROMUSCULAR APPARATUS OF LOWER DIGESTIVE TRACT

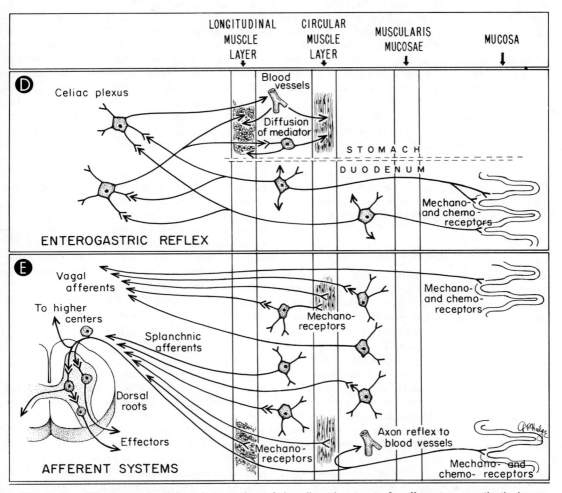

Fig 1–1.—Efferent and afferent innervation of the digestive tract. **A,** efferent sympathetic innervation. **B,** parasympathetic innervation. **C,** peripheral reflex system. **D,** enterogastric reflex. **E,** afferent systems. See Table 2–1 for detailed listing of the nerve tracts, their mediators, and their effects.

those having their cell bodies in centrally located ganglia and those having their cell bodies within the myenteric plexuses. Those of the first class have sensory endings in the mucosal epithelium, in the plexuses, and within the muscle layers. Their fibers pass centrally with the vagus and with the sympathetic rami to the dorsal roots. Cell bodies of afferent vagal fibers are in the nodose ganglion, and those of nerves traveling centrally with the sympathetics are in the

dorsal root ganglia. Among the latter are the dorsal root dilators, whose action on intestinal blood vessels is the same as that of cutaneous dorsal root dilators on the vessels of the skin. Of the 30,000 fibers contained in the vagus nerves of the cat, at least 80% are afferent, and so are 50% of the 30,000 fibers in sympathetic nerves to the gut.

Numerous vago-vagal reflexes, whose afferent and efferent fibers are both in vagal nerves, influence secretion and motility. Slowly adapting chemoreceptors of afferent fibers lie in the gastric and intestinal mucosa. These respond to perfusion of the lumen with solutions of high or low pH, hypotonic or hypertonic solutions of NaCl, and solutions of monosaccharides of which glucose is the most potent. Slowly adapting mechanoreceptors are in series with smooth muscle, and they respond to increased tension with increased rate of firing.

The sympathetic nerves contain afferent fibers whose cell bodies lie in the intramural plexuses. Some of their axons may pass through the celiac and other sympathetic ganglia and enter the central neuraxis before synapsing. Others appear to synapse in the celiac ganglion with postganglionic sympathetic fibers, and they are probably the afferent limb of reflex arcs that regulate gastrointestinal motility.

Efferent Nerve Endings

Each postganglionic fiber in the plexuses branches many times and ramifies extensively. Branching is the anatomical basis of divergence of nervous effects, and stimulation of only a few parasympathetic preganglionic fibers or sympathetic postganglionic fibers influences many effector cells. Bundles of terminal branches enter the muscular coats accompanying capillaries. Fibers leave the neurilemma and pass as naked filaments between muscle fibers. Terminal fibers containing vesicles do not penetrate the plasma membrane but intrude into cell pockets, where their tips and the cell membrane meet. The vesicles probably contain stores of chemical mediators that are liberated when the nerve is excited. Large numbers of naked nerve fibers pass close to all muscle cells, and these may also liberate transmitters at intervals along their length. Nerve fibers approaching one cell are derived from many cell bodies, and influences having widespread origins converge on a single cell.

Membrane Potentials

Contraction of a smooth muscle cell follows electrical changes in its membrane. The electrical changes, in turn, depend upon electrolyte distribution between the cell and the extracellular fluid, upon the permeability of the cell membrane, and upon electrogenic pumps within the membrane.

Potassium is at higher concentration in intracellular water than in extracellular water, and therefore it tends to diffuse across the cell membrane from its higher to its lower concentration. Within the bulk of the cell, K^+ ions are electrostatically balanced by an equal number of anions; some of the anions are high molecular weight organic compounds, and some are Cl^- ions. At the cell membrane, a fringe of K^+ ions diffuses outward, carrying positive charges across the membrane. This outward diffusion of K^+ ions establishes an electrical gradient across the membrane, positive outward and negative inward, which exists because positive K^+ ions are physically separated across the membrane from anions remaining within the cell. The electrical gradient itself opposes further diffusion of K^+ ions, and an equilibrium is quickly reached in which the outward diffusion force, created by the concentration difference, is balanced by an electrostatic force in the opposite direction. Thus, the potential difference across the cell membrane is maintained by a tendency for K^+ to diffuse, but only a very small separation of charges actually occurs. The general equation expressing the potential difference (E_K) if only K^+ ions diffuse is

$$E_K = \frac{RT}{F} \ln \frac{P_K[K^+]_o}{P_K[K^+]_i} \qquad (1.1)$$

where R is the gas constant (8.3 joule $\cdot$ degree^{-1} $\cdot$ mole^{-1}), T is the absolute temperature (311 K), and F is the Faraday constant of electrical charge carried by 1 mole of univalent ions (96,500 coulombs $\cdot$ equivalent^{-1}). The logarithmic term is to the base e, which is 2.3 times logarithms to the base 10. The potential E_K, expressed in millivolts, is called the *potassium equilibrium potential*. When the appropriate numerical values are substituted, the equation becomes

$$E_K = 61 \log_{10} \frac{P_K[K^+]_o}{P_K[K^+]_i} \qquad (1.2)$$

The values in brackets are K^+_i and K^+_o, the concentrations (more accurately, the activities) of K^+ ions inside and outside the cell. Each concentration is multiplied by the constant P_K, which expresses the permeability of the membrane to K^+ ions. This is an essential factor, for if the membrane were not at all permeable to K^+ ions, no fringe of K^+ ions could diffuse through it, and there could be no potential difference across the membrane, despite the existence of a concentration difference. When the contribution of K^+ ions alone is being considered, values of P_K greater than zero cancel out. In cat circular intestinal muscle, the electrolyte concentration of extracellular fluid is equal to that of an ultrafiltrate of plasma, and the K^+_o concentration is 4 mEq/kg H_2O. Intracellular concentrations are difficult to measure, but a good estimate of K^+_i is 164 mEq/kg of cell water. When these values are substituted in equation (1.2), a potassium equilibrium potential of -101 mV is obtained. This is the maximum potential difference K^+ could contribute.

The measured resting membrane potential of intestinal smooth muscle is always less than the potassium equilibrium potential, and consequently there must be additional determinants of the membrane potential.

Sodium and Chloride Potentials

The two other ions whose diffusion affects the resting membrane potential are sodium and chloride.

The concentration of Na^+ ions in extracellular fluid is about 153 mEq/kg H_2O. The best estimate of the concentration of Na^+ ions within intestinal smooth muscle cells is 19 mEq/kg H_2O, which is greater than the concentration in nerve or striated muscle. Because the diffusion gradient is inward, diffusion of Na^+ ions causes a potential difference across the cell membrane opposite to that of K^+ ions. The equation for the Na^+ equilibrium potential is

$$E_{Na} = 61 \log_{10} \frac{P_{Na}[Na^+]_o}{P_{Na}[Na^+]_i} \qquad (1.3)$$

Here, the factor P_{Na} is the permeability of the membrane to Na^+ ions. Using the values given for Na^+_o and Na^+_i, the maximum contribution of Na^+ to the membrane potential is calculated to be 55 mV.

Chloride concentration outside the cell is 132 mEq/kg H_2O, and its concentration inside intestinal smooth muscle cells is estimated to be 55 mEq/kg H_2O. The diffusion gradient, like that of sodium, is inward; but, because Cl^- ions carry a negative charge, the diffusion potential is, like that of K^+ ions, positive outward. The equation for the Cl^- equilibrium potential is

$$E_{Cl} = 61 \log_{10} \frac{P_{Cl}[Cl^-]_i}{P_{Cl}[Cl^-]_o} \qquad (1.4)$$

Here P_{Cl} is the permeability to Cl^- ions, and the difference in charge is allowed for by placing the inside concentration in the numerator. The calculated contribution of Cl^- ions to the membrane potential is -23 mV.

Because the actual resting membrane potential is more negative than the chloride equilibrium potential, chloride cannot be passively distributed between extracellular and intracellular fluids. It is probable that chloride ions are actively accumulated within the cells.

Resting Membrane Potential

The general expression for the membrane potential produced by the tendency of the three ions to diffuse is

$$E = 61 \log_{10} \frac{P_K[K^+]_o + P_{Na}[Na^+]_o + P_{Cl}[Cl^-]_i}{P_K[K^+]_i + P_{Na}[Na^+]_i + P_{Cl}[Cl^-]_o} \qquad (1.5)$$

This equation is clearly not simply the sum of the three equations for the individual ions; it is, rather, the equation relating the three ions to the membrane potential derived on the assumption that the electrical field within the membrane is constant with respect to the distance through the membrane. It is called the *constant field equation*. The assumption is equivalent to saying that the charge density within the membrane has negligible effects on movement of ions through it. If the equation is rewritten

$$E_m = 61 \log_{10} \frac{[K^+]_o + \frac{P_{Na}}{P_K}[Na^+]_o + \frac{P_{Cl}}{P_K}[Cl^-]_i}{[K^+]_i + \frac{P_{Na}}{P_K}[Na^+]_i + \frac{P_{Cl}}{P_K}[Cl^-]_o} \qquad (1.6)$$

the role of the permeability constants is obvious: the actual membrane potential depends on the ratios of the permeability of sodium and chloride to that of potassium.

The potassium permeability constant for guinea pig taenia coli has been calculated to be 11×10^{-8} cm $\cdot$ sec^{-1}; $P_{Na} = 1.8 \times 10^{-8}$ and $P_{Cl} = 6.7 \times 10^{-8}$ in the same units. The ratio P_{Na}/P_K is 0.16, and the ratio P_{Cl}/P_K is 0.61. Substitution of these values in equation (1.6) gives a calculated membrane potential of -37 mV. The calculated value is lower than that determined by inserting a microelectrode into smooth muscle cells. Potentials up to -80 mV have been measured in cat intestinal smooth muscle, and the average of a large number of measurements is -56 mV. In guinea pig taenia coli, the resting membrane potential is -60 mV.

There are probably two reasons for the discrepancy between calculated and observed membrane potentials. The first is that the other ions, particularly calcium, may contribute to the diffusion potential across the membrane. The other is that there is probably an electrogenic pump in the membrane that transfers sodium ions from the inside of the cell to the outside. If positively charged sodium ions are pumped out of the cell unaccompanied by anions, the act of pumping establishes a potential difference across the membrane, and this potential adds to that produced by the tendency of ions to diffuse through the membrane. Consequently, the membrane potential is higher than the diffusion potential. There is also an electrogenic pump which transfers calcium from the inside to the outside of the cell.

The membrane potential is reduced by stretching the muscle or by application of acetylcholine, and this brings the membrane nearer to threshold for action potentials. Epinephrine and norepinephrine, acting through alpha receptors, hyperpolarize the membrane by increasing potassium permeability. This stops action potentials and relaxes the muscle. Epinephrine, acting through beta receptors, also hyperpolarizes the membrane, perhaps by stimulating electrogenic extrusion of calcium.

Calcium

Calcium has at least three functions in intestinal smooth muscle: it controls the permeability of the membrane to sodium, it carries current inward during the action potential, and it couples electrical and mechanical activity.

The concentration of ionized calcium in extracellular fluid is $2-3$ mEq/L, and its concentration is controlled by the same factors controlling its concentration in plasma. Calcium is bound to the membrane of smooth muscle cells to the extent of 10 mEq/kg of cells. Although the total concentration of calcium within the cells is $2-3$ mEq/kg of cell water, calcium is almost entirely bound within the cells, and the concentration of ionized calcium is about 10^{-4} mM. The calcium equilibrium potential is, like the sodium

equilibrium potential, opposite to that of potassium, and when the calcium permeability of the membrane increases at the beginning of the action potential, the membrane potential approaches zero or overshoots slightly.

Potential Changes: The Membrane and Action Potentials

The resting membrane potential of smooth muscle cells—that is, the transmembrane potential recorded when there are no action potentials, no prepotentials, and no electrical control activity—is not constant. Spontaneous variations occur. Increased transmembrane potential, hyperpolarization, is associated with decreased tension, and decreased transmembrane potential, hypopolarization, is associated with increased tension. This is an important difference between striated and smooth muscle. The circular smooth muscle of the internal anal sphincter, for example, does not have action potentials, but it contracts following slow depolarization of its membrane.

The action potential is an abrupt fall in the membrane potential followed by a quick return to the resting level. The variation is known as the action potential; and because it appears sharp in records, it is often called a spike. In mammalian nerve and striated muscle, the spike is caused by a sudden increase of about 200-fold in the permeability of the membrane to sodium. The membrane potential, as expressed by equation (1.6), thereupon approaches the sodium equilibrium potential; and because the sodium equilibrium potential is positive inward, the membrane potential actually reverses, or overshoots.

The most thoroughly studied intestinal smooth muscle is the taenia coli of the guinea pig. In this tissue, sodium is not responsible for the action potential. If the taenia coli is bathed in a solution containing only 10 mN Na, its intracellular sodium concentration falls to about 24 mEq/kg H_2O. At these sodium concentrations the sodium equilibrium potential reverses to become 22 mV, outside positive. Nevertheless, if the muscle is electrically stimulated, an action potential with an overshoot of 20 mV, outside negative, occurs. The action potential and its overshoot depend upon the presence of calcium in the external medium and bound to the membrane. When the action potential is initiated, calcium is released from the membrane, and the permeability of the membrane to calcium increases. Calcium ions cross the mem-

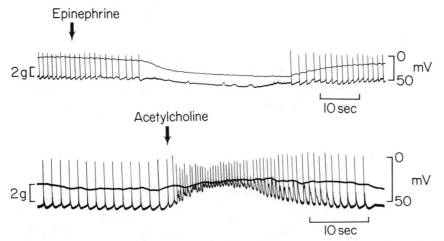

Fig 1–2.—Membrane potential and tension recorded from strips of guinea pig taenia coli. **Top,** the effect of adding epinephrine. **Bottom,** the effect of adding acetylcholine. (Adapted from Bulbring, E., Kuriyama H.: *J. Physiol.* 166:59, 1963.)

brane into the cells, and as ions they carry current into the cells. The entry of calcium is responsible for the rising phase of the action potential. As calcium is released from the membrane, the permeability of the membrane to other ions increases, and the falling phase of the action potential is probably the result of a large increase in the membrane's potassium conductance.

The amplitude of the action potential of intestinal smooth muscle is variable. In some instances, depolarization may be incomplete, so that at the height of the spike there may be residual polarization of the membrane, positive outward. Figures 1–2 through 1–5 show spikes of several sizes following one another.

Spike discharges occur spontaneously at frequencies ranging from 1 to 10/sec. Discharges are started by acetylcholine or, when spontaneously occurring, are increased in frequency (see Fig 1–2). Spike frequency is decreased, or spikes are abolished, by epinephrine. Figure 1–2 also shows that spike magnitude is not uniform and that the resting membrane potential is decreased by acetylcholine and increased by epinephrine.

Potential Changes: The Prepotential and Junction Potentials

The membrane potential of a smooth muscle cell often decreases spontaneously and rhythmically by a

few millivolts. Such changes may or may not be followed by spike discharges. In Figures 1–2 through 1–5 each spike is preceded by a small, slow depolarization, and in Figure 1–3 some of the small depolarizations are followed by spikes, whereas others are not. These prepotentials resemble those occurring in pacemaker tissue of the heart.

A single electrical shock to a parasympathetic nerve twig going to intestinal smooth muscle evokes, after a latent period of 400 msec, a slow depolarization, which may be followed by a spike. This slow potential, which resembles the end-plate potential of striated muscle, decays with a half-life of 150–300 msec. It is probably caused by acetylcholine liberated by the nerve endings. When repetitive shocks are given, the slow potentials sum. On the other hand, stimulation of sympathetic nerves to intestinal smooth muscle hyperpolarizes the membrane, suppresses spontaneous prepotentials, and abolishes action potentials. This effect is mediated by the action of norepinephrine on alpha receptors of smooth muscle cells and is mimicked by circulating epinephrine.

Conduction of Excitation

Rhythmic changes in length and tension move along intestinal smooth muscle. Ability to propagate excitation is a property of the muscle itself, for propagation occurs in sheets or tubes of completely dener-

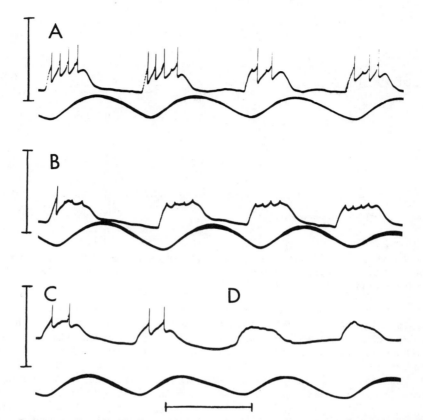

Fig 1–3.—Spontaneous electrical and mechanical activity recorded from sheets of longitudinal muscle of rabbit jejunum. The electrical record is the transmembrane potential of a single cell, but the tension record is that of a sheet of cells. Consequently, tension may increase as the result of contraction following action potentials in many cells other than the one from which the electrical record is made. **A** and **B** are continuous records; **C** and **D** are taken from two other preparations. Electrical waves are above and mechanical response is below in each frame. Each slow electric wave consists of prolonged depolarization, which usually gives rise to several generator-type potentials. Generator potentials can give rise to spikes, **A,** or decay, as in the last three slow waves of **B**. In some cases, where a single generator potential fails to elicit a spike, a second generator potential seems to sum with the first, and a spike follows as in the third slow wave of **A**. In some cases, miniature generator potentials occur at the top of slow waves, none being large enough to elicit a spike, as in **D**. All spikes are followed by undershoots. Vertical calibrations: 0 to −40 mV; horizontal calibrations; 5 sec. Mechanical records are uncalibrated, but upward deflection indicates increase in tension. (From Bortoff A.: *Am. J. Physiol.* 201:203, 1961.)

vated muscle and in nerve-free embryonic tissue. Mechanical stretch of a resting cell by active ones is not essential for propagation; excitation passes over immobilized strips of muscle, and it jumps the cut between two separate but closely apposed rings of muscle that are not mechanically linked.

Conduction is the coordinated activity of many cells. It is fast in tissue like pig esophageal muscularis mucosae, which has closely packed cells, and slow in cat longitudinal muscle, whose extracellular space is relatively large. Velocity of conduction is about ten times as fast in the longitudinal direction of muscle fibers as it is in the direction perpendicular to the fiber axis. The moving wave of activity is approximately 1

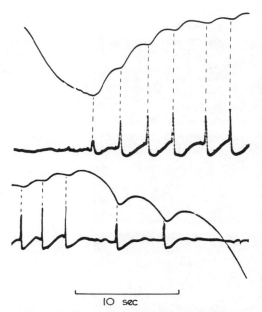

Fig 1–4.—Tension and electrical activity of a strip of taenia coli. Tension increases as spike frequency increases and falls as spike frequency decreases. (From Bulbring E.: *Lectures on the Scientific Basis of Medicine,* vol. 7. London: Athlone Press, 1959.)

IO sec

mm long in the longitudinal direction, the length of five or so muscle cells. The minimal diameter of a fiber bundle that can be excited to conduct is 0.1 mm, which includes 200–300 cells in cross section. Thus the wave passing through the tissue involves at least 1,000 cells at a time.

Conduction of excitation within the bundle is by electrotonic spread of current. There is a graded potential along each active cell at the leading edge of spreading activity (Fig 1–6). The depolarized part of the membrane is a current sink. Current flows from a neighboring resting cell which, because it is still polarized, has a positively charged membrane. The flow partially depolarizes the resting cell; and when the degree of depolarization reaches a critical value over a sufficient area of the cell membrane, abrupt depolarization of the resting cell occurs. The cell now becomes an active one, and it is a current sink for adjacent resting cells.

Because smooth muscle cells are very small, there is a very large surface-to-volume ratio in the tissue.

Current flowing across a depolarized area of one cell membrane spreads over a wide area of adjacent cell membranes. This means that the density of current flowing across one or only a few cells is not great enough to depolarize the critical area of the membrane of a resting cell. Converging electrical fields from many cells are needed to depolarize a sufficient area of the cell membrane so that excitation and conduction can occur. This is the reason that bundles of cells are the conducting unit in intestinal smooth muscle.

If current flows from a polarized membrane to a current sink in an active cell, there must be a return flow of current to complete the electric circuit. Such flow would be greatly facilitated were there low-resistance pathways between adjacent cells. Such a pathway could be provided by gap junctions, or nexuses, between cells. These provide low-resistance pathways between cells.

Gap junctions may be scarce or entirely absent in some tissues in which conduction nevertheless occurs. An example is the longitudinal muscle of the stomach. In that case, current must flow from the interior of one cell to the interior of its neighbor through high-resistance pathways.

Potential Changes: Basic Electrical Rhythm or Electrical Control Activity

An electrode placed in a cell of the longitudinal muscle of the stomach or small intestine records a conducted wave of depolarization distinct from the prepotentials or action potentials. The magnitude of the wave is 10–15 mV; its duration is 1–4 msec; and its frequency, which is characteristic of the tissue, is 3/min in the human stomach, 5/min in the dog stomach, and 17–19/min in the dog duodenum. Figure 1–3 shows these waves recorded from the rabbit jejunum; they are seen to occur with and without prepotentials and action potentials superimposed upon them. Bursts of action potentials, when they occur, are local and are followed within 0.5 sec by muscular contraction. Although the waves occur only in longitudinal muscle in the stomach and small intestine, the intimate association of the circular with the longitudinal layer is important for propagation of the wave in the longitudinal layer (Fig 1–7). As the wave of depolarization passes along the longitudinal layer, it is reinforced by current flow resulting from the partial depolarization of the cells of the circular muscle layer

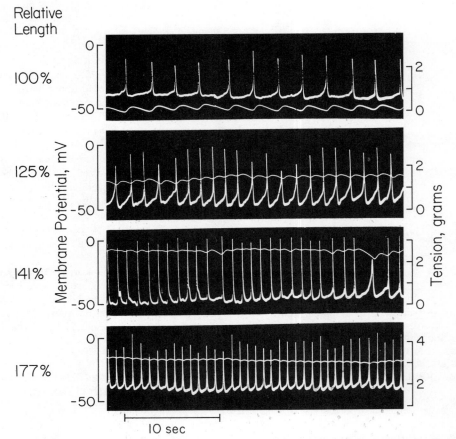

Relative
Length

Fig 1–5.—Tension and action potentials of a strip of taenia coli as it is stretched from its rest-ing length. (Adapted from Bulbring E., Kuriyama H.: *J. Physiol.* 169:198, 1963.)

lying beneath the partially depolarized cells of the longitudinal layer.

Longitudinal and circular muscles contract when action potentials occur during the phase of depolarization. Therefore, the frequency and velocity of contractions are governed by the frequency and velocity of the wave of depolarization in the longitudinal layer. For this reason, the wave is called the Basic Electrical Rhythm (BER), Pacesetter Potential (PSP), or Electrical Control Activity (ECA). There is no general agreement on the name, and the student must be prepared to recognize all of them. Because it is nonspecific, the term ''slow wave'' should not be used.

When the waves of depolarization are recorded us-

ing extracellular electrodes, they appear as the first derivative of the transmembrane potential difference changes. The BER recorded with extracellular electrodes is shown in Figure 1–8, and the figure also shows the relation between the control activity, spiking, and pressure development.

Local factors determine whether the positive deflection of the electrical control activity is followed by action potentials and contraction. The most important of these is the activity of the local intramural plexuses. When many impulses carried in the plexuses liberate acetylcholine, the excitability of smooth muscle cells is increased, and the cells respond with action potentials and contraction. Stimulation of the vagus nerve or administration of cholinergic drugs

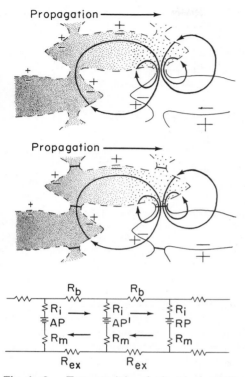

Fig 1–6.—Two models of electrical relations between smooth muscle cells. **Top,** cells connected by low-resistance gap junctions. **Middle,** cell membranes closely apposed but not fused to form low-resistance gap junctions. There is high resistance to current flow between cells. **Bottom,** the equivalent circuit. (Adapted from Barr L.: *J. Theor. Biol.* 4:73, 1963.)

also causes bursts of action potentials and contraction in phase with the electrical control activity. Sympathetic stimulation, by depressing plexus reflexes, reduces spiking and contraction without affecting the electrical control activity.

Mechanical Properties

Records in Figure 1–4 show that each spike is followed by an increase in tension. When a muscle is stimulated, spike frequency and tension increase. When the muscle is inhibited, spike frequency and tension both decrease.

Mechanical properties of smooth muscle are qualitatively similar to those of striated muscle but quantitatively different. Resting intestinal smooth muscle is greatly elongated by stretch, but a large fraction of the elongation results from reorientation of cells and fasciculi to a direction parallel with the direction of stretch, and from stretch of connective tissue components. A smaller part results from elongation of muscle fibers themselves. The tension-length diagram of smooth muscle is similar in shape to that of striated muscle; when tension developed is plotted against initial length, a curve, concave downward, is obtained with maximal tension at resting length. Initial velocity of shortening with a 2-gm load is of the order of 0.4 cm/sec, whereas that of frog sartorius is about 4 cm/sec with the same load.

Response to Stretch and Release of Stretch

Four properties of smooth muscle permit the viscera to accommodate to changes in volume with minimal changes in transmural pressure: (1) elastic behavior of the muscle, (2) stress relaxation, (3) active but intrinsic changes in muscle tension, and (4) reflexly mediated contraction and relaxation.

Pressure in a hollow organ is directly proportional to tension in its wall. For a sphere, which can serve as a model for the stomach, the relation is $P = 2T/r$, where T is the tension or total force in dynes per centimeter along a length of wall and r is the radius. For a cylinder such as the intestine, the relation between transmural pressure and tension in the circumferential direction is $P = T/r$. The relation among volume, pressure, and tension is much more complex. As volume increases, the wall becomes thinner, and the number of tension-producing elements per unit of cross-sectional area of the wall diminishes. Consequently, the number of elements exerting tension along the unit length of the wall also diminishes. Although tension along the length of the wall increases as the wall is stretched, the rise in tension, and therefore in transmural pressure, for an increment in volume is not so steep as would be expected if wall thickness remained constant. These relations among transmural pressure, tension in the wall, and contained volume hold for any distensible structure, living or dead. In life, the tension in the wall of the digestive tract is maintained by active contraction of its muscle, and adjustments in active tension are the means by which pressure and volume are regulated.

When smooth muscle is quickly stretched, its

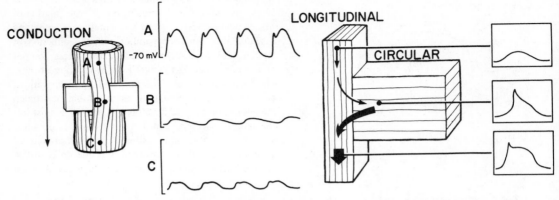

CONDUCTION

LONGITUDINAL

CIRCULAR

-70 mV

A

B

C

Fig 1–7.—Circular muscle layer amplification of the electrical control activity wave in the longitudinal muscle layer. **Left,** A plastic plate is slipped between the two muscle layers, and records are obtained from points A, B, and C. **Right,** The wave propagated in the longitudinal layer spreads to the circular layer through low-resistance pathways, and regenerative amplification occurs. The amplified wave returns to the longitudinal layer. (Adapted from Connor J.A. et al. *J. Physiol.* 273:665, 1977.)

length and tension increase, and it obeys the law that increase in length is directly proportional to tension. However, if the muscle is held at constant length, its tension falls; or if it is held at constant tension, its length slowly increases (Fig 1–9). This slow response is stress relaxation; it is the result either of sliding of myofibrils past one another or of inhibition of active contraction. After an increase in volume has raised tension in the wall of the stomach or intestine, stress relaxation reduces tension, and transmural pressure

DOG DUODENUM, ELECTRODES 68 mm APART

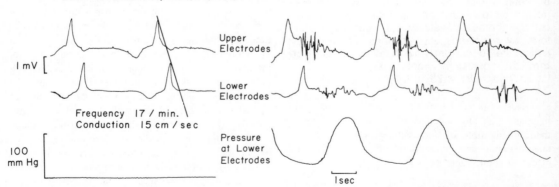

1 mV

Frequency 17 / min.
Conduction 15 cm / sec

Upper
Electrodes

Lower
Electrodes

Pressure
at Lower
Electrodes

100
mm Hg

1 sec

Fig 1–8.—Electrical activity and pressure recorded from duodenum of an unanesthetized dog. At an earlier operation, the duodenum was brought out through the abdominal wall and covered by a skin flap so that if formed a permanent, handle-like loop. Electrical recording was done by leading from two pairs of needle electrodes; pressure was recorded from a transducer inserted by mouth to the site of the lower electrodes. **Left,** control waves without spikes sweep over the duodenum at a frequency of 17/min and with conduction velocity of 15 cm/sec. There are no spikes (the small deflections on the upper record are electrocardiogram), and intraluminal pressure is atmospheric. **Right,** spontaneous smooth muscle spikes follow the positive phase of each control wave; contraction and pressure increases at lower electrode coincide with spiking and are proportional to spike frequency. (Courtesy of P. Bass and C. F. Code.)

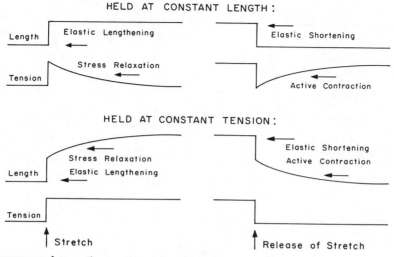

HELD AT CONSTANT LENGTH:

Length — Elastic Lengthening — Elastic Shortening

Tension — Stress Relaxation — Active Contraction

HELD AT CONSTANT TENSION:

Length — Stress Relaxation / Elastic Lengthening — Elastic Shortening / Active Contraction

Tension

Stretch — Release of Stretch

Fig 1–9.—Response of smooth muscle to stretch and release of stretch.

falls toward the level prevailing before volume increased.

Reflexly regulated receptive relaxation is described in chapter 4.

If tension is being maintained by active and regular spike discharge, stretch may inhibit spiking for about 30 sec, during which the muscle relaxes to accommodate itself to the new length. On release of tension, smooth muscle frequently contracts, so that tension is restored at shorter length. On the other hand, stretch of intestinal muscle may be followed not by relaxation but by a burst of spikes and increased tension (see Fig 1–5).

REFERENCES

Bennett M.R., Burnstock G.: Electrophysiology of the innervation of intestinal smooth muscle, in Code C.F. (ed.): *Handbook of Physiology: Sec. 6. Alimentary Canal*, vol. 4. Washington, D.C., American Physiological Society, 1968, pp. 1709–1732.

Bennett A., Stockley H.L.: The intrinsic innervation of the human alimentary tract and its relation to function. *Gut* 16:443, 1975.

Bohr D.F., Somlyo A.P., Sparks H.V., Jr. (eds.): *Handbook of Physiology: Sec. 2. The Cardiovascular System*, vol. 2, *Vascular Smooth Muscle*. Bethesda, Md., American Physiological Society, 1980.

Bolton T.B.: Mechanisms of action of transmitters and other substances on smooth muscle. *Physiol. Rev.* 59:606, 1979.

Bortoff A.: Myogenic control of intestinal motility. *Physiol. Rev.* 56:418, 1976.

Bulbring E., Brading A.F., Jones A.W., et al (eds.): *Smooth Muscle*. Baltimore, Williams & Wilkins Co., 1970.

Bulbring E., Bolton T.B. (eds.): Smooth muscle. *Br. Med. Bull.* 35:209, 1979.

Burnstock G.: Evolution of the autonomic innervation of visceral and cardiovascular systems in vertebrates. *Pharmacol. Rev.* 21:247, 1969.

Burnstock G.: Neurotransmitters and trophic factors in the autonomic nervous system. *J. Physiol.* 313:1, 1981.

Burnstock G., Costa M.: Inhibitory innervation of the gut. *Gastroenterology* 64:141, 1973.

Casteel R., Godfraind T., Ruegg, J.C. (eds.): *Excitation-Contraction Coupling in Smooth Muscle*. Amsterdam, Elsevier, 1977.

Chamley-Campbell J., Campbell G.R., Ross, R.: Smooth muscle cells in culture. *Physiol. Rev.* 59:1–61, 1979.

Connor J.A., Prosser C.L., Weems, W.A.: A study of pacemaker activity in intestinal smooth muscle. *J. Physiol.* 240:671, 1974.

Daniel E.E., Daniel V.P., Duchon G., et al.: Is the nexus necessary for cell-to-cell coupling in smooth muscle? *J. Memb. Biol.* 28:207, 1976.

Duthie H.L. (ed.): *Gastrointestinal Motility in Health and Disease*. Lancaster, MTP Press, 1978.

El-Sharkawy T.Y., Morgan K.G., Szurszewski J.H.: Intracellular electrical activity of canine and human gastric smooth muscle. *J. Physiol.* 279:291, 1978.

Furness J.B., Costa M.: The adrenergic innervation of the gastrointestinal tract. *Ergeb. Physiol.* 69:1, 1974.

Gabella G.: Fine structure of the myenteric plexus in the guinea-pig ileum. *J. Anat.* 11:69, 1972.

Gabella G.: *Structure of the Autonomic Nervous System*. London, Chapman and Hall, 1976.

Gershon M.D., Erde, S.M.: The nervous system of the gut. *Gastroenterology* 80:1571, 1981

Goodford P.J.: Distribution and exchanges of electrolytes in intestinal smooth muscle, in Code C.F. (ed.): *Handbook of Physiology:* Sec. 6. *Alimentary Canal,* vol. 4. Washington, D.C., American Physiological Society, 1968, pp. 1743–1766.

Hillarp N.-A.: Peripheral autonomic mechanisms, in Field J. (ed.): *Handbook of Neurophysiology,* vol. 2. Washington, D.C., American Physiological Society, 1960, pp. 979–1006.

Holman M.E.: An introduction to electrophysiology of visceral smooth muscle, in Code C.F. (ed.): *Handbook of Physiology:* Sec. 6. *Alimentary Canal,* vol. 4. Washington, D.C., American Physiological Society, 1968, pp. 1665–1708.

Kuriyama H.: Ionic basis of smooth muscle action potentials, in Code C.F. (ed.): *Handbook of Physiology:* Sec. 6. *Alimentary Canal,* vol. 4. Washington, D.C.: American Physiological Society, 1968, pp. 1767–1792.

Prosser C.L.: Rhythmic potentials in intestinal muscle. *Fed. Proc.* 37:2153, 1978.

Ruegg J.C.: Smooth muscle tone. *Physiol. Rev.* 51:201, 1971.

Sarna S.K.: Gastrointestinal electrical activity: Terminology. *Gastroenterology* 68:1631, 1975.

Schofield G.C.: Anatomy of muscular and neural tissue in the alimentary canal, in Code C.F. (ed.): *Handbook of Physiology:* Sec. 6. *Alimentary Canal,* vol. 4. Washington, D.C., American Physiological Society, 1968, pp. 1579–1628.

Shoenberg C.F., Needham D.M.: A study of the mechanism of contraction in vertebrate smooth muscle. *Biol. Rev.* 51:53–104, 1976.

Stephens N.L. (ed.): *Biochemistry of Smooth Muscle.* Baltimore, University Park Press, 1977.

Wood J.D.: Neurophysiology of Auerbach's plexus and control of intestinal motility. *Physiol. Rev.* 55:307, 1975.

Wood J.D., Mayer C.J.: Adrenergic inhibition of serotonin release from neurones in guinea pig Auerbach's plexus. *J. Neurophysiol.* 42:594, 1979.

Yokoyama S., Ozaki T., Kajisuka T.: Excitation conduction in Meissner's plexus of rabbit small intestine. *Am. J. Physiol.* 232:E109, 1977.

2

Chemical Messengers of the Digestive Tract

BECAUSE THE PROPERTIES of the chemical messengers* of the digestive tract are complex, those properties must be understood before the messengers' even more complex functions can be understood. The general properties of the messengers will be cataloged in this chapter. In subsequent chapters, when the behavior of each of the organs of the digestive tract is described, the way in which the messengers, alone or in conjunction, influence the functions of the organ will be set forth in detail.

Chemical messengers of the digestive tract may act in four ways:

1. A messenger may be a *neurotransmitter,* released by the terminals of one nerve fiber and acting across a very short distance upon another nerve fiber, a muscle cell, or a gland cell. Acetylcholine and norepinephrine are familiar neurotransmitters, but several amines, purines, and peptides are now known to be, or are thought to be, neurotransmitters in the digestive tract.

2. The messenger may be part of an *endocrine system* in which a stimulus acting upon a receptor cell causes that cell to liberate a hormone into the blood so that it can act upon a distant target cell. The gastrin system is the best understood of those in the digestive tract.

3. In a *neuroendocrine system* the messenger is liberated from the terminals of a nerve fiber by an action potential, and, instead of acting locally, the messenger is carried to a distant target cell. The antidiuretic hormone is a member of a neuroendocrine system. In the digestive tract peptides are thought to be the messengers of a neuroendocrine system.

4. A messenger liberated by one cell may act upon an immediately adjacent cell. In this case the messenger is part of a *paracrine* system. Somatostatin may be a paracrine messenger in the digestive tract.

Pharmacological versus Physiological Actions

A solution containing the hormone in question is injected into an animal, and the responses are observed. The question whether a response is caused by the hormone or by some contaminant is eventually solved when the pure hormone becomes available. Then the pharmacological actions of the pure hormone are cataloged.

Several criteria must be rigorously satisfied before the physiological role of the hormone is determined.

1. The natural stimuli for the hormone's release must be identified. In the case of secretin, acid in contact with the mucosa of the upper small intestine releases the hormone into the blood. Do nervous stimuli or other components of the chyme release it as well?

2. The natural stimulus for release of the hormone must have the correct magnitude and time course to account for the magnitude and time course of the response. Tenth normal hydrochloric acid in the duodenum and jejunum releases secretin, but tenth normal hydrochloric acid is a wildly unphysiological stimulus. When chyme is gradually emptied from the stomach, does it contain enough acid at the right time

*Some years after *secretin,* the first authentic messenger of the digestive tract, was discovered, W.B. Hardy suggested that it be called a *hormone.* The word means *I excite,* and secretin excites the pancreas to produce a bicarbonate-containing fluid. When inhibitory messengers were discovered, they were for a while called *chalones,* from the Greek verb *to slacken.* However, a single messenger may excite under some circumstances and inhibit under others. Consequently, the word *hormone* as it is now used to denote a messenger which may excite or inhibit has, like the word *democracy,* lost its original meaning.

to release enough secretin to stimulate the pancreas as the pancreas is observed to be stimulated?

3. Do the amount and kind of hormone released account for the observed response? To answer this question, one must know the magnitude and time course of the increase of hormone in the plasma following the physiological stimulus. In the case of a hormone having multiple forms, the time course of each must be known. If a similar increment in amount and kind is produced by infusion, is the same response reproduced?

4. Do other circumstances determine or influence the response? Perhaps concomitant excitatory or inhibitory nervous influences are important, or those of other hormones. In the case of secretin, the naturally occurring stimulus does not, in fact, release enough secretin to stimulate the pancreas if it were acting alone. However, at the same time, the pancreas is being excited by acetylcholine released from vagal nerves during the cephalic phase of digestion, and, more important, cholecystokinin, released from the intestinal mucosa by protein and fat digestion products, acts synergistically upon pancreatic cells to permit a small amount of secretin to become an adequate stimulus.

In a few instances the physiological action, as distinct from the pharmacological action, of a hormone in governing its primary response is known with some confidence. Secondary actions uncovered pharmacologically are often found to be unphysiological. For example, cholecystokinin inhibits acid secretion. It is released from the intestinal mucosa by fat digestion products, and when fat is in the lumen of the upper small intestine, acid secretion is inhibited. However, in both man and the dog, infused cholecystokinin inhibits gastrin-stimulated acid secretion only in a dose which is supramaximal for the physiological actions of the hormone, and, therefore, its effect upon gastric secretion is pharmacological, not physiological. An intravenous (IV) bolus dose of 50 ng kg^{-1} of little gastrin is half-maximal for acid secretion in man, whereas 250 ng kg^{-1} is half-maximal for contraction of the lower esophageal sphincter. On this basis, the action of little gastrin upon the lower esophageal sphincter has been judged to be pharmacological.

Acetylcholine and the Catecholamines

In addition to their roles in controlling motility and blood flow in the digestive tract, acetylcholine and catecholamines have direct and indirect effects upon secretion.

Acetylcholine released from postganglionic fibers of the parasympathetic system or from fibers of the intrinsic plexuses directly stimulates secretion by cells of the salivary, gastric, and pancreatic glands. In addition, acetylcholine acts synergistically with other messengers. Nevertheless, acetylcholine is not exclusively excitatory, for as a mediator of vagal fibers to the antral mucosa it inhibits release of gastrin. Acetylcholine also reversibly increases the permeability of tight junctions between cells of the gastric mucosa with the result that the mucosa leaks plasma proteins into its lumen. The anticholinergic drug, atropine, reduces gastric shedding of plasma proteins which occurs in the protein-losing gastropathy of Ménétrier's disease.

Norepinephrine released from sympathetic postganglionic fibers stimulates salivary acinar cells to secrete and salivary duct cells to reabsorb. If catecholamines directly affect other digestive glands, those effects have not been shown to be important. However, catecholamines may have an indirect effect upon gastric secretion. Vagotomy interrupts preganglionic fibers to gastric glands and to gastrin-secreting cells and thereby abolishes the cephalic phase of gastric secretion. To test for completeness of vagotomy the blood concentration of glucose is reduced below 50 mg dl^{-1} by administration of insulin. Under normal conditions the glucose-sensing cells of the hypothalamus relay impulses to the vagal nuclei, which in turn send impulses by vagal fibers to the stomach, stimulating acid secretion through acetylcholine and releasing gastrin. If vagotomy is complete, the path is interrupted, and gastric secretion should not increase during insulin hypoglycemia. Nevertheless, plasma gastrin concentration does increase during insulin hypoglycemia even if vagotomy has been accurately performed. During insulin hypoglycemia epinephrine is secreted by the adrenal medulla, and epinephrine then releases gastrin, which in turn stimulates acid secretion. Consequently, an increase in acid secretion during insulin hypoglycemia does not necessarily prove that vagotomy is incomplete or that vagal fibers have regenerated.

The familiar classification of fibers of the autonomic nervous system is that all preganglionic fibers are cholinergic, that all parasympathetic postganglionic fibers are cholinergic, and that most sympathetic postganglionic fibers are adrenergic, the rest being cholinergic. This classification is no longer true or useful. Many preganglionic and postganglionic fibers of both divisions contain and liberate other mediators: purines, 5-hydroxytryptamine, and a host of diverse peptides. To make matters more complicated, it appears that a single fiber may liberate two mediators, acetylcholine and a peptide, during the passage of a single impulse. This branch of neuroendocrinology is in a primitive state, and anything written about it now is sure to be obsolete soon.

Gastrin

Gastrin was identified by the classical methods of endocrinology. In an experimental animal the endocrine organ, the gastric antrum, and the receptor organ, the body of the stomach, were transplanted so that there could be no nervous connection between the two. Irrigation of the mucosa of the antrum with liver extract discharged a messenger into the blood, and the messenger upon reaching the mucosa of the body of the stomach stimulated its oxyntic cells to secrete acid.

Further analysis showed that some constituents of liver extract or of a meal are more effective liberators of gastrin than others and that when the pH of chyme bathing the antral mucosa is low, none is effective. At first it was thought that impulses in vagal nerves to the stomach do not stimulate release of gastrin, but after errors in the experiments were eliminated it was found that vagal impulses both stimulate and inhibit gastrin release. Then other hormones, some familiar and some still poorly characterized, were shown to release gastrin or to inhibit its release. Eventually radioimmunological assay of gastrin in plasma permitted its measurement under physiological conditions.

Gastrin is a polypeptide, and 60 years elapsed between gastrin's discovery in 1905 and the time when peptide chemistry had sufficiently advanced to permit chemical characterization of the hormone. Then it was found that there are at least six distinct gastrins in one animal species, each with a characteristic structure, potency, and physiological half-life. When one of these gastrins, or its synthetic derivative, pentagastrin, was administered to an animal, a multiplicity of responses by many organs was revealed, and great effort was required to determine which was a physiological and which a pharmacological response. At the target organ, the action of gastrin was found to be affected by at least three other hormones. Finally,

important species differences were uncovered, and something true of gastrin in the dog turned out to be false in man.

The chemical and pharmacological problem is to discover all the properties and functions of a messenger. The physiological problem is to determine which of these is important in governing the behavior of the digestive tract. The pedagogical problem is to decide what fraction of our knowledge is appropriately described in an introductory text. For gastrin and a few other messengers these problems are approaching solution in 1982. For many other messengers the answers are far less definite, and the student is warned that for gastrointestinal endocrinology, as for the rest of physiology, the questions remain the same, but the answers change.

The Gastrins

Gastrins are a family of straight-chain peptides. The structures of three are given in Table 2–1. The hormone synthesized in the G cells contains 34 amino acids, and it is called *big gastrin*. The amino acid at its N-terminus is pyroglutamic acid (Glp), and the phenylalanine at its C-terminus is amidated. These terminal substitutions prevent destruction of the hormone by aminopeptidases and carboxypeptidases. Big gastrin is only one of the forms of gastrin in plasma. Another form is a heptadecapeptide, *little gastrin,* which is derived from big gastrin by cleavage after two basic amino acids. Little gastrin is the most abundant form of gastrin stored in the G cells, and it too is released into plasma. A smaller molecule containing 14 amino acids called *mini-gastrin* is found in some samples of plasma. A very small amount of a much larger and still uncharacterized molecule called *big big gastrin* occurs in the void volume of plasma samples run through a Sephadex column, and it may be present in some gastrinomas. There is some as yet inconclusive evidence that the C-terminal tetrapeptide of gastrin is present in the mucosa of the antrum and duodenum and that it is absent from systemic plasma because it is cleared from portal blood by the liver. In all forms of gastrin, the sixth amino acid from the C-terminus is tyrosine, and this tyrosine may or may not be sulfated. Sulfated and nonsulfated gastrins occur in approximately equal quantities, and they are equally potent as hormones.

Gastrins are conventionally named according to their species source, their amino acid content and

TABLE 2–1.—STRUCTURE OF SOME GASTRINS, CHOLECYSTOKININ, AND CAERULEIN

Human big gastrin, HG–34–II*

```
  1     2     3     4     5     6     7     8     9    10    11    12    13    14    15    16    17
Glp — Leu — Gly — Pro — Gln — Gly — His — Pro — Ser — Leu — Val — Ala — Asp — Pro — Ser — Lys — Lys —
 18    19    20    21    22    23    24    25    26    27    28    29    30    31    32    33    34
Gln — Gly — Pro — Trp — Leu — Glu — Glu — Glu — Glu — Glu — Ala — Tyr — Gly — Trp — Met — Asp — Phe — NH₂
                                                                  |
                                                                 HSO₃
```

Human little gastrin, HG–17–II

```
  1     2     3     4     5     6     7     8     9    10    11    12    13    14    15    16    17
Glp — Gly — Pro — Trp — Leu — Glu — Glu — Glu — Glu — Glu — Ala — Tyr — Gly — Trp — Met — Asp — Phe — NH₂
                                                                  |
                                                                 HSO₃
```

Human mini-gastrin, HG–14–II

```
  1     2     3     4     5     6     7     8     9    10    11    12    13    14
Trp — Leu — Glu — Glu — Glu — Glu — Glu — Ala — Tyr — Gly — Trp — Met — Asp — Phe — NH₂
                                          |
                                         HSO₃
```

Porcine little gastrin, PG–17–II   5 —Met—

Feline little gastrin, FG–17–II   8 —Ala—

Canine little gastrin, CG–17–II   5 —Met—

Ovine little gastrin, OG–17–II   5 —Val—   10 —Ala—

Pentagastrin

```
C(CH₃)₃—OCO—NH—CH₂—CH₂—CO——Trp—Met—Asp—Phe—NH₂
```

Cholecystokinin and its C-terminal octapeptide

```
              26    27    28    29    30    31    32    33
(25 more) — Asp — Tyr — Met — Gly — Trp — Met — Asp — Phe — NH₂
                   |
                  HSO₃
```

Caerulein

```
Glp — Glu — Asp — Tyr — Thr — Gly — Trp — Met — Asp — Phe — NH₂
                   |
                  HSO₃
```

*Gastrin Is are not sulfated.

their sulfation. Thus: HG–17–II is *H*uman heptadecapeptide *G*astrin, sulfated. The unsulfated form is numbered I. A more immediately intelligible nomenclature, not generally acceptable, is to omit designation of the unsulfated form and to label the sulfated one = S. Because sulfation has no effect upon the potency of any gastrin, designation of sulfation is unimportant and will be ignored here.

The entire spectrum of physiological activity is exhibited by the last four amino acids at the C-terminus of the gastrins. Unfortunately, biochemical convention is to number peptides from the N-terminus, and therefore the active fragment of big gastrin resides in amino acids 31–34, of little gastrin in amino acids 14–17, and of mini-gastrin in amino acids 11–14, all the same amino acids in the same sequence.

Little gastrin obtained from the tissues of pig, cat, dog, and sheep differs from human little gastrin in having one or two amino acid substitutions in the middle of the peptide chain (see Table 2–1). Each substitution is the result of a single base change in the codon triplet, and because the substitutions occur in the nonspecific part of the molecule, they have no evolutionary significance.

A synthetic product called *pentagastrin* consists of the C-terminal tetrapeptide to which a substituted beta-alanine has been added. This compound has the advantages of being stable, water-soluble, and patentable. It is the commercially available product having all the physiological properties of the natural gastrins.

Only 10% of the gastrin contained in G cells of the antrum is big gastrin; the rest is little gastrin. Big gastrin is the larger fraction of gastrin extractable from the proximal duodenum. There is only a minute amount of any gastrin in the oxyntic glandular mucosa, and the amount extractable from the normal pancreas is so small as to be debatable. Big gastrin is twice as abundant as little gastrin in most plasma samples; after feeding, it rises more rapidly and remains higher than does little gastrin.

Big gastrin has a half-life in man of 42 min, and little gastrin has a half-life of 7 min. The difference in half-lives accounts for the fact that the plasma concentration of big gastrin is higher than that of little gastrin. Although big gastrin composes approximately three fourths of the total molar concentration of circulating gastrin after a protein meal, it contributes less than half of the acid-stimulating activity.

Injection of equimolar amounts of big gastrin and little gastrin causes equal secretions of acid. However, big gastrin, having a much longer half-life, rises to a higher plasma concentration under this circumstance, and it is therefore much less potent, at the molecular level, than is little gastrin in stimulating acid secretion. Little gastrin is removed from blood to the extent of 21% to 30% by a single passage through the vascular beds of the brain, leg, kidney, and intestine. Hepatic removal is 40%.

Gastrin concentrations are measured by radioimmunoassay, and the normal fasting plasma concentration is 50–100 pg/ml (a pg or picogram is 10^{-12} gm). The upper limit of normal is about 200 pg/ml. Except in the most expert hands, the assay does not differentiate among the various forms of the hormone or between active and inactive forms. Consequently, most reported assay results tell only whether the concentration is high, medium, or low, and reported values are most useful when they show changes occurring as the result of some procedure. A peptide consisting of the first 13 amino acids of little gastrin is sometimes present in plasma. Because it lacks the four amino acids at the C-terminus, it does not have the physiological actions of gastrin. However, it reacts with antibodies raised against the N-terminus of gastrin, and therefore it interferes with the radioimmunological assay of gastrin.

Actions of Gastrin

The major short-term action of gastrin is to stimulate secretion of acid. (The major long-term action is a trophic one discussed below.) In man, when the plasma concentration of gastrin rises in response to a meal, acid secretion rises by 1.8% of the stomach's capacity to secrete for every picogram per milliliter rise in plasma gastrin concentration.

Gastric blood flow increases when gastrin stimulates acid secretion. Pepsinogen secretion also accompanies acid secretion, but secretion of pepsinogen is not entirely a direct consequence of the action of gastrin upon chief cells. When acid, whose secretion has been stimulated by gastrin, flows over the surface of the mucosa, it excites receptors, which, through a cholinergic reflex, stimulate pepsinogen secretion. Many other responses can be detected when gastrin is infused IV (Table 2–2). Most are probably pharmacological.

TABLE 2–2.—ACTIONS OF GASTRINS

Physiological actions:
  Gastrins stimulate
    Acid secretion
    Pepsinogen secretion
    Gastric blood flow
    Contraction of the circular muscle of the stomach
    Growth of gastric and small intestinal mucosa and the
      pancreas
Pharmacological (and possibly physiological) actions:
  Gastrins stimulate
    Water, bicarbonate, and electrolyte secretion by
      pancreas, liver, and small intestinal mucosa
    Enzyme secretion by pancreas and small intestinal
      mucosa
    Contraction of the lower esophageal sphincter,
      gallbladder, small intestine, and colon
  Gastrins inhibit
    Contraction of the pyloric sphincter, ileocecal
      sphincter, and sphincter of the hepatopancreatic
      ampulla
    Gastric emptying
    Absorption of glucose and electrolytes in the small
      intestine
  Gastrins release
    Insulin, glucagon, and calcitonin

Control of Gastrin Release

Gastrin is synthesized and stored in G cells in the pyloric glandular and duodenal mucosa. Gastrin is released from basal storage granules into interstitial fluid and thence into portal blood. Only insignificant amounts are carried in gastric and intestinal lymph. The G cells also liberate gastrin into the gastric lumen. G cells are flask-shaped; their necks project into the lumen of the glands, and the exposed surface is covered with short microvilli which, one supposes, receive stimuli for release of gastrin. The G cells also are innervated by fibers from the intrinsic plexuses.

Impulses reaching the antral mucosa through vagal nerves and the plexuses both stimulate and inhibit gastrin release. The mediator for stimulation is probably a peptide, perhaps vasoactive intestinal peptide (VIP). The mediator for inhibition is acetylcholine, and therefore atropine enhances the response of antral G cells to positive stimuli.

The major stimulus for gastrin release is a neutral solution of some L-amino acids. The effective ones are phenylalanine, cystine, tryptophan, hydroxyproline, and tyrosine. Peptides, but not native proteins, also release gastrin. Solutions of calcium salts, including milk, release gastrin. Consequently, the use

of calcium-containing antacids is inappropriate. Decalcified milk has only a small effect in releasing gastrin. Hypercalcemia resulting from IV infusion of calcium salts releases gastrin, but a patient with hypercalcemia caused by hyperparathyroidism does not have hypergastrinemia unless he has a gastrinoma, a tumor that secretes gastrin.

In the dog, gastrin is also released from a surgically prepared pouch of the antrum by distention, but it is not released by distention of the whole, intact stomach. Distention of the stomach is not an important releaser of gastrin in man.

Epinephrine releases gastrin. In man, the effect is antagonized by a beta-adrenergic blocking agent.

Aliphatic alcohols, of which ethanol is the most potent, release gastrin in the dog, a confirmed teetotaler, but in man, ethanol, if it releases gastrin at all, has only a trivial effect. Caffeine does not release gastrin, but decaffeinated coffee does so and stimulates acid secretion, an effect attributable to peptides in the brew.

Peptides extracted from the duodenal mucosa, when injected IV, stimulate release of gastrin from the antral mucosa. The effective peptide is similar to or identical with the peptide bombesin. However, perfusion of the small intestine of the dog with liver extract does not cause a rise in plasma gastrin concentration, although the perfusion stimulates acid secretion. Consequently, the physiological role of intestinal gastrin releasers is in doubt.

Gastrin is released into plasma by peripheral stimulation of the sciatic and other motor nerves, but the significance of this fact is obscure. In the colon, gastrin (and cholecystokinin) is confined to neurones, and there it may be a neurotransmitter.

Gastrin release is inhibited by acid in contact with the pyloric glandular mucosa. Release is completely suppressed when the pH of antral contents is 1.0. When a solution of amino acids is placed in the stomach of a normal man, the increment in plasma gastrin is the same when the pH of the solution is 3.0, 4.0, or 5.5, but when the pH of the solution is 2.5, the rise in plasma gastrin is very small. A patient with pernicious anemia has achlorhydria, and his fasting plasma gastrin concentration is high. If such a patient drinks acid, preferably through a straw so that he does not dissolve his teeth, his plasma gastrin concentration falls steeply within 15 min. When he eats a meal, his plasma gastrin rises to several times the fasting level, because gastrin release is not inhibited by acid.

Inhibition of gastrin release by acid in contact with the antral mucosa is abolished by a small dose of atropine. Thus, the inhibition of gastrin release by acid appears to be mediated by a cholinergic pathway.

Cholecystokinin

The fact that fat- and protein-digestion products in the duodenum cause contraction of the gallbladder by means of a hormone was demonstrated by the classical means of endocrinology in 1928, and the hormone responsible was named cholecystokinin. Years later, the fact that fat- and protein-digestion products in the duodenum stimulate enzyme secretion by the pancreas through mediation by a hormone was established by similar means, and the hormone responsible was named pancreozymin. Still later, a polypeptide isolated from the duodenal mucosa proved to have the properties of both hormones; cholecystokinin and pancreozymin are the same molecule. For convenience, the molecule is given its earlier name, but its property of stimulating the pancreas is still called pancreozymin. The hormone is often referred to as CCK or CCK-PZ, but here the name will be spelled out.

Cholecystokinin is a polypeptide containing 33 amino acids. Its partial structure is given in Table 2–1. The last five amino acids in its C-terminal sequence are identical with those in gastrin. Since the last four of these constitute the active fragment of gastrin, both hormones have the same physiological actions. They differ quantitatively, not qualitatively. In gastrin, the sulfated tyrosyl residue is separated from the active C-terminal sequence by a single amino acid, and whether or not the tyrosyl residue is sulfated makes no difference to the hormone's physiological activity. In cholecystokinin, the sulfated tyrosyl residue is separated from the C-terminal sequence by two amino acids, and this tyrosyl residue must be sulfated for the molecule to have cholecystokinin activity upon the gallbladder. Sulfation is not necessary for action upon the stomach.

The C-terminal octapeptide of cholecystokinin (see Table 2–1) has all of the hormone's actions. As a synthetic product it is a convenient substitute for cholecystokinin in experimental or clinical work. As a natural product it occurs in blood and the brain. It may account for as much as 30% of the cholecystokinin immunoreactivity in the blood of a man who has been given a fatty meal.

Cholecystokinin is present in cells in the mucosa of the duodenum and jejunum but not the ileum. It is released by the L-forms of essential amino acids. The D-forms and the nonessential amino acids are ineffective. The most potent stimulus is L-tryptophan, and the optimal concentration of tryptophan is 50 mM. In decreasing order of potency are phenylalanine, lysine, methionine, arginine, valine, isoleucine, histidine, leucine, and threonine. The maximal response to tryptophan is 80% of that to a maximal dose of cholecystokinin, and the rate of secretion of pancreatic enzymes when the duodenum is irrigated with a mixture of essential amino acids is nearly the same as that following IV administration of a maximally tolerated dose of cholecystokinin.

Perfusion of the duodenum with acid at a physiological rate releases cholecystokinin as well as secretin, but the amount of cholecystokinin released is only one fifth that of secretin.

The actions of cholecystokinin are listed in Table 2–3.

In the dog, gastric emptying is slowed when cholecystokinin or its octapeptide is injected or when cholecystokinin is released by perfusion of the duodenum with tryptophan. The amount of cholecystokinin required to inhibit gastric emptying is the same as that required for contraction of the gallbladder and secretion of pancreatic enzymes. The effect of cholecystokinin in slowing gastric emptying is a physiolog-

TABLE 2–3.—ACTIONS OF CHOLECYSTOKININ

Physiological actions
  Cholecystokinin
    Stimulates contraction of the gallbladder and relaxation of the sphincter of the hepatopancreatic ampulla
    Stimulates secretion of enzymes by the pancreas
    Weakly stimulates secretion of bicarbonate-containing juice by the pancreas and liver but is synergistic with secretin
    Slows gastric emptying
    Is a trophic hormone for the pancreas
Pharmacological (and possibly physiological) actions
  Cholecystokinin
    Weakly stimulates acid secretion in absence of gastrin
    Competitively inhibits gastrin-stimulated acid secretion
    Stimulates pepsinogen secretion
    Inhibits the lower esophageal sphincter
    Stimulates motility of the sigmoid colon
    Stimulates secretion by the duodenal glands
    Releases insulin and glucagon
    May be a neurotransmitter in the central nervous system
    Terminates or stimulates feeding behavior

ical one in the dog, and it is probably also physiological in man.

Because cholecystokinin has the same C-terminal sequence of amino acids as do the gastrins, cholecystokinin weakly stimulates acid secretion. It competitively inhibits gastrin-stimulated acid secretion by occupying gastrin receptors and denying them to the more potent hormone. Both these effects occur only at a dose of cholecystokinin supramaximal for pancreatic secretion, and they are therefore pharmacological.

Cholecystokinin and its octapeptide occur in the brain of man and many other mammalian species. The hormones are present in nerve terminals isolated from the cerebral cortex, and they may be neurotransmitters.

When an animal feeds, many mechanisms are activated which eventually result in satiety, that is, in cessation of eating and in characteristic postprandial behavior such as grooming or somnolence. Infusion of impure preparations of cholecystokinin or of its octapeptide in pharmacological doses suppresses or reduces feeding in rats and monkeys and perhaps in man, and it is possible that cholecystokinin from the intestine or cerebral nerve terminals is one of the satiety signals.

Caerulein

The skin of the Australian frog *Hyla caerulea* contains a high concentration of a decapeptide, caerulein, whose structure is given in Table 2–1. It has the same C-terminal tetrapeptide as gastrin and cholecystokinin and has a sulfated tyrosyl residue in a position corresponding to cholecystokinin's. Consequently, it shares the properties and potency of the other two polypeptides. In fact, its ability to cause contraction of the gallbladder is so great that it has been used in nanogram per kilogram doses during cholecystography.

Secretin

Secretin (correctly pronounced sĕ-krē′ tin) holds pride of place as the first gastrointestinal hormone to be discovered. In the afternoon of January 16, 1902,*

---

*For an eyewitness account of the discovery of secretin on Jan. 16, 1902, see Martin, C.: Obituary notice: E.H. Starling. *Br. Med. J.* 1:900–904, 1927.

Bayliss and Starling found that when acid was placed in an extrinsically denervated loop of upper small intestine of an anesthetized dog, the pancreas responded by secreting. This could not be a nervous reflex, and Starling exclaimed: "Then it must be a chemical reflex!" He then showed that a crude extract of the intestinal mucosa, when given IV, also stimulated pancreatic secretion. Although in retrospect the experiment was far from conclusive, Starling's exclamation announced the birth of the science of endocrinology; he saw that a specific stimulus acting on a specific receptor organ releases a specific messenger, which, traveling by the blood to a distant, specific target organ, elicits a specific response. In addition, the experiment of Bayliss and Starling demonstrated a negative feedback loop in which the stimulus evokes a response which ends in the elimination of the stimulus.

The structure of the polypeptide secretin is given in Table 2–4. Of the 27 amino acids in secretin, 14 occupy the same position, counting from the N-terminus, as do those in glucagon. Consequently, secretin in pharmacological doses mimics glucagon, and glucagon in pharmacological doses mimics secretin.

The chief action of secretin is to stimulate secretion of bicarbonate-containing fluid by pancreas and liver. The concentration of secretin in the plasma rises after a meal, and administration of antibodies to secretin profoundly depresses pancreatic output of both bicarbonate and enzymes. The half-life of circulating secretin is about 4 min.

Secretin is present in S cells throughout the duodenum, but it is most concentrated in the duodenal bulb. Over the range of pH from 0 to 3.0, release of secretin is independent of the pH of fluid entering the duodenum; the amount of secretin released is proportional to the amount of acid entering the duodenum, not to the concentration of acid. Above pH 3.0 there is a sharp decline in the amount of secretin released as the pH rises until an absolute threshold is reached at pH 4.5–5.0. When the contents of the stomach are acid, the duodenal bulb is acidified only briefly when a spurt of chyme enters it (see Fig 4–1), and acid is completely neutralized in the first portion of the duodenum. Therefore, only a small quantity of secretin is released into the blood during digestion of a meal, and this amount of secretin *in itself* is not enough to stimulate the actually occurring pancreatic and biliary secretion of bicarbonate.

Synergism of secretin and cholecystokinin accounts

TABLE 2–4.—STRUCTURE OF SECRETIN*

1	2	3	4	5	6	7	8	9	10	11	12	13	14
His	*Ser*	Asp	*Gly*	*Thr*	*Phe*	*Thr*	*Ser*	Glu	Leu	*Ser*	Arg	Leu	Arg

15	16	17	18	19	20	21	22	23	24	25	26	27
Asp	*Ser*	Ala	*Arg*	Leu	*Gln*	Arg	Leu	Leu	*Gln*	Gly	*Leu*	Val—NH_2

*The amino acids in italics occur in corresponding positions in glucagon.

for the quantity of bicarbonate and enzymes secreted. The two hormones have the following relations:

1. Cholecystokinin is released by acid in the duodenum at a rate one fifth that at which secretin is released by the same amount of acid.

2. Secretin is released by acid but not by any other component of chyme.

3. Cholecystokinin alone weakly stimulates secretion of bicarbonate-containing juice by pancreas and liver, but it amplifies the action of secretin.

4. Secretin alone weakly stimulates secretion of enzymes by the pancreas, but it amplifies the action of cholecystokinin.

5. Secretin and cholecystokinin stimulate pepsinogen secretion and inhibit acid secretion.

Because the structure of secretin has nothing in common with the structure of gastrin, secretin might be expected to be a noncompetitive inhibitor of gastrin-stimulated acid secretion. It is so in the dog, but it is a competitive inhibitor in man.

Secretin's inhibition of acid secretion is probably only a pharmacological property; so are its effects in stimulating contraction of the gallbladder, inhibiting contraction of the lower esophageal sphincter and gastric and intestinal motility, releasing insulin, and causing lipolysis in adipose tissue.

Gastric Inhibitory Polypeptide

Gastric inhibitory polypeptide (GIP) occurs at highest concentration in the duodenal and jejunal mucosa and in lower concentration in the mucosa of the ileum. The peptide isolated from pig mucosa is a linear polypeptide of 43 amino acids, 15 of the first 26 in the same position as in porcine glucagon. The most potent stimuli for its release into blood are intraluminal glucose and fat digestion products, but acid and amino acids are also effective. Therefore, its concentration in plasma rises to a peak about 60 min after ingestion of a meal. Its identifying action is inhibition of acid secretion. However, when GIP was infused into human subjects in a dose resulting in a plasma concentration 19 times greater than that occurring after a meal, GIP did not inhibit pentagastrin-stimulated secretion of acid or of pepsin. In pharmacological doses GIP slows gastric emptying, causes mesenteric vasodilatation, and augments insulin release by glucose.

Vasoactive Intestinal Peptide

Vasoactive intestinal peptide (VIP) was first identified as a peptide extractable from digestive organs which, upon infusion, causes vasodilatation. In pharmacological doses it inhibits acid and pepsin secretin and stimulates ventilation and pancreatic, hepatic, and intestinal secretion. Its concentration in blood does not rise during a meal. On the other hand, VIP is found by immunohistochemical methods in cell bodies and nerve terminals throughout the digestive tract, particularly in the wall of the gallbladder and in the sphincters where it may mediate relaxation. In addition, VIP is present in high concentration in sympathetic ganglia and in the central nervous system, particularly in cerebral grey matter. VIP binds to membranes derived from the cerebral cortex, hippocampus, striatum, and thalamus. These facts suggest that VIP is a neurotransmitter and not a hormone. Cells of neuroblastomas contain a large amount of VIP, and release of VIP from a neuroma is probably one of the causes of the profuse watery diarrhea known as pancreatic cholera.

Bombesin

At least 19 biologically active peptides have been extracted from amphibian skin, and similar peptides are found in the digestive tract of birds and mammals.

In addition to caerulein, the best known of these is bombesin, the tetradecapeptide first obtained from the skin of the European frog *Bombina bombina*. Bombesin stimulates acid secretion by liberating gastrin and pancreatic secretion by liberating cholecystokinin. A heterogeneous group of bombesin-like peptides, apparently derived from a large probombesin molecule, is in the gastrointestinal mucosa of chickens, turkeys, rats, and men. Present methods cannot detect it in blood. Bombesin could participate in the intestinal phase of gastric secretion, but unfortunately for this idea gastric secretion occurring when chyme is in the intestine is not mediated by bombesin-released gastrin. Therefore, the hormonal status of bombesin remains to be established.

Pancreatic Polypeptide

Pancreatic polypeptide is a linear peptide of 36 amino acids which, like glucagon, was first identified as a minor impurity in preparations of insulin. It can be measured by radioimmunoassay. Eating a meal causes a prompt increase in its plasma concentration. The response is sustained for several hours, and because its half-life is only 5.5 min, it must be continuously released.

The rise in pancreatic polypeptide concentration is attenuated by vagotomy, and release effected by peripheral stimulation of the vagus is blocked by atropine. However, the natural stimulus for its release from endocrine cells scattered throughout the pancreas remains to be identified. The candidate hormone, bombesin, in a dose equal to that which releases gastrin, also releases pancreatic polypeptide, and the effect of bombesin is blocked by atropine. When pancreatic polypeptide is given IV in a dose resulting in plasma concentrations similar to those found after a meal, it inhibits pancreatic secretion of bicarbonate and enzymes which had been stimulated by secretin and caerulein. It also causes relaxation of the gallbladder and contraction of the choledochal sphincter, but the role of these actions in the integrated response to a meal is unknown. It has no role in regulation of gastric secretion.

Other Peptides

On the one hand many peptides isolated from the digestive tract are, like VIP and bombesin, only candidates for membership in the endocrine society, and on the other hand some functions of the digestive tract are thought to be controlled by hormones which are as yet no more than names. Table 2–5 lists some of both. Statements therein are tentative, for, as in a Pirandello play, the reality behind appearance is uncertain.

Origin of Gut Endocrine Cells

The fact that peptide hormones of the digestive tract occur in the brain and in nerves of the gut as well as in many kinds of endocrine cells has been explained as the result of the common origin of the cells synthesizing and secreting chemical messengers. The most vigorously argued view is that early in embryological development, cells break from the neuroectoderm and, migrating to the gut and its appendages, become parents of hormone-secreting cells. These cells synthesize biologically active amines as well as peptides, and, because they have similar histochemical properties, they are called members of the APUD (*Amine Precursor Uptake and Decarboxylation*) series, a name that could well be discarded with no loss. A dissenting opinion is that some of the gut endocrine cells are of endodermal origin. Whatever its origin, a gut endocrine cell may give rise to an adenoma or a neuroma releasing one or more hormones. A gastrinoma secretes a family of similar hormones, and a neuroma may secrete vasoactive intestinal peptide. Several hormone-secreting tumors may develop simultaneously, producing one or another form of multiple endocrine adenomatosis. Thus, gastrinomas and hyperplasia or tumors of the parathyroid glands are often associated.

Trophic Actions*

Trophic actions of gastrin upon the oxyntic glandular mucosa, the mucosa of the small intestine, and the pancreas have been demonstrated in experimental animals. Synthesis, storage, and release of gastrin de-

---

*The adjective *trophic* comes from the Greek word *trophikos,* meaning nursing. The adjective *tropic* comes from *tropos,* a turn. Hence, a *trophic* hormone nourishes a tissue, whereas a *tropic* hormone turns, or changes, the response of its target organ. Some hormones of other endocrine systems, which were first called trophic hormones, are now called tropic ones, e.g., the adrenocorticotropic hormone. There has been some misguided effort, based on analogy. to call the trophic function of gut hormones a tropic one.

TABLE 2–5.—SOME CANDIDATE HORMONES OF THE DIGESTIVE TRACT

NAME	SUPPOSED FUNCTION
Bulbogastrone	In duodenal mucosa; postulated to account for inhibition of gastric secretion and motility by acid in the duodenum
Chymodenin	In duodenal mucosa; selectively stimulates pancreatic secretion of chymotrypsinogen
Duocrinin and enterocrinin	In intestinal mucosa; stimulate intestinal secretion; may be GIP and other known hormones
Enkephalin	In nerves of intrinsic plexuses; infusions stimulate acid secretion and increase gastric blood flow
Enterogastrone	A name applied to agent in extracts of intestinal mucosa inhibiting gastric motility and secretion; agent thought to be hormone meditation inhibitory action of fat; name now obsolete
Enteroglucagon	In duodenal mucosa; mimics glucagon; may be secretin, GIP, and other hormones
Gastrone	In gastric juice; inhibits gastric secretion and causes pyrexia; perhaps a bacterial toxin
Incretin	Postulated to account for insulin release during glucose absorption; may be GIP
Motilin	In gut enterochromaffin cells; released by lumenal alkalinization; initiates interdigestive myoelectric complex
Somatostatin	In hypothalamus and digestive tract; inhibits gastrin-stimulated and cholecystokinin-stimulated secretions and gastrin release (by paracrine action?)
Urogastrone	In human urine; almost identical with epidermal growth factor; inhibits gastric secretion
Villikinin	In duodenal mucosa; stimulates contraction of intestinal mucosal villi

pend upon frequent use of the digestive tract. Gastrin concentrations in plasma and antrum fall by more than two thirds when a rat is fasted, and they return to normal on refeeding. Normal levels are maintained on a liquid but not on a bulky, nonnutritive diet. When a rat or dog is completely supported by parenteral nutrition, the weights of the oxyntic mucosa, the small intestinal mucosa, and the pancreas fall, but the fall can be prevented by administration of gastrin, cholecystokinin, and secretin. On the other hand, administration of gastrin to the rat, dog, or man stimulates DNA and protein synthesis in the three tissues named but not in liver or muscle.

In man, there is an association between average plasma gastrin concentrations and acid secretory capacity of the stomach. There is a decrease in secretory capacity after antrectomy, which removes a major source of gastrin. On the other hand, a patient with duodenal ulcer often has a secretory capacity greater than normal and a supranormal gastrin release during digestion. A patient with chronic hypergastrinemia resulting from a gastrinoma usually has an acid secretory capacity far above normal. It is a reasonable but as yet unproved hypothesis that the trophic action of gastrin accounts for these facts.

Cholecystokinin, secretin, and caerulein are trophic agents for the pancreas. Hormones of the anterior pituitary also control gastrointestinal mucosal growth. The mucosa atrophies in hypophysectomized animals, and administration of hormones of the anterior pituitary partially restores the mucosa. It is likely that hormones of the anterior pituitary and releasing or release-inhibiting hormones of the hypothalamus influence gastrointestinal function in an as yet undetermined fashion.

REFERENCES
Bryant M.G., Bloom S.R.: Distribution of the gut hormones in the primate intestinal tract. *Gut* 20:653, 1979.

Dockray G.J.: Molecular evolution of gut hormones: Application of comparative studies on the regulation of digestion. *Gastroenterology* 72:344, 1977.

Erspamer V., Falconieri Erspamer G., Melchiorri P., Negri L.: Occurrence and polymorphism of bombesin-like peptides in the gastrointestinal tract of birds and mammals. *Gut* 20:1047, 1979.

Falasco J.D., Smith G.P., Gibbs J.: Cholecystokinin suppresses sham feeding in the rhesus monkey. *Physiol. Behav.* 23:887, 1979.

Grube D., Forssmann W.G.: Morphology and function of enteroendocrine cells. *Horm. Metab. Res.* 11:589, 1979.

Hubel K.A.: Secretin: A long progress note. *Gastroenterology* 62:318, 1972.

Johnson L.R.: Gastrointestinal hormones and their functions. *Ann. Rev. Physiol.* 39:135, 1977.

Miller L.J., Gorman C.A., Go V.L.W.: Gut-thyroid interrelationships. *Gastroenterology* 75:901, 1978.

Pearse A.G.E., Polak J.M., Bloom, S.R.: The newer gut hormones; cellular sources, physiology, pathology, and clinical aspects. *Gastroenterology* 72:746, 1977.

Rehfeld J.F.: Immunochemical studies on cholecystokinin. II. Distribution and molecular heterogeneity in the central nervous system and small intestine of man and dog. *J. Biol. Chem.* 253:4022, 1978.

Ryan G.P., Johnson, L.R.: Role of gastrin in mucosal DNA

synthesis after vagotomy and atropine in the rat. *Am. J. Physiol.* 235:E565, 1978.

Sidhu G.S.: The endodermal origin of digestive and respiratory tract APUD cells: Histopathologic evidence and a review of the literature. *Am. J. Pathol.* 96:5, 1979.

Smith G.: Satiety effect of gastrointestinal hormones, in Beers R.F., Jr., Bassett E.G. (eds.): *Polypeptide Hormones.* New York, Raven Press, 1980, pp. 413–420.

Straus E.: Radioimmunoassay of gastrointestinal hormones. *Gastroenterology* 74:141, 1978.

Walsh J.H., Grossman M.I.: Gastrin. *N. Engl. J. Med.* 292:1324, 1377, 1975.

Zimmerman E.G. (ed.): Conference on peptides of the brain and gut. *Fed. Proc.* 38:2286, 1978.

Zollinger R.M., Coleman D.W.: *The Influences of Pancreatic Tumors on the Stomach.* Springfield, Ill., Charles C Thomas, Publisher, 1974.

PART II

Motility

3

Chewing and Swallowing

FOOD IS TAKEN into the mouth in bites ranging from a few cubic millimeters upward; there it is broken up, mixed with saliva, and lubricated. The act of chewing is partly voluntary and partly reflex, but the act of swallowing, once initiated, is entirely reflex. From the mass in the mouth, a bolus of 5–15 cu cm is separated and projected into the pharynx, where it is engulfed by the muscles controlled by the stereotyped swallowing reflex. Pressure generated by muscular contraction moves the bolus through the hypopharynx past the hypopharyngeal sphincter into the esophagus. During swallowing, the passages from the pharynx into the nose and trachea are closed, and respiration is briefly inhibited. In man in the upright position, a liquid or slippery bolus moves rapidly down the esophagus under the influence of gravity; a solid or sticky bolus is slowly propelled by a peristaltic wave. The lower esophageal sphincter opens before the bolus, and the sphincter closes after it, preventing regurgitation.

Chewing

The extent to which a mouthful of food is chewed varies among species; in some, such as the dog and the cat, food is reduced in size only enough to permit swallowing. In man, particles are usually reduced to a few cubic millimeters, but the amount a mouthful is chewed depends on the nature of the food, habit, incidental conversation, and early training. The extent of chewing has a negligible effect on the chemical processes of digestion; but the bolting of unchewed food, especially when nervous tension is high, often causes epigastric distress.

The pattern of chewing is centrally determined. The jaw and face areas of the motor cortex occupy a large part of the precentral and postcentral gyri, and a large number of cortical efferent neurones passing through the internal capsule are responsible for initiating voluntary chewing movements. The chewing rhythm is generated in many sites in the lower brain stem, and the final efferent motoneurones are in the trigeminal nucleus.

The jaw is usually closed in opposition to gravity by tonic contraction of the masseter, the medial pterygoid, and the temporalis muscles. Gravity, pulling the jaw down, activates muscle spindle receptors, which, by a monosynaptic reflex through the trigeminal nucleus, stimulate contraction of the antigravity muscles of the jaw.

When food is in the mouth, pressure against the gums, the teeth, and the anterior part of the hard palate stimulates receptors and generates afferent impulses which inhibit motoneurones to the jaw-closing muscles and stimulate contraction of the digastric and lateral pterygoid muscles, opening the jaw. During the period of jaw opening there is a brisk afferent discharge from muscle spindles in the temporalis and masseter muscle which are being stretched. This discharge stops abruptly during the initial phase of rapid rebound contraction of these muscles, but it begins again and continues during grinding movements of the jaw. As food is broken by the teeth, Golgi tendon organs of the jaw-closing muscles send afferent impulses, and there is immediate inhibition of those muscles. This prevents teeth from banging suddenly together and terminates the chewing stroke. Tooth contact occurs in about 20% of the cycle, which lasts two thirds of a second.

Most persons chew a bolus on one side of the mouth at a time, but some divide the bolus to chew it on both sides simultaneously. Lateral movements of the jaw vary extensively in natural chewing. The intercuspal position of the teeth is passed at each stroke, but the side from which it is approached varies at random. Consequently, there is a random shift of the bolus in the mouth.

The pressure exerted on a lower molar during reflex chewing of cooked meat is 3.9–15.7 newtons/mm², and the vertical load taken by a single tooth is 70–120 newtons. Very much higher pressures are exerted when hard foods are chewed. Efficiency of chewing depends upon the brittleness and consistency of the food chewed. Ninety-four percent of a mouthful of pecans will pass through a 10-mesh screen after 20 chews, but nearly 200 chews are required to reduce pickled cauliflower to the same fineness.

Mouth Movements in Swallowing

The digestive tract moves its contents by creating a pressure gradient. At the beginning of a swallow, the tip of the tongue separates a bolus from the rest of the material in the mouth and brings it into the midline between the anterior portion of the tongue and the hard palate. The jaw shuts, and the soft palate is elevated. The forepart of the tongue is pressed firmly against the roof of the mouth and, together with the closed lips, seals off the anterior portion of the mouth. The palate and the contracted palatopharyngeal muscles form a partition between the mouth and the nasal cavity, which prevents pressure generated in the mouth from being dissipated through the nose. When selective paralysis of motor components occurs, as in poliomyelitis, the bolus is regurgitated into the nasopharynx. The bolus, which lies in a groove in the midportion of the tongue, is propelled into the oral pharynx as the tongue rolls backward on the hyoid bone, pressing against the palate progressively more posteriorly. Respiration is now briefly inhibited, the larynx is abruptly raised, and closure of the glottis cuts off the laryngeal airway. As the larynx rises, the epiglottis is thrust into the path of the oncoming bolus. The bolus tilts the epiglottis backward; and in response to further upward movement, the epiglottis is retroverted until it hangs slantwise over the closed glottis. Although the epiglottis acts as a lid over the closed glottis, its presence is not necessary to prevent food from entering the trachea. Firm closure of the glottis seals the airway, and this prevents aspiration of food even if the epiglottis is surgically removed. Driven by a pressure difference of 4–10 mm Hg generated by movement of the tongue, the bolus pours over and around the epiglottis. The hypopharyngeal sphincter, formed by the cricopharyngeal muscles, is opened by upward movement of the structures to which it is attached, and the bolus moves with a wedge-shaped leading edge through the hypopharynx into the upper esophagus. After it passes the

level of the clavicle, the larynx descends, the upper respiratory passage and the glottis open, the tongue moves forward, and respiration resumes (Fig 3–1). The whole of the pharyngeal part of swallowing occurs within 1 sec.

Mouth Movements in Drinking

When a liquid is drunk, by sucking or through a straw, subatmospheric pressure is generated in the mouth by retraction of the tongue without breaking the lingual-palatine seal. This fills the mouth; then, suddenly, the posterior portion of the tongue is depressed, allowing fluid to run into the pharynx. The mandible is raised, closing the jaw firmly, and the swallowing movement begins with a sweeping posterior movement of the tongue.

The champion beer drinker, who steadily pours the contents of a bottle into his mouth, does not, contrary to appearances, down the beer by opening a passage directly into his esophagus. He elevates and extends his jaw to increase the capacity of his mouth and pharynx, and he keeps the level of liquid in his mouth even with his lower lip. Before any beer enters the pharynx, the hyoid bone, larynx, and pharynx drop down as in yawning, and the dorsum of the tongue is wedged against the descended soft palate. The glottis is closed, breathing is suspended, and then the dorsum of the tongue moves forward and the soft palate elevates, allowing a flood of beer to flow into the pharynx, where it is held by the closed hypopharyngeal sphincter. The tongue now moves like a cam at the rate of once a second to force fluid through the sphincter and into the esophagus. This pumping process continues until the mouth is empty; then an ordinary swallow travels from mouth to stomach to sweep away the foam (Fig 3–2).

Air Swallowing

Air normally present in the pharynx is trapped when a bolus is delivered into the pharynx by the tongue. Most of it passes into the trachea the instant before the glottis closes, but often a small amount of air precedes the bolus down the esophagus and into the stomach. Most swallowed air passes no farther than the stomach and is soon expelled by belching, but some does pass through the intestines. Air in frothy saliva or in food, particularly meringues and soufflés, is swallowed. In fact, the "dry swallow"

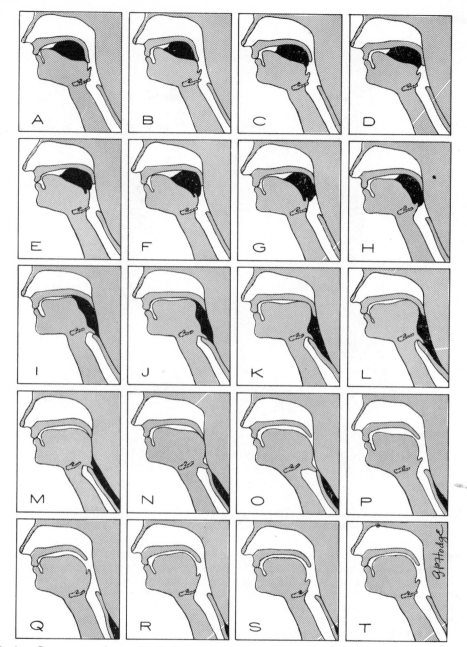

Fig 3–1.—Sequence of events during swallowing. **A** and **B,** the soft palate forms a partition extending to the base of the tongue. **C, D, E,** and **F,** the soft palate is elevated to obstruct the nasopharynx as the bolus moves backward over the tongue. **G, H,** and **I,** the bolus tilts the epiglottis backward. From **H** through **R** the glottis, not shown, cuts off the laryngeal airway. **J** and **K,** the bolus passes smoothly over the convex epiglottis, and the tongue moves backward as a piston. **K,** the bolus is slightly delayed at the hypopharyngeal sphincter. **O** and **P,** the soft palate relaxes, and the epiglottis ascends. **Q–T,** the bolus moves down the esophagus. The entire sequence occupies 1.3 sec. (Adapted from Rushmer R.F., Hendron J.A.: *J. Appl. Physiol.* 3:622, 1951.)

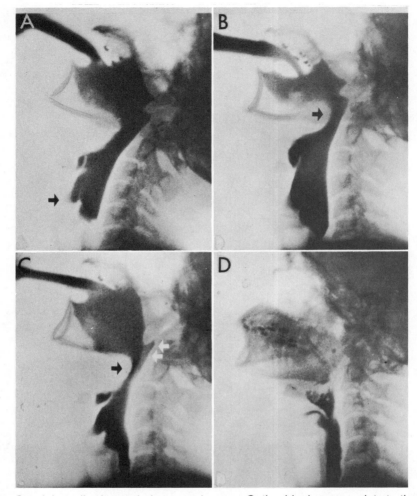

Fig 3–2.—Special swallowing technique used by competitive beer drinker. **A,** fluid is being poured into the wide-open mouth. The oropharynx is open, permitting the laryngeal pharynx to fill; the laryngeal vestibule is distended *(black arrow).* **B,** the dorsum of the tongue bulges backward *(black arrow);* fluid in the mouth is forced down into the already distended laryngeal pharynx. **C,** the *black arrow* points to the nearly completed, cam-like action of the tongue, the hypopharyngeal sphincter is open, and fluid has passed into the esophagus. **D,** the final swallow occurs with the mouth closed, the larynx elevated, and the epiglottis folded over. (From Ramsey G.H., et al.: *Radiology* 64:498, 1955.)

consists of air and saliva. During periods of excessive salivation, such as those accompanying nausea, the large volume of air swallowed with saliva may cause distress. As much as 500 cc may be swallowed with a meal. Air may also be voluntarily swallowed by trapping it with the tongue while the lips are closed.

Persons lacking a larynx may learn to fill the esophagus with air, which is then expelled slowly as the basis of a form of speech. On the other hand, those with respiratory paralysis can be trained to swallow air with the glottis open, a form of respiration called *glossopharyngeal* or *frog breathing*. The first breath of life may be swallowed, for in five of eight newborn babies observed by successive radiophotographs, the initial inflation of the lungs was found to follow first distention and then compression of the pharyngeal cavity.

The Swallowing Reflex

Swallowing may always be a reflex act. Voluntary efforts to swallow are ineffective unless there is something, if only a few milliliters of saliva, to swallow. Then voluntary movement of the tongue initiates the reflex by throwing saliva against areas of the mouth and pharynx fitted with a large number of receptors from which impulses travel to the swallowing center in the glossopharyngeal nerve and the superior laryngeal branch of the vagus. When afferent impulses arrive at the swallowing center in the medulla they evoke a complex, stereotyped act of swallowing lasting as long as 9 sec in which there is efferent discharge through the six nuclei and the motoneurones listed in Figure 3–3. In addition, afferent impulses in the superior laryngeal nerve travel through relay stations in the pons and thalamus to the frontal cortex, and the swallowing reflex can be facilitated through efferent pathways from the frontal cortex to the medullary swallowing center.

When the swallowing center is activated, interneurones within it fire in an ordered sequence through the efferent pathways listed in Figure 3–3 so that there is a pattern of sequential contraction of buccopharyngeal muscles lasting about 0.5 sec. The muscles involved are poorly provided with proprioceptors, and they contain no gamma-efferent system. Feedback from the muscles plays no significant part in determining the sequence or strength of their contraction.

At the end of the pharyngeal phase other interneurones in the swallowing center begin to discharge, some immediately and some after long latent periods. Their output, lasting many seconds, is responsible for the peristaltic wave moving slowly down the esophagus. Messages reaching the swallowing center by way of afferent nerves from the esophagus quantitatively modify the peristaltic wave; the larger the bolus swallowed, the stronger is the peristaltic contraction. At each stage of the interneuronal activity responsible for buccopharyngeal contraction and esophageal peristalsis neurones which are to be responsible for a subsequent phase of activity are inhibited.

The rigidly ordered pattern of discharge is illustrated in Figure 3–4. Five muscle groups in the pharynx contract concurrently for 250–500 msec. This generates high pressure behind the bolus and propels it toward the hypopharyngeal sphincter. At rest, the esophagus is closed at the pharyngoesophageal junction by passive elasticity of surrounding structures; the cricopharyngeal muscle is not usually contracted. This muscle, however, is easily activated reflexly by manual displacement or by the presence of a foreign object in the lumen. If the muscles surrounding the junction are initially contracted, they are reflexly inhibited during the early part of swallowing as the bolus approaches and passes through the sphincter. The sphincter is opened by contraction of the muscles that raise the larynx and cricoid cartilage. After the bolus has passed through the sphincter, the muscles belonging to it contract, firmly shutting the sphincter. Pressure within the sphincter reaches 120 mm Hg in the anterior-posterior direction and 60 mm Hg in the lateral direction. The sphincter remains firmly closed until the peristaltic wave in the esophagus, beginning just below the sphincter, has carried the bolus onward, thus preventing regurgitation.

Structure of the Esophagus

In man, the upper third of the muscular wall of the esophagus consists of striated muscle arranged as a layer of longitudinal fiber bundles surrounding a layer of circular fiber bundles. Between these muscle layers and the stratified squamous epithelium forming the lining of the esophagus is a thick submucous elastic and collagenous network. This, together with the thick muscularis mucosae, throws the epithelium into folds whose surfaces appose one another, obliterating the lumen. When swallowing occurs, the folds are smoothed out in the part of the esophagus occupied by the bolus.

The muscular wall of the lower two thirds of the esophagus is smooth muscle arranged in two layers, the inner being roughly circular but containing many helical, elliptical, or oblique bundles and the outer being roughly but irregularly longitudinal. Transition between striated muscle above and smooth muscle below generally occurs in the middle third, and it is higher for the inner layer than for the outer. In some animals—dogs, mice, and elephants, for example—the whole esophageal muscle is striated. There is no muscular structure that forms an anatomical sphincter separating the esophagus from the stomach; neverthe-

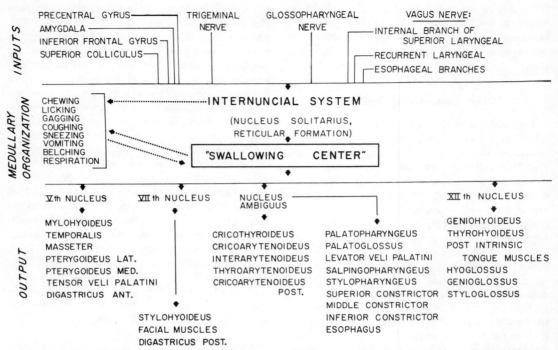

Fig 3–3.—Outline of afferent and efferent systems involved in swallowing, and the requirement of interaction between swallowing and other synergies. (From Doty R.W.: Neural organization of deglutition, in Code C.F. (ed.): *Handbook of Physiology: Sec. 6. Alimentary Canal,* vol. 4. Washington, D.C.: American Physiological Society, 1968.)

less, the lower end of the esophagus does behave like a physiological sphincter.

At rest, the upper seven eighths of the esophagus, below the pharyngoesophageal junction, is relaxed. When an esophagoscope is passed downward, the esophagus is found to be closed about 2 cm above the diaphragmatic hiatus and 3–4 cm above the cardia. There is sometimes a slight thickening of the esophageal muscle at this point, which marks the upper boundary of the physiological lower esophageal sphincter. The lower one eighth of the esophagus, between the constriction and the cardia, is sometimes called the esophageal vestibule.

Motor Innervation of the Esophagus

Striated muscle forming the upper third of the esophagus is innervated by motoneurones in the glossopharyngeal and vagus nerves. These motoneurones end in motor endplates upon striated muscle fibers.

The smooth muscle of the rest of the esophagus is supplied by the vagus nerve in characteristic autonomic fashion: preganglionic fibers end in synapse with ganglion cells in the intrinsic plexuses. Postganglionic fibers from the plexuses innervate the smooth muscle cells. Preganglionic and postganglionic fibers are cholinergic and excitatory; parasympathomimetic drugs stimulate contraction, and atropine prevents it. When swallowing occurs, the swallowing center sends out through the motoneurones and the vagal fibers a sequence of efferent discharges to progressively distal segments of both striated and smooth muscle segments of the esophagus, which result in peristalsis. However, there is a mechanism of intrinsic control within the plexuses which can produce a peristaltic wave independent of central control.

Between swallows, there is pressure in the lower esophageal sphincter as the result of contraction of its smooth muscle. Contraction is in part intrinisic, occurring in the absence of any extrinsic nervous influ-

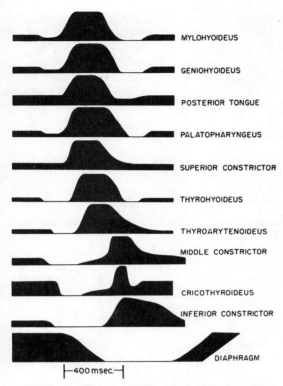

MYLOHYOIDEUS

GENIOHYOIDEUS

POSTERIOR TONGUE

PALATOPHARYNGEUS

SUPERIOR CONSTRICTOR

THYROHYOIDEUS

THYROARYTENOIDEUS

MIDDLE CONSTRICTOR

CRICOTHYROIDEUS

INFERIOR CONSTRICTOR

DIAPHRAGM

├─400 msec.─┤

Fig 3–4.—Summary of outflow of the swallowing center of the dog, recorded electromyographically. Height of line for each muscle indicates relative intensity of activity, but contours of rise and fall are schematic. The upper five muscles represented are part of the pharynx; these contract almost simultaneously, creating high pressure above the bolus. At the same time, the lower five muscles shown, which belong to the hypopharynx and hypopharyngeal sphincter, relax. After the bolus has passed, these muscles contract, firmly shutting the sphincter. The diaphragm relaxes while the bolus is passing through the hypopharynx. Pressure changes caused by this muscular activity are shown in Figure 3–7. (From Doty R.W., Bosma J.F.: *J. Neurophysiol.* 19:44, 1956.)

ence. In addition, contraction is mediated by cholinergic postganglionic impulses from the intrinsic plexuses. Excitatory fibers are also contained in the sympathetic innervation of the lower esophageal sphincter. About one third of the resting contraction can be attributed to the action of norepinephrine upon alpha receptors in the sphincter's smooth muscle

cells. Furthermore, sympathetic impulses facilitate contraction of the lower esophageal sphincter by stimulating cholinergic myenteric neurones and by inhibiting intrinsic inhibitory neurones.

During swallowing the lower esophageal sphincter relaxes as the result of both the suppression of excitation and of inhibitory impulses carried in cervical vagal fibers and in fibers of the intrinsic plexuses innervating the sphincter. The inhibitory action of the plexuses can be independent of vagal innervation.

The nature of the inhibitory mediators has not been unequivocally established, and there may be many of them. It is likely that vagal inhibitory fibers release 5-hydroxytryptamine (5-HT) at their synapses with inhibitory postganglionic cells and that terminal postganglionic fibers inhibit smooth muscle of the sphincter by releasing adenosine triphosphate (ATP) or a closely related purine. Vagal fibers to the sphincter contain vasoactive intestinal peptide (VIP). That peptide inhibits the sphincter, and the inhibitory nerves may be, in the jargon of the day, VIP-ergic.

Both sympathetic and parasympathetic stimulation and their mediators, norepinephrine and acetylcholine, cause the muscularis mucosae of the esophagus to contract.

Methods of Measuring Intraluminal Pressure

Pressure in the digestive tract can be measured by passing into it a small tube connected to a pressure-sensing device, such as a strain gauge. The tube, which is usually made of polyethylene or polyvinyl plastic, has an outside diameter of less than 2.5 mm and an inside diameter of about 1.5 mm. The tube has an opening 1.5 mm in diameter on its side just proximal to its tip, and the tip is sealed. The tube is filled with fluid, and fluid is forced through it at a rate on the order of 50 μl/min. Continuous flow of fluid through the tube is necessary to keep the distal opening of the tube unobstructed and to permit accurate recording of the pressure in the sphincter. Often several tubes are bound into a bundle with their openings radially arrayed at the same distance from the tip or separated by known distances. Four such tubes with openings separated by 5 cm were used to obtain the record shown in Figure 3–7. In some instances, such as the arrangement shown in Figure 3–5, the miniature transducer itself is placed at the tip of the tube. In other instances, the miniature gauges are arranged

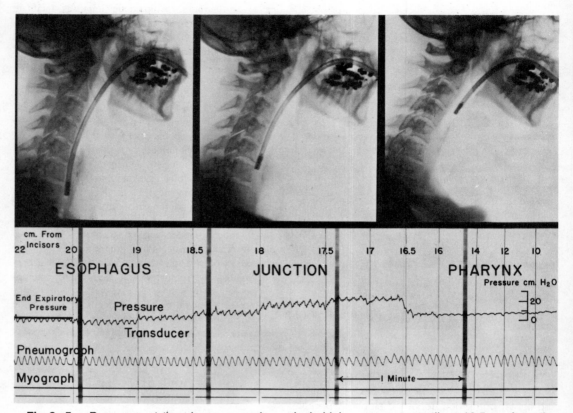

Fig 3–5.—Pressures at the pharyngoesophageal junction recorded by means of a transducer whose position at the tip of the tube is shown in the roentgenograms. As the transducer is withdrawn from the esophagus, it records increasingly higher pressures until, at 16.5 cm from the incisors, the upper limit of the superior esophageal sphincter is passed. (From Fyke F.E., Code C.F.: *Gastroenterology* 29:24, 1955.)

radially in order to measure lateral forces from several directions. When multiple tubes or many transducers are used, the diameter of the complete device may be 6 mm or more in diameter, and the process of measurement, particularly within narrow sphincters, may itself disturb the function to be measured. For example, the presence of the tube and transducer shown in Figure 3–5 within the hypopharyngeal sphincter probably stimulated reflex contraction of the muscles surrounding the sphincter, with the result that the sphincter was not truly at rest.

Reflex stimulation occurring during measurement may be responsible for differences in magnitude of pressure for the same sphincter reported from different laboratories, and the absolute magnitudes quoted in this book are not to be taken seriously.

Resting Pressure in the Esophagus

Pressure in the resting midesophagus is equal to intrathoracic pressure because the esophageal muscle is flaccid. In fact, respiratory physiologists use esophageal pressure as a measure of intrathoracic pressure. During quiet breathing the pressure is 5–10 mm Hg below ambient pressure. It falls on inspiration and rises on expiration.

Pressure at each end of the esophagus is higher than that within the esophagus. Pressure in the mouth and pharynx, when these are at rest, is approximately atmospheric; and were not the upper end of the esophagus closed by the upper esophageal sphincter, air would flow from the mouth into the esophagus until esophageal pressure became equal to atmospheric pressure.

Intragastric pressure prevails beyond the lower end of the esophagus. This is usually 5–10 mm Hg above ambient pressure; it rises in inspiration and falls on expiration. Because intragastric pressure is higher than intraesophageal pressure, gastric contents would flow from stomach to esophagus unless their junction were closed by the lower esophageal sphincter.

The Lower Esophageal Sphincter at Rest

If the open tip of the pressure-recording tube is passed all the way through the esophagus into the stomach, it registers the basic intragastric pressure of 5–10 mm Hg above ambient pressure (Fig 3–6). Superimposed on this basic pressure are variations caused by respiratory movements: quiet inspiration causes an increase in intragastric pressure of about 5 mm Hg, and quiet expiration causes a fall toward ambient pressure.

As the tip is withdrawn upward from the stomach into the subdiaphragmatic esophagus, which is 2 cm long in a normal man, the baseline pressure rises until, at a point close to the hiatus of the diaphragm through which the esophagus passes into the thorax, it is higher than the intragastric pressure. Above this point it falls until it is below ambient (and therefore below intragastric) pressure. When the subject is in end-expiration, the elevation in pressure is first detected 3 cm below the diaphragm and ends 3 cm above it. When the subject is in end-inspiration, the zone of increased pressure begins 3 cm below the diaphragm and ends 1 cm above it. Thus, a zone of elevated pressure 4–6 cm long separates the stomach and esophagus and acts as a barrier between the two. This is the physiological lower esophageal sphincter, and it is not identical with any anatomical structure at the cardia or with the hiatus.

In normal persons, the resting pressure in the lower esophageal sphincter varies from minute to minute; it averages 15–35 mm Hg and remains within that range in repeated observations on the same subject over a 6-month period. In persons with esophageal reflux, pressure in the sphincter is between 1 and 10 mm Hg.

When the tip of a pressure-sensing catheter is in the stomach, changes in pressure coincident with respiration are superimposed upon the baseline pressure. These are positive upon inspiration and negative upon expiration. As the tube is withdrawn into the esophagus, the respiratory excursions invert, becoming negative upon inspiration and positive upon expiration. The respiratory-pressure inversion point occurs at variable points along the lower esophageal sphincter, and it is not a reliable landmark of the hiatus of the diaphragm.

Chemical Mediators and the Lower Esophageal Sphincter

Gastrointestinal hormones are believed to share control of the lower esophageal sphincter, but it is impossible to give a definitive account of how control is exerted. Almost every known mediator has been tested for its effect upon the sphincter in man or an experimental animal. Many have been found, in pharmacological doses, to stimulate or to inhibit by direct action upon the sphincter muscle or to stimulate or inhibit through neural mechanisms. For example, 5-HT directly stimulates sphincter muscle, indirectly stimulates by activating cholinergic neurones, and indirectly inhibits by stimulating nonadrenergic inhibitory neurones. In normal circumstances, cholecystokinin causes relaxation of the sphincter by stimulating inhibitory, nonadrenergic, noncholinergic nerves. However, cholecystokinin directly stimulates muscle of the sphincter to contract. Consequently, when the inhibitory neurones are absent, as in achalasia, cholecystokinin paradoxically increases sphincter pressure.

The question of whether gastrin is a normal determinant of sphincter pressure is unsettled. Pharmacological doses of gastrin definitely increase sphincter pressure. Intravenous infusion of gastrin at a rate which produces submaximal acid secretion, and is therefore presumably physiological, causes a dose-related rise in sphincter pressure and enhances the sphincter's response to a cholinergic drug. Sphincter pressure rises 4–8 mm Hg during digestion of a meal, but if the meal is acidified to pH 1.2 before ingestion, a process that inhibits release of gastrin, the rise does not occur. On the other hand, injection of somatostat-

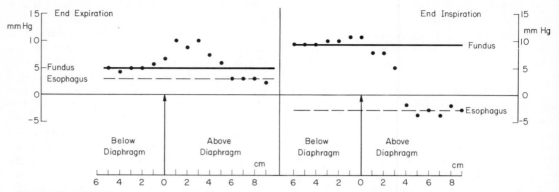

Fig 3–6.—Resting pressures at the gastroesophageal junction in an adult man, recorded by passing an open-tip catheter connected with a transducer into the stomach and slowly withdrawing it. The position of the tip was observed fluoroscopically, and the position of the diaphragm was determined by inversion of respiratory excursions on the record. End-expiratory pressures are on the left; the esophageal pressure is above atmospheric. End-inspiratory pressures are on the right; esophageal pressures are subatmospheric. At both stages of respiration, there is a zone of pressure higher than that in the body, or fundus, of the stomach, and the zone moves upward in expiration. (Adapted from Vantrappen G., et al.: *Gastroenterology* 35:592, 1958.)

in inhibits a rise in pressure following ingestion of glycine or alkalinization of the stomach, both of which cause release of gastrin, but somatostatin has no effect upon basal sphincter pressure or its response to pentagastrin. This suggests that the rise in sphincter pressure following a meal is only partially, if at all, mediated by gastrin. Finally, many investigators have failed to find a correlation between plasma gastrin concentration within physiological limits and sphincter pressure.

Pharyngeal Pressures During Swallowing

The upper esophageal sphincter, which separates the pharynx from the esophagus, is closed by the elasticity of its tissues and by minimal contraction of the cricopharyngeus and the inferior pharyngeal constrictor muscles. Closing pressure when there is nothing in the lumen is unknown. When a probe 6 mm in diameter is pulled through the sphincter, anterior-posterior pressure is found to be about 120 mm Hg and left-right pressure about 60 mm Hg. These pressures are artifactually high.

Pressure in the pharynx rises abruptly during the act of swallowing to a maximum of about 100 mm Hg as the bolus passes through it (Fig 3–7). The pressure elevation lasts only about 0.5 sec. A fraction

of a second before the wave of high pressure reaches the pharyngoesophageal junction, the sphincter is opened by contraction of muscles raising the cricoid cartilage. If muscles of the sphincter are contracted, they are reflexly inhibited as the pressure wave approaches the sphincter (see Fig 3–4). The bolus passes through the open sphincter into the esophagus. The sphincter remains open only during the half-second that high pressure exists in the pharynx above it. Then the sphincter closes, and within another half-second its pressure rises to 90–100 mm Hg, or about twice its resting pressure. While pressure in the sphincter is highest, a peristaltic wave begins at the upper end of the esophagus, generating a pressure of about 30 mm Hg. The higher pressure in the sphincter prevents reflux from esophagus to pharynx.

After the peristaltic wave has passed farther down the esophagus and pressure in the upper esophagus is again low, the sphincter's pressure subsides, over 2 or more sec, to its resting level. Thus, the sphincter is closed before swallowing begins, relaxes, and opens as the bolus is propelled toward it, remains open as the bolus passes, snaps shut again, and maintains a high pressure, preventing reflux until esophageal peristalsis has cleared the upper esophagus. Pressure in the upper esophageal sphincter is reflexly augmented by acid in the lumen or by distention of the upper 5 cm of the esophagus.

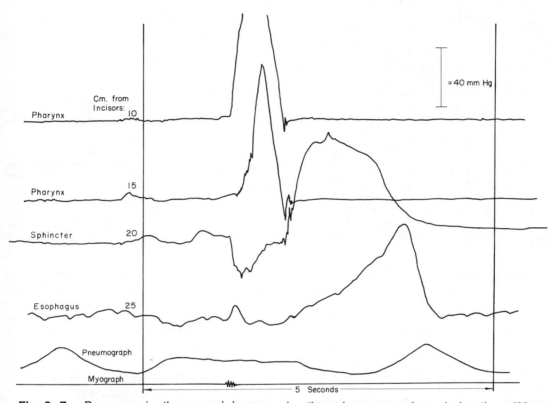

Fig 3–7.—Pressures in the normal human pharynx and upper esophagus during a swallow. Four transducers are in place at distances from incisors given on the left (see Fig 3–5). The beginning of the swallow is signaled by the myograph, which records action potentials in the jaw muscles. Respiration is interrupted. Pressure in the pharynx (10-cm trace) rises to a high level, and this increased pressure travels quickly down the pharynx (15-cm trace). The third transducer, in the pharyngoesophageal junction (20-cm trace), shows that the sphincter relaxes at the time the pharyngeal pressure is high. As soon as the pressure in the pharynx has fallen to baseline level, pressure in the sphincter rises and remains high for more than a second. While the sphincter is tightly closed, the peristaltic wave (25-cm trace) begins to move slowly down the esophagus. (Courtesy of C.F. Code.)

Esophageal Peristalsis

There are three kinds of esophageal peristalsis. *Primary* peristalsis is the wave of contraction moving down the esophagus which usually, but not invariably, follows the oral and pharyngeal swallowing movements. When the pharyngeal phase of swallowing is rapidly repeated, as in drinking a glass of water, a primary peristaltic wave occurs only after the last swallowing movement. *Secondary* peristalsis is a similar wave originating just below the upper esophageal sphincter without any antecedent mouth or pha-ryngeal movements. *Tertiary* peristalsis is the wave occurring only in the smooth muscle part of an esophagus that has been vagotomized.

The peristaltic wave beginning just below the upper esophageal sphincter pushes a bolus ahead of it. The pressure generated ranges from 30 to 120 mm Hg. The force acting on a bolus varies with bolus size; when the bolus is 6 mm in diameter, the force exerted will lift 30 gm, and when it is 13 mm in diameter, the force will lift 125 gm. The pressure generated by a peristaltic wave rises to a peak in 1 sec, lasts at the

Fig 3–8.—Pressures in the lower part of the normal human esophagus during a swallow. Three transducers are in place at distances from the incisors given on the left. The beginning of the swallow is signaled by the myograph, which records action potentials in the jaw muscles. Almost immediately, the pressure in the lower esophageal sphincter (39-cm trace) falls, and it remains low for about 8 sec. In the meantime, the peristaltic wave is passing down the lower esophagus (29- and 34-cm traces). After the pressure in the lower esophagus has fallen to resting level, the pressure in the sphincter rises, and it remains elevated for about 10 sec. (Adapted from Fyke F.E., Jr., Code C.F., Schlegel, J.F.: *Gastroenterologia* 86:135, 1956.)

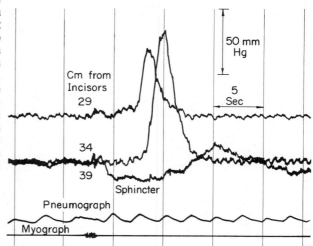

peak 0.5 sec, and subsides in about 1 sec (Fig 3–8). The whole course of rise and fall may occupy one point in the esophagus for 3–7 sec. The length of esophagus contracting at any one instant is 10–30 cm. The mean velocity of peristalsis in the upper esophagus is 2.9–3.3 cm sec^{-1}, and it falls to 2.2 cm sec^{-1} just above the lower esophageal sphincter. Therefore, in an adult man the peristaltic wave is completed about 9 sec after swallowing begins. Because the longitudinal muscle contracts, the esophagus shortens during peristalsis.

The peristaltic wave is governed by both the sequential activation of its muscle by efferent nerves firing in a pattern determined by the swallowing center and by the program built into the nerves of its intrinsic plexuses. Continuity of the esophagus is not essential, for if its muscle, but not its nerve supply, is cut across, the wave begins at the end of the esophagus below the cut as it dies out above the cut. Afferent impulses from receptors within the esophageal wall are not essential for progress of the coordinated wave of peristalsis. Nevertheless, impulses carried to higher centers by afferent nerves from the esophagus modify the reflex. This accounts for the adjustment of peristaltic force to the size of the bolus.

Secondary peristalsis is the result of afferent impulses from the esophagus. If the normal esophagus is distended at any point, as by a piece of peanut butter sandwich left behind from previous swallows, afferent nerves are stimulated, and contraction begins with forcible closure of the hypopharyngeal sphincter

and sweeps downward. Secondary peristalsis occurs without any movements of the mouth or pharynx and without the subject's awareness of it.

In old age, the pressure generated by esophageal peristalsis is diminished, but its velocity and duration are the same as in a young person.

Lower Esophageal Sphincter During Swallowing

When a normal man swallows, the lower esophageal sphincter relaxes. Relaxation begins at the upper edge of the sphincter 1.3 sec after the start of swallowing; at this time, esophageal peristalsis is just beginning at the top of the esophagus. A wave of relaxation moves down the sphincter and continues into the upper part of the stomach. The pressure barrier between stomach and esophagus is not completely abolished by relaxation of the sphincter, for the lowest pressure occurring at the hiatus during swallowing is slightly above intragastric pressure. This residual barrier prevents reflux during inspiration, when the pressure difference between stomach and esophagus is greatest. Then, about 6 sec after swallowing has begun, a wave of contraction starts at the top of the sphincter and reaches its maximal force 2–3 sec later, as esophageal peristalsis dies out in the muscle just above it. Contraction moves downward through the sphincter at the rate of 0.6 cm/sec, and its force diminishes as it moves. Pressure generated by contraction is greatest above the hiatus, where 20–30 mm

Hg above resting pressure may be reached. This high pressure dies out over 5–10 sec. The 39-cm trace in Figure 3–8 was recorded from a part of the sphincter showing this behavior: early and prolonged relaxation followed by late and equally prolonged contraction. A record made from the sphincter below the hiatus would show only relaxation followed by return to resting pressure. Thus, the barrier between stomach and esophagus is reduced but not obliterated very early in swallowing, remains low until the esophagus has completed its propulsive wave, is raised high after the wave has died out, and slowly falls to its resting level.

During rapidly repeated swallowing, the lower esophageal sphincter relaxes at the beginning of swallowing movements, and it remains relaxed until the single peristaltic wave that follows the last swallow has reached it.

Bolus Movement During Swallowing

The actual movement of the bolus effected by swallowing depends on its consistency and the position of the subject. When a man in the upright position swallows water, it is shot rapidly into the esophagus by buccopharyngeal movements, and in the esophagus the head of the column reaches the lower esophageal sphincter within 1 sec (Fig 3–9). Descent of the bolus is aided by gravity. It is sometimes held up briefly if the sphincter has not yet completely relaxed. If a stethoscope is placed on the abdomen over the lower esophageal sphincter of an erect subject, a gurgle is heard 5–8 sec after he swallows water; water waiting at the bottom of the esophagus is pushed through the sphincter by the slow peristaltic wave. Water also moves along the esophagus of a man in the horizontal position much faster than the peristaltic wave, apparently as the result of momentum derived from the force of pharyngeal contraction. When a man in the head-down position swallows water, the esophagus fills, the column of water being supported by the closed pharyngoesophageal sphincter. The weak propulsive force of esophageal peristalsis is incapable of raising the whole column of fluid into the stomach. On repeated swallowing in this position, more water is forced into the esophagus by presssure developed in the pharynx, while an equal volume at the top of the column is delivered into the stomach.

With the subject in any position, a pasty mass moves more slowly through the pharynx and esopha-

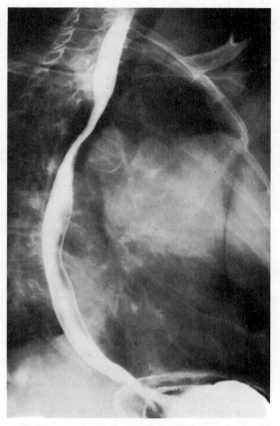

Fig 3–9.—Lateral oblique view of the normal human esophagus during swallowing of a thin mixture of barium sulfate and water. The subject is upright; and the mixture, falling by gravity, fills the whole esophagus well ahead of the peristaltic wave. (Courtesy of F.J. Hodges and J.N. Correa.)

gus, taking, on the average, about 5 sec to reach the stomach. Solid masses move still more slowly, and small particles may be left behind in the esophagus to be gathered up by a succeeding peristaltic wave.

Drinking water at 0.5–3 C slows or abolishes esophageal peristalsis. Nevertheless, the lower esophageal sphincter relaxes and contracts with each swallow, although no peristalsis occurs above it. Sphincteric relaxation is delayed, and water often pools above the sphincter. Relaxation, when it occurs, is prolonged and is followed by exaggerated contraction. In contrast, drinking water at 58–61 C accelerates movements of the esophagus as compared with those occurring when water at 23–26 C is drunk. Rate of propagation of peristalsis is increased, the waves have a shorter duration, relaxation of the lower esophageal sphincter is briefer, and the subsequent contraction is feebler.

Reflux into the Esophagus

During quiet respiration, the intragastric pressure is about 10 mm Hg higher than intraesophageal pressure, but reflux does not occur. This is because pressure in the lower esophageal sphincter is higher than intragastric pressure and because the sphincter resists distention. However, reflux does not occur when the lower esophageal sphincter relaxes as part of the orderly peristalsis characteristic of swallowing.

As the stomach fills during a meal, resting pressure in the lower esophageal sphincter rises. The concurrent rise in plasma gastrin or in an unidentified hormone may be in part responsible. Pressure in the sphincter rises when the gastric mucosa of the cardia just distal to the sphincter is irrigated with acid, and this local reflex may contribute as gastric contents gradually become acid.

External abdominal compression raises intragastric pressure, but it does not alter the gradient between the subhiatal portion of the sphincter and the stomach. The reason is that intra-abdominal pressure is transmitted equally to the subdiaphragmatic portion of the gastroesophageal sphincter and to the stomach. Although the absolute pressure in the two areas rises, the pressure in the sphincter remains higher than that in the stomach. The same mechanism prevents reflux when the pressure difference between stomach and esophagus is raised by forced inspiration on a closed glottis. In deep inspiration there is also a steep rise in pressure in a narrow subhiatal zone of the sphincter, probably caused by pinch-cock action of the contracted diaphragm. Contraction of the diaphragm may also increase pressure in the sphincter during quiet inspiration.

Some persons believe that when intra-abdominal pressure rises, there is, in addition to the increment in pressure resulting from mechanical transmission of intra-abdominal pressure to the sphincter, a further increment as the result of a cholinergic reflex. This mechanism is said to be effective if the sphincter is entirely in the thorax, as it is in hiatus hernia. Consequently, patients with hiatus hernia who have a normal response to an increase in intra-abdominal pressure do not have reflux when intra-abdominal pressure

rises. On the other hand, in some persons the increment in lower esophageal pressure is less than the increment in intra-abdominal pressure; in them, reflux occurs whether the sphincter is in its normal position or is in the thorax.

Prevention of reflux has also been attributed to valve-like mucosal folds at the cardia, to the acute angle the stomach sometimes makes with the esophagus and to contraction of oblique sling-like fibers of smooth muscle at the cardia. None of these is important in man.

The stomach can accept a large volume of food without an appreciable rise in pressure; but a heavy meal, large amounts of swallowed gas, or carbon dioxide generated by the taking of sodium bicarbonate may raise intragastric pressure and bring the lower edge of the gas bubble normally present in the gastric fundus below the level of the cardia. Then, when the sphincter relaxes during swallowing, eructation into the esophagus follows, and esophageal pressure rises abruptly. Carminatives, the food seasoners, and ingredients of after-dinner liqueurs that produce the sensation of warmth and facilitate eructation reduce intrasphincteric pressure and allow reflux. If the subject is in the supine position, gastric contents quickly run into the esophagus. Afferent nerves are stimulated, with the result that a secondary peristaltic wave begins at the pharyngoesophageal sphincter and sweeps down the esophagus to empty it again.

To study belching, large volumes of air were run through a tube into the stomachs of normal human subjects. Intragastric pressure rose 4–7 mm Hg during entry of the first 200–600 cc, but it did not rise further until the total volume was about 1,600 cc. Sooner or later (5–157 sec) after the pressure plateau was reached, there was sudden reflux of air into the esophagus, and intragastric and esophageal pressures became equal. Then, if the subject did not belch, secondary peristalsis stripped the esophagus and returned the air to the stomach. Belching occurred in 18 subjects during reflux of air into the esophagus. At the moment of belching, pressure in the esophagus, stomach, and abdomen rose abruptly, apparently following contraction of somatic muscles, and gas was expelled from the common gastroesophageal cavity. Pressure in the esophagus remained equal to intragastric pressure 6–10 sec after the belch, until secondary peristalsis emptied the esophagus.

During reflux of gastric contents, the pH within the esophagus may be as low as 2. In this environment pepsin of gastric contents can attack the esophageal mucosa. In some men and women the burning sensation known as heartburn is felt if the pH is below 4. It is not known whether the sensation is aroused by acid stimulation of receptors within the mucosa or by increased tension in the esophageal muscle. Perfusion with 0.1N HCl of the lower esophagus of patients subject to heartburn reproduces the symptoms; these are invariably accompanied by motor abnormalities, including nonpropulsive synchronous contractions of the esophagus, prolonged peristalsis, and raised intraluminal pressure. No pain or motor abnormalities occur in normal subjects similarly perfused. In some persons, heartburn is an important symptom associated with coffee-drinking. In them, resting lower esophageal sphincter pressure is half the normal value, and if they have acid in the stomach, easy reflux results in heartburn. Heartburn attributable to acid reflux occurs in about 50% of pregnant women in the last five months of pregnancy. Secondary peristalsis has been found in 45% of pregnant women examined who complained of heartburn but never in those free from it. The major cause of heartburn in pregnancy is diminished resistance at the lower esophageal sphincter. There is circumstantial evidence that the high level of progesterone prevailing during pregnancy inhibits the lower esophageal sphincter. Mechanical factors are important. No intra-abdominal esophageal segment can be identified. Whenever abdominal pressure increases during bending, stooping, or lying, pressure is transmitted to the stomach; because there is no subdiaphragmatic portion of the esophagus, the pressure cannot at the same time reinforce the sphincter. Heartburn subsides in the last weeks of pregnancy as the uterus descends, and it ceases immediately after delivery.

When a person goes to sleep with a full stomach, reflux may occur as many as 20 times in 12 hours, but it is restricted to periods of arousal or wakefulness. In other persons, such reflux never occurs.

If acid is present in the lower esophagus of a normal person, it is quickly cleared by 4–12 swallows or secondary peristaltic waves. Swallowed saliva dilutes and neutralizes regurgitated acid. A person with symptomatic reflux may require 28 or more swallows to clear the same amount of acid from his esophagus. A normal person has 15 or so episodes of reflux in 15 hours, and in each the refluxed gastric contents are returned to the stomach in 5–15 min. In a patient with abnormal reflux, the pH of the lower esophagus may be less than 3.0 for 30 min to an hour after each episode of reflux, because he has difficulty clearing

his esophagus. Two-fifths of patients with reflux also have delayed gastric emptying.

Mechanisms controlling adult esophageal function are incomplete at birth. The infant's superior laryngeal sphincter is not strongly closed; peristalsis does not occur in his lower esophagus; there is little or no abdominal esophagus; and the lower esophageal sphincter does not close between swallows. On the other hand, the crural muscle fibers of the diaphragm are well developed, and they act as an accessory closing mechanism during expiration as well as during inspiration. Reflux into the esophagus is infrequent, but when it does occur, absence of secondary esophageal peristalsis and weakness of the superior laryngeal sphincter allow gastric contents to bubble out the mouth in the manner of a fumarole overflowing.

Reverse Peristalsis and Rumination

Reverse peristalsis does not occur in the esophagus of man or other nonruminating animals. A ruminator, such as the cow, aspirates rumenal contents into its esophagus by relaxing its lower esophageal sphincter and inspiring against a closed glottis. This steepens the gradient between rumen and esophagus by dropping the intraesophageal pressure 30–40 mm Hg. Neither rumen nor abdominal muscle contracts, and intra-abdominal pressure does not rise. The cud is then carried to the mouth by reverse peristalsis of the esophagus, traveling 1 m/sec.

Fermentation in the rumen generates a large volume of gas, which must be expelled. A thousand-pound cow produces 1–2 L/min, chiefly carbon dioxide and methane. Eructation begins with movements of the rumenoreticulum, which force gas into the area of the cardia. Relaxation of the lower esophageal sphincter allows gas to fill the esophagus. Then, after the sphincter has closed again, a powerful antiperistaltic wave, sweeping upward at 1.6 m/sec, forces gas into the pharynx. Because the glottis is open and the nasopharyngeal airway is closed, more than half the gas is forced into the lungs, where carbon dioxide and methane are absorbed.

Achalasia of the Esophagus

In achalasia, or failure of the lower esophageal sphincter to relax, the swallowed bolus may have difficulty passing into the stomach (Fig 3–10). The pharyngeal swallowing reflex and the action of the pha-

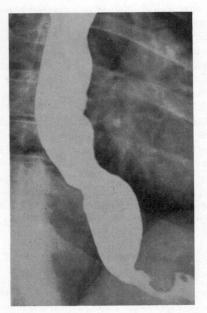

Fig 3–10.—Frontal view of an abnormal human esophagus; achalasia of the lower esophageal sphincter, with dilation of the esophagus. (Courtesy of F.J. Hodges, and J.N. Correa.)

ryngoesophageal sphincter are normal, but in the esophagus itself, peristalsis does not occur. If the esophagus does contract, the contractions, lasting 2–8 sec, engage the whole muscle at once. Pressure generated by such massive contractions is 10–20 mm Hg, and four or five of them may occur repetitively after each swallow.

Resting pressure in the lower esophageal sphincter of a person with achalasia is twice normal. When swallowing occurs, the sphincter relaxes only partially, if at all, leaving a residual gradient of pressure between stomach and esophagus. Therefore, the bolus fails to pass into the stomach, and the esophagus above the sphincter is enlarged. One or several meals may lodge above the sphincter and pass slowly into the stomach over a period of hours. When the person lies down to sleep, aspiration of esophageal contents into the lungs may be the cause of pneumonia. Treatment is mechanical dilatation or surgical weakening of the sphincter.

In achalasia, lesions have been found in the dorsal vagal nucleus, in vagal trunks, and in the myenteric plexuses of the esophagus. These lesions may be caused by a neurotropic virus. The trypanosomes of Chagas' disease attack postganglionic fibers in the esophagus as well as in the heart. The result of the lesions is that esophageal muscle is denervated and therefore more sensitive to acetylcholine released by the remaining preganglionic vagal fibers. In a normal person, injection of 5–10 mg of the parasympathomimetic drug methacholine has no effect upon the esophagus. In a person with achalasia, that dose causes prolonged, vigorous, and painful contractions of the esophagus.

Other Causes of Delayed Emptying

In some patients who complain of inability to swallow or of pain on swallowing, the esophagus may appear anatomically and physiologically normal during x-ray examination while a suspension of barium sulfate is being swallowed. Some patients have been considered neurotic on the basis of such negative findings. Conscious rejection is frequently expressed in the phrase "I can't swallow that," and states of rejection below the level of consciousness can be expressed by inability to swallow. Therefore, esophageal dysfunction may be a neurotic symptom, and in patients with true achalasia, exacerbation and remission of symptoms are often related to emotional stress and the return of relative security. However, in some persons who can swallow water without difficulty, swallowing a bite of meat or a peanut butter sandwich may evoke abnormal contraction of the distal esophagus. The solid bolus, instead of falling through the esophagus as does a liquid, is moved by the peristaltic wave, and in these individuals, afferent impulses aroused by the bolus reflexly arouse premature and exaggerated contraction of the lower esophageal sphincter.

Esophageal emptying is also delayed by pain. The esophagus of a normal subject that requires 12 sec or less to empty may take 25–102 sec when the subject's hand is plunged in ice water.

Sensations from the Esophagus

Afferent fibers mediating esophageal sensations travel centrally with the sympathetic nerves; the only afferents from the esophagus traveling centrally with the vagus nerve are those concerned with secondary peristalsis. Distention of the esophagus at any level causes pain in the midsternum. Pain is aroused by strong and continued contractions. The mean amplitude of peristaltic waves in persons complaining of chest pain is 170 mm Hg as compared with 81 mm Hg in persons free of pain. In some persons, normal peristalsis of the lower half of the esophagus is replaced by simultaneous contraction of the whole esophageal muscle, which is prolonged and often repetitive. This is called "diffuse spasm of the esophagus." These contractions develop pressures of 150–200 mm Hg, and they are accompanied by severe, aching substernal pain, which may extend between the tips of the clavicles and upward into the throat and jaws. The pain subsides as esophageal pressure falls. A sense of fullness or substernal constriction is caused by dilatation of the upper and middle two thirds of the esophagus. Pain caused by abnormal esophageal contractions may be precipitated by exercise or emotional upsets; and because it closely resembles cardiac pain with identical distribution, it is often confused, particularly by laymen, with pain arising in the heart. Differential diagnosis is made difficult by the fact that in many patients such pain arises both from the esophagus and from coexistent ischemic heart disease.

REFERENCES

Anderson D.J.: Mastication, in Code C.F. (ed.): *Handbook of Physiology:* Sec. 6. *Alimentary Canal,* vol. 4. Washington, D.C.: American Physiological Society, 1968, pp. 1811–1820.

Bennett J.R., Atkinson M.: The differentiation between oesophageal and cardiac pain. *Lancet* 2:1123, 1966.

Car A., Jean A., Roman C.: A pontine primary relay for ascending projections of the superior laryngeal nerve. *Exp. Brain Res.* 22:197, 1975.

Code C.F., Schlegel J.F.: Motor action of the esophagus and its sphincters, in Code C.F. (ed.): *Handbook of Physiology:* Sec. 6. *Alimentary Canal,* vol. 4. Washington, D.C.: American Physiological Society, 1968, pp. 1821–1840.

Cohen B.R., Wolf B.S.: Cineradiographic and intraluminal pressure correlations in the pharynx and esophagus, in Code C.F.(ed.): *Handbook of Physiology:* Sec. 6. *Alimentary Canal,* vol. 4. Washington, D.C.: American Physiological Society, 1968, pp. 1841–1860.

Cohen S.: Motor disorders of the esophagus. *N. Engl. J. Med.* 301:184, 1979.

Diamant N.E., El-Sharkawy T.Y.: Neural control of esophageal peristalsis. *Gastroenterology* 72:546, 1977.

DiDion L.J.A., Anderson M.C.: *The "Sphincters" of the Digestive System.* Baltimore, Williams & Wilkins Co., 1968.

Doty R.W.: Neural organization of deglutition, in Code C.F. (ed.): *Handbook of Physiology:* Sec. 6. *Alimentary Canal,* vol. 4. Washington, D.C.: American Physiological Society, 1968, pp. 1861–1902.

Dougherty R.W.: Physiology of eructation in ruminants, in Code C.F. (ed.): *Handbook of Physiology:* Sec. 6. *Alimentary Canal,* vol. 5. Washington, D.C.: American Physiological Society, 1968, pp. 2673–2694.

Goodwin G.M., Luschei E.S.: Discharge of spindle afferents from jaw-closing muscles during chewing in alert monkeys. *J. Neurophysiol.* 38:560, 1975.

Goyal R.K., Rattan S.: Neurohumoral, hormonal, and drug receptors for the lower esophageal sphincter. *Gastroenterology* 74:598, 1978.

Grand R.J., Watkins J.B., Torti F.M.: Development of the human gastrointestinal tract. *Gastroenterology* 70:790, 1976.

Henderson R.D.: *Motor Disorders of the Esophagus.* Baltimore, Williams & Wilkins Co., 1976.

Jean A.: Control bulbaire de la dégultition et de la motricité oesophagienne. Thèses, Université de Droit, d'Economie et des Sciences d'Aix Marseille, 1978.

Kawamura Y. (ed.): *Physiology of Mastication.* Basel, S. Karger A.G., 1974.

Palmer E.D.: Disorders of the cricopharyngeus muscle: A review. *Gastroenterology* 71:510, 1976.

Stevens C.E., Sellers A.F.: Rumination, in Code C.F. (ed.): *Handbook of Physiology:* Sec. 6. *Alimentary Canal,* vol. 5. Washington, D.C.: American Physiological Society, 1968, pp. 2699–2704.

Sturdevant R.A.L.: Is gastrin the major regulator of lower esophageal sphincter pressure? *Gastroenterology* 67:551, 1974.

4

Gastric Motility and Emptying

THE EMPTY STOMACH is small and relaxed, and most of the time feeble contractions passing over its muscle have negligible effects upon intragastric pressure. During the interdigestive period, when a person is hungry, pressure in the lower esophageal sphincter rises, and strong peristaltic waves, whose frequency and rate of progress are governed by the basic electrical rhythm originating in a pacemaker high on the greater curvature, sweep over the stomach at a frequency of 3/min. Residual contents are emptied into the duodenum, where similar intense activity soon begins and continues to the terminal ileum. Such bouts of gastric contraction continue for 10–20 min, are repeated at approximately hourly intervals, and are abruptly terminated by eating.

As the stomach fills, its fundus and body relax to accommodate its contents, and pressure increase in the stomach is small. Food swallowed tends to form layers in the body of the stomach, and gastric contents are well mixed with gastric secretions only in the antrum. Immediately after the stomach is filled, its peristaltic waves diminish in force; they gradually become stronger, the more quickly the less fat in the meal. As they engage the strong muscle of the antrum, they may deepen sufficiently to mix the stomach's contents and to propel small portions into the duodenum. Emptying is effected by these waves in the antrum and pylorus, and the rate of emptying is chiefly regulated by factors controlling the vigor of antral contraction. In addition to adventitious influences such as pain, nausea, and emotional disturbances, the major ones affecting gastric motility are the composition and volume of gastric contents delivered to the duodenum. These, operating through nervous and humoral pathways, regulate the rate of emptying so that delivery of chyme is adjusted to duodenal motility and emptying.

Structure and Innervation of the Stomach

The stomach acts as a unit because the longitudinal and circular muscles of its wall are continuous sheets. The excitatory wave, which can be recorded as the basic electrical rhythm of the stomach, begins high on the greater curvature and moves over the longitudinal layer to the pylorus, setting the rhythm of gastric contraction. Its velocity, which is low in the body and high in the terminal antral segment, determines the rate at which gastric peristaltic waves move. The strength of contraction is governed by both anatomical and physiological characteristics of the muscle. Contractions are usually weak in the body, whose muscle is thin, and strong in the antrum, whose muscle is thick. The myenteric and submucous plexuses are continuous throughout the stomach, and in some unknown way they regulate the response of the muscle to excitation. The extrinsic innervation, sympathetic via the celiac plexus and parasympathetic via the vagus, is distributed to the whole of the stomach. In man, the effect of sympathetic stimulation is to depress all movement of the stomach. Stimulation of some fibers in the vagus nerves enhances gastric motility, and stimulation of other fibers inhibits it. Both sets of nerves have small effects upon the frequency and rate of progress of the basic electrical rhythm, and both, acting through the intrinsic plexuses, reduce or enhance contraction of gastric muscle.

A dense network of fibers within the circular and longitudinal muscle and in the myenteric plexus of the pylorus of the cat give the immunohistochemical reaction for enkephalin-like substances. Many cell bodies in the plexus give the same reaction, but the antrum and body of the stomach are only sparsely supplied with similar nerves. These enkephalin-containing nerves probably mediate contraction of the pylorus. Local intra-arterial injection of morphine or enkephalinamide stimulates contraction of the sphincter, and vagally induced contraction is blocked by the opiate antagonist, naloxone.

There is a partial discontinuity between stomach and duodenum. At the pylorus, the circular muscle of the antrum ends in a thick ring, which is anatomically and physiologically identified as the pyloric sphincter. Beyond the muscle, there is a thick ring of connective tissue, the hypomuscular segment, which almost completely separates the stomach and the duodenum. The longitudinal muscle layer dips down into this barrier, and a few bundles pass through it to the duodenum. The myenteric plexus of the stomach is continuous with that of the duodenum. The basic electrical rhythm of the stomach can be recorded 2–3 mm beyond the pylorus, but the rhythm of the more distal duodenum is independent of it. Duodenal contractions are influenced by gastric peristalsis, for about 85% of duodenal contractions occur when the gastric terminal antral segment is contracting. This rough coordination is mediated by the few longitudinal fibers passing the barrier, by the plexuses and by the mechanical effect of chyme entering the duodenum. The connective tissue ring at the pylorus extends into the submucosa, and there is no connection between the submucous lymphatic systems of stomach and duodenum.

Blood supply of the stomach is derived from small arteries lying on the serosal surface of the stomach and arising from large arteries along the greater and lesser curvatures. These vessels penetrate the gastric muscle and divide into arterioles, capillaries, and venules, which serve the muscle and mucosa. Veins draining both muscle and mucosa return through the muscle to join large veins accompanying the large arteries. The consequence of this arrangement is that when gastric muscle contracts, veins within it are occluded, and upstream venules and capillaries become engorged. When there is strong contraction, as in peristalsis, blood flow in the small arteries is cut off.

There is abundant collateral circulation in the stomach; almost all major arteries must be tied off before any part of the stomach is deprived of blood.

Types of Movements of the Stomach

The volume of the chamber of the stomach of a fasting man is 50 ml or less. There is little or no tension in its wall, and intraluminal pressure is equal to intra-abdominal pressure. No movement of the muscle of the body can be seen by x-ray examination. When either a small balloon or an open-tip manometer is placed in the antrum, pressure changes reflecting contraction of the antral muscle at the rate of 3/min are recorded (Fig 4–1). Although the sequence may appear to be nonrhythmic, waves are separated by some multiple of 20–22 sec.

Pressure waves in the digestive tract are empirically classified by size: Type I waves are simple monophasic waves of low amplitude, 3–10 mm Hg, lasting 5–10 sec. Their rhythm is characteristic of that part of the bowel from which they are recorded. Type II waves are simple ones of greater amplitude, 8–40 mm Hg, lasting 12–60 sec. Type III are complex waves whose characteristic is a rise in baseline pressure, which lasts for a few seconds to some minutes.

As fasting proceeds and man or any other mammal enters the postabsorptive state, the *interdigestive myoelectric complex* appears. This is more fully described in the chapter on intestinal motility; here the participation of the stomach is sketched. Pressure in the lower esophageal sphincter abruptly rises. Action potentials in the longitudinal and circular muscles appear, and the muscle contracts. The action potentials and contraction accompany the basic electrical rhythm as it sweeps from high on the greater curvature over the stomach to the pyloric sphincter. Soon action potentials and contraction become more intense, with the result that strong peristaltic waves at the frequency and velocity of the basic electrical rhythm pass down the stomach for 10 or more min. Residual gastric contents are emptied into the duodenum, and even large, solid objects, such as swallowed coins, leave the stomach. Then frequency of action potentials and strength of contraction die out, and the stomach becomes quiet until another interdigestive myoelectric complex appears approximately one hour after the beginning of the first one.

The pattern of the interdigestive myoelectric complex is set within the neuromuscular apparatus of the stomach and intestine; it is aroused by external stimuli. The chief stimulus seems to be a rise in plasma concentration of the hormone motilin, though what causes this rise is unknown. The vagus is unimportant in initiating and controlling the complex, for in a vagotomized animal the complex is only slightly disturbed. A rise in plasma gastrin upon feeding is one, but not the only, factor terminating the occurrence of the complex.

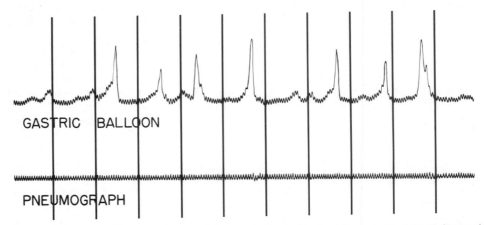

Fig 4–1.—Sequence of large and small waves passing over the antrum of a normal fasting human subject. The record was made using a small balloon. Vertical lines mark minute intervals. The waves occur at intervals which are multiples of 21 sec. (Courtesy of C. F. Code.)

A fasting person often experiences the pangs of hunger during a strong gastric contraction. The nature of the sensations of hunger, appetite, and satiety is a problem of physiological psychology too esoteric to be discussed here.* Many persons, when hungry, feel intermittent pangs or sharp gnawing sensations referred to the lower midchest or epigastrium. Afferent impulses from the stomach signal tension in its wall; impulses in these fibers could be the input to the central nervous system arousing the sensation. However, the correlation between contractions of the stomach and presence or intensity of the hunger sensation is very poor, and we may reject the naive equation of gastric contractions with hunger pangs.

Pressure in the Full Stomach

The body of the stomach relaxes with each swallow, and food is accommodated with little rise in intragastric pressure. The efferent pathway is the vagus nerve, and the fibers responsible are nonadrenergic,

noncholinergic. The neurotransmitter may be dopamine. Relaxation is sustained during repetitive swallows.

Figure 4–2 shows the pressure recorded in a vagally innervated pouch of the fundus of a dog's stomach, the fundus being defined as that portion of the stomach lying orad to the entrance of the esophagus. In this instance, when the dog swallowed, the fundus relaxed along with the lower esophageal sphincter, but there was no change in antral motility. Such receptive relaxation, which was observed in 26 of 40 trials, provides some additional volume for swallowed food; on eating, the volume of the body increases as well. The drinking of 225 ml of water raises pressure in the human stomach less than 2 mm Hg. When air is rapidly injected into the human stomach by way of a tube, the pressure rises by 4–7 mm Hg for the first 200–600 cc; after this, there is no further rise in pressure until a volume of 1,600 cc is reached. Intra-abdominal pressure rises at the same time, so that gastric transmural pressure remains approximately constant. Tension in the stomach wall must rise, for in a hollow organ at constant transmural pressure, tension in the wall is directly proportional to the radius.

Afferent impulses from the mouth and esophagus are relayed through the hypothalamus to the vagal nucleus. Sham feeding and distention of the esophagus cause relaxation of the body of the stomach. Afferent impulses from the heart and chemoreceptor trigger

---

*The older literature is treated authoritatively and at length in Code C.F. (ed.): *Handbook of Physiology:* Sec. 6. *Alimentary Canal,* vol. 1. Washington, D.C., American Physiological Society, 1967. More recent work is described in Tepperman J.: *Metabolic and Endocrine Physiology,* ed. 4. Chicago, Year Book Medical Publishers, 1980. A useful review is Smith G.P., Gibbs J.: Postprandial satiety, in *Progress in Psychobiology and Physiological Psychology,* vol. 8. New York, Academic Press, 1979, pp. 179–242.

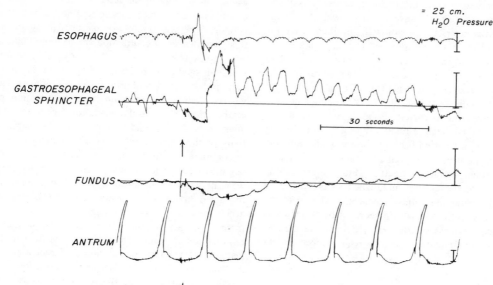

Fig 4–2.—Simultaneous records made from a dog's lower esophagus, lower esophageal sphincter, a vagally innervated pouch of the fundus of the stomach, and the gastric antrum. At the arrow, the dog swallowed; 5 sec later the peristaltic wave reached the lower esophagus. The lower esophageal sphincter relaxed before the peristaltic wave reached it, and it subsequently closed at high pressure for a long period. The fundus of the stomach relaxed as soon as the swallow began and remained relaxed for 25 sec. Antral contractions did not change. (From Lind J.F., et al.: *Am. J. Physiol.* 201:197, 1961.)

zone (see chap. 7) also result in nonadrenergic, noncholinergic impulses in vagal fibers to cause the relaxation of the stomach that precedes vomiting. Distention of the stomach itself, either body or antrum, stimulates slowly adapting tension receptors, and afferent impulses from these receptors traveling centrally in the vagus nerve are reflected in efferent vagal impulses, which cause gastric relaxation. However, the stomach may also respond to distention with contractions. In an experiment on a fistulous human subject whose stomach was distended by a balloon at a pressure of 8 mm Hg, a small increase in volume sufficient to double intragastric pressure was followed 2 of 10 times by strong contraction.

The abdominal contents are relatively mobile, and their specific gravity is approximately that of water. The physical result is that the viscera apply hydrostatic pressure to the serosal surface of the stomach, proportional to the vertical height of the viscera. The contents of the stomach, with the exception of the gas bubble, have nearly the same specific gravity as the viscera. Although the gastric contents are acted upon by gravity pulling them downward, this force is balanced by an equal and opposite force of the abdominal viscera applied to the serosal surface of the stomach, pushing them up. The stomach behaves like a water-filled bag suspended in water. Gravity can affect only contents whose specific gravity differs from that of the surrounding viscera.* The gas bubble rises, and dense materials sink. Because the density of most food is close to that of water, the contents of the stomach resemble a thick Irish stew with small lumps of meat and vegetables submerged in gravy. However, the density of the common x-ray contrast medium, barium sulfate, is 4.50, and it rapidly sinks

---

*The same physical principle applies to the esophagus, but the specific gravity of the thoracic contents averages about half that of water. Consequently, the force of gravity does help to move material through the esophagus.

to the lowest part of the stomach unless it is finely dispersed as a stable suspension in water. Viscous food forms layers in the stomach, stratified in the order in which it was swallowed. The strata are sometimes vertical and sometimes oblique. Figure 4–3 shows that the relative position of two layers of fried meatballs remained constant in the stomach of a Swedish medical student for 65 min, although considerable emptying occurred. Solid lumps stay in the body of the stomach; a relatively large lump entering the antrum is squirted back by the retropulsive action of antral peristalsis, and small lumps are ground by the terminal antral contraction. Fluid moves more quickly into the antrum and, when swallowed on a stomach containing solids, seems to pass around the solid mass. Acid and pepsin secreted by the mucosa penetrate the mass slowly; but, judging from the low pH of antral contents, secretions readily reach the antrum.

Movements of the Full Stomach

Three kinds of movements occur in the full stomach: peristaltic waves, systolic contraction of the terminal antrum and diminution of size of the stomach. The basic electrical rhythm begins at a pacemaker high on the greater curvature of the stomach; there are no pacesetter potentials on the fundus. A peristaltic wave follows the depolarization as it moves over the longitudinal muscle of the body, antrum, and pylorus. In man, the frequency of gastric peristalsis is 3/min, and the frequency and rate of progress are affected by nervous and humoral factors, to be described below. The electrical wave consists of a moving band of partial depolarization of the smooth muscle cells of the longitudinal layer (see Fig 1–3). When recorded by means of extracellular electrodes, it appears as a diphasic or triphasic complex that is the derivative of the transmembrane potential variations (see Fig 1–5). The wave moves over the muscle of the body of the stomach at less than 1 cm/sec, more rapidly over the greater than over the lesser curvature. It lasts about 1.5 sec at any one point and consequently occupies about 1–2 cm of muscle. The wave accelerates in the antrum to 3–4 cm/sec.

Excitation spreads from the longitudinal to the circular muscle, and the latter contracts with variable intensity. Consequently, a circular band of contraction, about 2 cm wide, moves as a peristaltic wave over the body and the antrum, following the electrical

wave. On account of the relation between the size of the stomach and the frequency and rate of peristalsis, two or three waves are usually present at any one time. The waves of contraction are no more than shallow ripples as they pass over the body, and during the first hour of digestion they usually remain shallow as they pass over the antrum. The hormone gastrin, whose plasma concentration rises as digestion proceeds, stimulates antral contraction by direct action upon antral muscle and by enhancing cholinergic activity of the intrinsic plexuses. Consequently, waves of contraction deepen as they near the angle, and they begin to push gastric contents ahead of them. As the wave moves over the antrum, the lumen is constricted but not obliterated. The wave of electrical excitation accelerates as it passes over the terminal antral segment (a length of about 5 cm in man), so that the terminal antral segment and the pylorus, which is the last part of the terminal antrum, contract almost simultaneously. This is the systolic contraction of the terminal antral segment. Because the pylorus is narrow, it closes early (Figs 4–4 and 4–5).

Hypoglycemia induced by insulin administration speeds the rate of propagation of the wave of excitation over the antrum, and it deepens the peristaltic wave by increasing the frequency of action potentials. Vagotomy completely abolishes this effect of insulin hypoglycemia. Distention of the stomach by filling it with water slows the rate of propagation but increases the frequency of action potentials. Vagotomy does not change the consequences of distension. Administration of cottonseed oil not only shows the rate of propagation, but it reduces the strength of peristaltic contraction by reducing the incidence of action potentials.

Gastric contents are viscous; and as they are pushed into the terminal antrum by the peristaltic wave, pressure in the antrum rises. About 60% of the time, the rise in pressure is enough to overcome the pressure barrier at the pylorus, and a small fraction of antral contents passes into the duodenum. Then, as the pylorus contracts, passage of chyme into the duodenum is abruptly stopped. Contraction of the terminal antral segment continues, and pressure within it rises by about 10–25 mm Hg. The contents of the segment, prevented from passing through the pylorus, are forcibly squirted by the terminal antral contraction back through the patent lumen into the proximal antrum. This retropulsion of chyme is highly effective in mixing food and digestive juices, and it creates strong

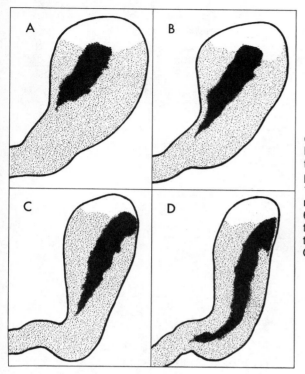

Fig 4–3.—Tracings of x-ray photographs of the stomach of a young man. Immediately before **A,** he ate 200 gm of fried meatballs, then 50 gm of meatballs mixed with powdered barium sulfate, and finally another 100 gm of meatballs without contrast medium. **B,** 10 min later. **C,** 35 min later. **D,** 64 min later. The layers remain stratified in the body of the stomach, but mixing occurs in the antrum. (Adapted from Nielsen N.A., Christiansen H.: *Acta Radiol.* 13:678, 1932.)

shearing forces in the chyme which break up friable masses of food. Then the terminal antral segment and pylorus relax, and both remain relaxed until the next peristaltic wave comes down the antrum.

The fraction of gastric contents passing through the pylorus at each cycle is small. The rate of emptying is a function of the volume of gastric contents, and the maximal rate for a 750-ml liquid meal is 20 ml/min. With three type II waves per min, this is 7 ml per wave, or 1% of gastric contents. When the stomach contains a smaller volume, about 1–2 ml is emptied per wave.

Sounds above 500 cps are produced by passage of chyme through the pylorus. They do not occur during the 3–4 sec preceding pyloric contraction, but they begin and continue during the 4–5 sec the pylorus is contracting. They are loudest when contraction is most vigorous.

As the stomach empties, its volume becomes smaller. Intragastric pressure does not rise, and this means that tension in the wall of the stomach must diminish with volume and radius of curvature.

Control of Gastric Motility in Man

The foregoing description of gastric motility is based on studies on the dog in which cineradiography can be combined with electrical recording. Observations on man are necessarily more limited. Electrical recording from the mucosal surface of the stomach is made by means of electrodes carried at the end of a nasogastric tube. Short-term electrical records are obtained from the serosal surface of the stomach by means of electrodes placed during peritoneoscopy or laparotomy. In some instances, electrodes have been sewn to the surface of the antrum and duodenum during cholecystectomy; they have been left in place for 6 days, to be removed at the same time that the T tube draining the bile duct is taken out. Observations made by these means show that there are minor quantitative differences between gastric motility in dog and in man.

In man, the frequency of the wave of excitation passing over the antrum ranges from 2.1 to 3.7/min, with a mean of about 3.1. In any one person at rest,

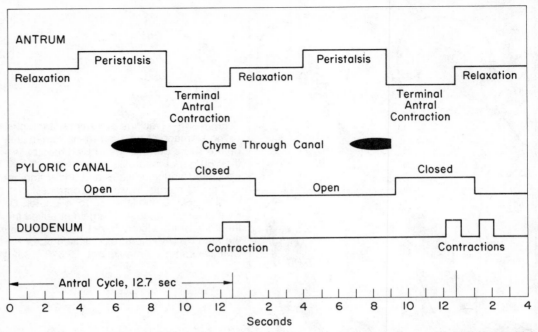

Fig 4–4.—Cycle of events occurring in the antrum, pyloric canal, and duodenum of a dog during digestion of a meal. The representative cycles were derived from study of many individual ones, such as that shown in Figure 4–5. The designation of "open" or "closed" applied to the pyloric sphincter refers not to force of contraction, which was not measured, but to whether the canal could be seen to be patent, i.e., to contain barium-impregnated chyme. (From Carlson H.C.: Ph.D. Thesis, University of Minnesota, 1962.)

the frequency varies by no more than 10%. Conduction velocity over the antrum is about 0.5 cm/sec, with a range of 0.3–1.4. This means that the wave of excitation is about 10 cm long. In man as in the dog, the velocity of propagation increases by about four times at the terminal antrum, and that segment contracts as a unit. The frequency found in the duodenum 10–12 cm from the pylorus is 12/min, but in the more proximal duodenum the gastric rhythm of 3/min is superimposed on an occasional rhythm of 12/min.

In a group of patients scheduled for vagotomy, regular antral electrical activity was found 96% of the time. Immediately after truncal vagotomy, it was present only 30% of the time, and after selective vagotomy 38% of the time. After highly selective vagotomy, the regular antral frequency was present in 74% of the records. When such subjects were restudied several weeks later, normal rhythm was found

98% of the time. When antral electrical activity is abolished or disorganized by vagotomy, gastric retention may be complete. Consequently, retention is more frequent in patients with truncal vagotomy, and this is the reason surgeons combine pyloroplasty with that operation. In patients with all types of vagotomy, normal rhythm is usually restored as time passes.

The Gastroduodenal Junction

The pylorus divides the gastroduodenal junction into a proximal part composed of the terminal antrum and a distal part composed of the duodenal bulb and the first part of the duodenum proper. The terminal antrum and the pylorus are part of the gastric muscle, and they form the peristaltic pump and funnel through which gastric contents are emptied into the duodenum. The duodenal cap, the first 2–5 cm of the

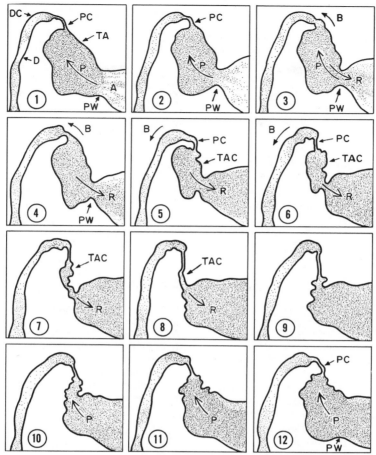

Fig 4–5.—Tracings made from a cineradiographic study of the motions of a dog's antrum during digestion of a meal made opaque with barium sulfate. Pictures were taken at 15 frames per sec, and every 15th frame was traced; consequently, each picture represents the shape of the antrum 1 sec later than the preceding one. **1,** A peristaltic wave of type II intensity *(PW)* indents the antrum. The barium mixture has been propelled *(P)* into the terminal antrum *(TA),* leaving only a small amount of barium along the folds of the antral mucosa *(A).* The pyloric canal *(PC)* is empty. Some barium is in the duodenal cap *(DC)* and duodenum *(D).* **2,** The pyloric canal *(PC)* is open but nearly empty, and the peristaltic wave *(PW)* has moved farther along the terminal antrum. **3,** The peristaltic wave *(PW)* is still farther along. The pyloric canal is open, and a small bolus of chyme *(B)* is in it. The largest part of the barium has begun to move backward *(R)* into the antrum. **4,** The bolus *(B)* continues to move through the open pyloric canal, and more of the content of the terminal antrum returns to the antrum *(R).* **5,** The pyloric canal *(PC)* begins to close and is nearly empty. The terminal antrum begins a systolic contraction *(TAC),* which increases retropulsion of its contents. **6,** Contraction of the terminal antrum continues *(TAC)* with more retropulsion *(R),* and some chyme moves down from the duodenal cap *(B).* **7** and **8,** The terminal antral contraction is almost complete, and most of the pyloric canal is closed. **9,** Terminal antral contraction is complete. **10,** The terminal antrum begins to fill again as the result of the propulsive force of another peristaltic wave in the antrum. **11,** Filling of the terminal antrum continues. **12,** The terminal antrum is full, and the pyloric canal *(PC)* begins to open. The whole cycle has taken 12 sec. (From Carlson H.C.: Ph.D. Thesis, University of Minnesota, 1962.)

duodenum, is a nearly spherical or ovoid chamber having a capacity of 5–10 ml, which receives the chyme as it is driven through the pylorus. Beyond the duodenal cap, the postbulbar part of the duodenum is a relatively narrow tube whose mucosa is thrown into interdigitating circular folds.

When an infused catheter is withdrawn from duodenum to stomach, a zone of high pressure is found between the two organs. This zone, which averages 1.5 cm in length, has a pressure 5 mm Hg above intra-abdominal pressure. It is a physiological sphincter dividing antrum and duodenum. A brief decrease in pressure in the sphincter occurs when antral peristalsis reaches the terminal antral segment, and at that time contents of the antrum flow through the gastroduodenal junction into the duodenal bulb. Then pressure rises briefly by about 40 mm Hg as the terminal antral segment contracts vigorously. When the duodenum is acidified with 0.1N HCl, sphincter pressure rises to 25 mm Hg, and there is no regurgitation into the stomach. Olive oil, amino acid solutions in the duodenum also cause an increase in sphincter pressure. This rise in pressure is mitigated or abolished when pentagastrin is infused or gastrin is endogenously released (Fig 4–6).

It is only fair to report that for at least a generation English investigators have been unable to find a pressure barrier at the pyloric sphincter. In the most recent study in which a catheter 2 mm in diameter was used, no pressure profile was found at the pyloric sphincter; and there was no response to irrigation of the duodenum with 0.1 N hydrochloric acid. If these observations are correct, the sphincter is always open at least 2 mm; and flow through the sphincter is entirely governed by the pressure gradient between stomach and duodenum.

In normal persons, there is little or no regurgitation through the gastroduodenal junction. The resting pressure in the junction in patients with gastric ulcers is normal, but the pressure fails to rise when the duodenum is acidified or is irrigated with olive oil or amino acid solutions. In these patients, infusion of cholecystokinin or secretin does not increase sphincter pressure, as it does in normal persons. Regurgitation of duodenal contents is frequent in patients with gastric ulcers, and the high and sustained concentration of bile acids and lysolecithin in the stomach may contribute to their mucosal damage.

In man, the antral peristaltic pump empties chyme into the duodenum in spurts at the gastric frequency

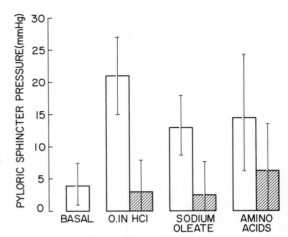

Fig 4–6.—Pressures in the human pyloric sphincter recorded when a tube of 5.4 mm outside diameter was pulled through. The open rectangles show the mean values in the basal condition and when the duodenum was irrigated with 0.1N HC1, sodium oleate, or amino acids. The bars show the extreme range of pressures found. The hatched rectangles show the results obtained when the subjects were intravenously infused with pentagastrin at a dosage of 2 µg/kg^{-1}/hr^{-1}. Essentially the same results were obtained when gastrin was endogenously released by intragastric irrigation with 1.5% glycine in 0.1N NaOH. (Adapted from Fisher R.S., Boden G.: *Am. J. Dig. Dis.* 21:468, 1976.)

of 3/min. Contractions of the postbulbar duodenum are governed by the duodenal basic electrical rhythm. This has a frequency of about 11/min in man. The duodenal cap lies anatomically and physiologically between two structures, which contract at different frequencies. It is filled and distended following antral contraction, and chyme spills from it into the postbulbar part of the duodenum. The cap itself contracts in an apparently irregular, eccentric manner, thereby displacing its contents into the postbulbar duodenum. The cap seldom empties completely.

Two mechanisms are responsible for the fact that 85% of proximal duodenal contractions occur just after the gastric peristaltic wave has delivered chyme to the cap. The first is simply the response of the duodenum to distention. The second is the conduction of the gastric basic electrical rhythm through the pylorus into the duodenum (Fig 4–7). Gastric longitudinal

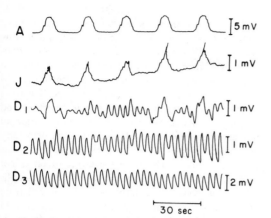

A

J

D₁

D₂

D₃

30 sec

Fig 4–7.—Monophasic potentials recorded simultaneously from points on the antrum, junction, and proximal duodenum of a rhesus monkey in situ. The antral electrode, *A,* was 14 mm orad to the junction; *J* was on the junction; D_1, D_2, and D_3 were 4, 10, and 17 mm aboral to the junction, respectively. The antral augmentation of duodenal slow waves is evident as far as 17 mm from the junction. (From Bortoff A., Davis R.S.: *Am. J. Physiol.* 215:889, 1968.)

muscle bands, particularly those on the lesser curvature, are continuous with those of the duodenal cap. The potential changes occurring at the gastric frequency mix with those occurring at the duodenal frequency for many millimeters beyond the pylorus, augmenting depolarization of duodenal muscle and therefore increasing the likelihood of excitation of duodenal circular muscle.

Gastric Emptying of Liquids

The rate at which the stomach empties is governed by the volume of gastric contents and the influence of chemical and physical properties of chyme in the duodenum and jejunum. If chyme is drained from the intestine by a fistula close to the pylorus, the stomach empties far more rapidly than if chyme is allowed to remain in the duodenum. Reduction in gastric motility mediated through the duodenum ranges from slight depression to complete cessation of all contraction. The terminal antrum, as an integral part of gastric muscle, is affected the same way: when it is relaxed, there is no pressure gradient to drive gastric contents into the duodenum. Therefore, emptying is minimal

despite relaxation of the pyloric sphincter. The only exception to this occurs when a very strong stimulus, such as 100 mN HCl, is placed in the duodenum; then the pylorus may contract from 1 to several min. After this brief spasm, which has no lasting influence on gastric emptying, the pylorus relaxes and remains flaccid. The duodenum, however, contracts vigorously at its basic rhythm until the acid is removed. In infants, hypertrophy and hyperplasia of the circular muscle of the pylorus sufficient to cause obstruction is sometimes called pylorospasm. Since it is not spasm in the sense that the muscle is abnormally contracted, pyloric stenosis is a better name.

The three major determinants of the rate of emptying of the stomach are the volume of the meal, its osmotic pressure and its chemical composition. The most nearly standardized data describing the time course of emptying of a liquid meal have been obtained by use of the *serial test meal.* In this test, a subject, after his stomach is washed out, is given a known volume of the meal, either by mouth or by tube. At some time, say 15 min later, the stomach is emptied as thoroughly as possible, and the volume and composition of the material recovered are measured.

If a substance not absorbed by the stomach (e.g., phenol red) is present in the meal at a known concentration, the volume of secretions added to the meal can be calculated from its dilution. Then, at some later time, at which the subject's stomach is assumed to be in the same functional state, a similar meal is given. This time, a different interval, say 30 min, is allowed to elapse before the gastric contents are recovered. A large number of replications of the procedure gives a description of the course of emptying. The relation of the volume of gastric contents to emptying has been measured this way, using a meal composed of 20 gm of citrus pectin, 35 gm of sucrose, 60–70 mg of phenol red, NaOH to adjust pH to 6.5, and distilled water to make 1 L. The osmotic pressure was adjusted by adding sucrose.

The shape of the full stomach is roughly that of a cylinder on top of a cone; and for a cylinder or cone, the transmural pressure P is equal to the tension in the wall T divided by the radius R. Rearranging, we have $T = P \cdot R$. The radius is proportional to the square root of the volume contained: $T \propto P\sqrt{V}$. Since tension in the wall of the stomach is the ultimate driving force for emptying, we expect the rate of emptying to be a function of the square root of the volume within the stomach. The data in Figure 4–8 show that the square root of the volume of gastric contents decreases in a linear fashion with increasing time. The slope of the line, or the rate of emptying, depends upon the composition of the meal. These data show that the concept of "emptying time" or the time taken for the last traces of a meal to leave the stomach is meaningless, for the bulk leaves early.

Although the square root of volume gives the best fit of the data, the data are almost as well fitted if the logarithm of the volume remaining in the stomach is plotted against time. In either case, the data show that with liquid meals, the rate of emptying is greatest when the volume is greatest, and that means that emptying is fastest at the beginning of digestion of a meal. Consequently, the greatest bulk of gastric contents is delivered to the duodenum before much gastric digestion or acidification has occurred.

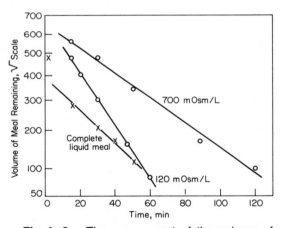

Fig 4–8.—The square root of the volume of gastric contents of a normal man plotted against time, showing that rate of emptying is a linear function of the square root of the volume. Data for two liquid meals are given. Each meal contained 20 gm of citrus pectin per liter and either 35 or 200 gm of sucrose per liter. The osmotic pressures are those of the meals determined by freezing-point lowering as the meals were ingested. The complete liquid meal was a mixture of milk product, cream, and sugar containing 1.28 cal per ml: fat 40%, protein 15%, and carbohydrate 45%. (Adapted from Hunt J.N., Spurrell W.R.: *J. Physiol.* 113:157, 1951; and Hunt J.N.: *Gastroenterology* 45:149, 1963. The square root plot was proposed by Hopkins A.: *J. Physiol.* 182:144, 1966.)

Fat in any digestible form (triglycerides or phospholipids) in the presence of bile and pancreatic juice is the most powerful of the chemical agents that slow gastric motility (Fig 4–9). Fatty acids are more effective than their corresponding glycerides. In the first 15 min after a man has eaten a meal containing 100 gm of fat, strong antral contractions (type II waves) are entirely absent, and thereafter their frequency and strength are about half those of the waves after an ordinary meal. However, during digestion of a fatty meal, small ripples (type I waves) continue to pass over the stomach. Therefore the effect of fat is not to abolish gastric peristalsis but to reduce the amplitude of the waves. If a fatty meal is removed from the stomach, its inhibitory influence disappears in 3–5 min. When a mixed meal is eaten, its fat is always emptied more slowly than its water.

Fat digestion products liberate cholecystokinin from the intestinal mucosa; that hormone inhibits gastric motility and is largely responsible for the effect of fat in delaying gastric emptying.

Protein digestion products also liberate cholecystokinin, and they delay gastric emptying. Tryptophan is the most potent amino acid in slowing gastric emptying, but on a molar basis dipeptides are almost twice as effective. The products of carbohydrate digestion (dextrins, oligosaccharides, and monosaccharides) have an inhibitory effect less powerful than that of fat.

Control of gastric emptying is not confined to the duodenum. If the duodenum is bypassed in perfusion experiments, irrigation of the jejunum or ileum with solutions of glucose or protein hydrolysate also inhibits gastric emptying.

Heavy exercise up to 71% of maximal oxygen uptake has little or no effect upon the emptying of liquids and none upon intestinal absorption of glucose, sodium, potassium, chloride, bicarbonate, or water. There has been no systematic study of the effect of exercise upon the digestion of regular meals. Emperor Frederick II (1215–50), *stupor mundi*, is said to have compared the full stomach of one of his servants, killed after a hard, postprandial ride, with the empty one of another, similarly fed servant, killed after a delightful conversation with the emperor himself, but the experiment is inadequately documented. Severe pain, fear, or other stimuli of massive sympathetic discharge delay gastric emptying. Patients with duodenal ulcers empty liquid meals more rapidly than do normal persons.

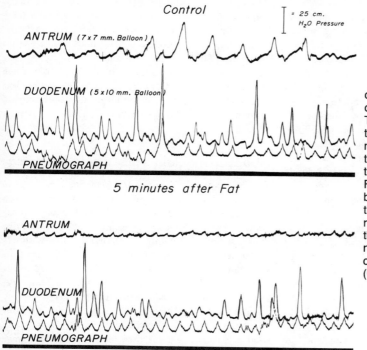

Control

ANTRUM (7 x 7 mm. Balloon)

$] = 25$ cm.
H_2O Pressure

DUODENUM (5 x 10 mm. Balloon)

PNEUMOGRAPH

5 minutes after Fat

ANTRUM

DUODENUM

PNEUMOGRAPH

Fig 4–9.—Effect of fat in the duodenum on antral and duodenal motility of a normal dog. The top record shows type I and type II waves in the antrum as measured by a small balloon; there are vigorous contractions in the duodenum similarly recorded. Five minutes after olive oil has been placed in the duodenum, the only pressure changes recorded by the antral balloon are those corresponding to respiratory movements; duodenal contractions are less vigorous. (Courtesy of C. F. Code.)

Water drunk with meals, aside from its effect on volume and osmotic pressure, has no special influence on emptying. The Olympic record for gastric emptying is held jointly by subjects of whole intestinal perfusion; each has had physiological salt solution pumped into his stomach through a tube at the rate of 75 ml/min for well over an hour.

Osmotic Pressure and Gastric Emptying

Gastric contents may have any osmotic pressure, and the osmotic pressure of gastric contents has little or no effect upon gastric function. On the other hand, a major task of the duodenum is to adjust the osmotic pressure of its contents to isotonicity, and to assist in accomplishing this, gastric emptying is regulated by the osmotic pressure of duodenal contents.

A comparison of the rates of emptying of two liquid meals of differing osmotic pressures is given in Figure 4–8. Each meal contained 15 gm of citrus pectin in 750 ml of water. One contained 35 gm of sucrose per liter, giving it a freezing-point depression equal to that of a solution of 120 milliosmol (mOsm).

The other contained 200 gm of sucrose per liter; it had a freezing point of −1.3 C, equivalent to an osmotic pressure of 700 mOsm. The fluid of higher osmotic pressure emptied more slowly, requiring about twice as long for an equal volume to leave the stomach. Nevertheless, at the beginning, about 350 ml left the stomach within 40 min; emptying was slowed, but not stopped, by high osmotic pressure. Variations in the osmotic pressure of gastric contents caused by sodium chloride also affect emptying. Emptying is fastest at 200 mOsm NaCl, and solutions of lower concentration, including pure water, empty more slowly. Above 200 mOsm NaCl, the rate of emptying is inversely proportional to the salt content.

The signal regulating emptying is probably given by receptors in the duodenal mucosa sensitive to an effective osmotic pressure difference across their walls. Such a difference, with the higher pressure outside, would make them shrink, thereby initiating nerve impulses to be carried along the afferent limb of the enterogastric reflex path. Nonpenetrating solutes—glucose, sorbitol, and sodium sulfate—are, on an osmolar basis, approximately equally effective.

The rapidly penetrating solution ammonium chloride has no influence upon the rate of gastric emptying. Very high concentrations of the penetrating solutes urea and glycerol slow gastric emptying, but their effect is much less per osmol than that of glucose. Ethyl alcohol solutions up to 1,720 mOsm (about 8%, or between strong beer and light wine) have almost no effect on duodenal osmoreceptors, and hence wine may be expected to flow through the pylorus like water. Strong drinks may delay gastric emptying. In one study of eight men, drinking 120 ml of 100 proof bourbon whiskey (50% ethanol v/v) just before eating almost doubled the time required for half of a liquid meal to leave the stomach. The fatty canapés often eaten during social drinking slow the emptying of cocktails into the duodenum, but because ethanol is rapidly absorbed through the gastric mucosa, they do little to diminish the effects of the drinks.

The osmotic pressure of gastric contents may be high or low, and adjustment of osmotic pressure is not a necessary prelude to emptying. The osmotic pressure of two glazed doughnuts and a pint of milk is 630 mOsm, whereas that a meal of steak, bread and butter, tossed salad, and tea is 232 mOsm. Gastric contents are diluted in the stomach by the addition of 500–1,000 ml of essentially isotonic gastric secre-

tions. Gastric contents that empty early are only slightly diluted, and they have nearly their original osmotic pressure. Thirty minutes after being eaten, a steak meal has been diluted 2–6 times, and its remnants after 90 min have been diluted 5–14 times. The final osmotic pressure attained by the mixture of food and secretions lies between 160 and 250 mOsm; the reason the mixture is hypotonic is that buffering of each hydrogen ion removes an osmotically active particle from solution.

Duodenal Acidity and Gastric Emptying

The acidity of gastric and intestinal contents is measured by passing a tube to which one or more miniature glass electrodes are attached, and the position of the electrodes is determined fluoroscopically. Immediately after eating, when antral contents are almost neutral, duodenal contents are likewise neutral, and fluctuations in duodenal pH are small and irregular. The pH of gastric contents falls progressively, so that 30 min after a meat meal the pH of chyme within the antrum lies between 2.5 and 3.4. An hour later the pH of antral contents is slightly below 2.0, and it is steady. At this time the pH recorded immediately beyond the pyloric sphincter fluctuates widely

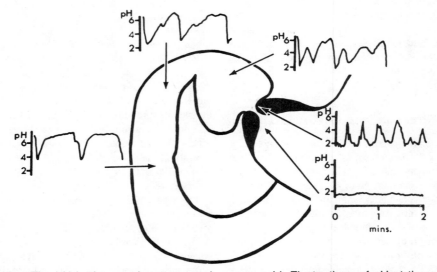

Fig 4–10.—The pH in the gastric antrum and at various sites in the duodenum recorded by means of a glass electrode. The records were obtained relatively late in the digestion and emptying of a meal when the contents of the antrum were acid. Fluctuations of pH at the pylorus and beyond occur at multiples of 3/min, which is the frequency of terminal antral contraction. (From Rhodes J., Prestich, C. J.: *Gut* 7:509, 1966.)

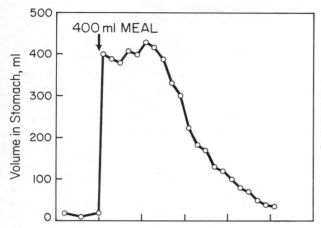

Fig 4–11.—Volume of gastric contents in a subject eating a 400-ml meal of steak, bread and butter, and vanilla ice cream. (Adapted from Malagelada J-R., Longstreth G. F., Summerskill W. H. J., et al.: *Gastroenterology* 70:203, 1976.)

(Fig 4–10) as chyme is delivered to the duodenal bulb in spurts. At the base of the duodenal bulb the prevailing acid condition is interrupted by neutrality, whereas farther along the prevailing neutral condition is interrupted by acidity. In normal subjects the average pH of contents of the middle of the second part of the duodenum lies between 5.4 and 7.8, and the pH is below 3.0 less than 1% of this time.

The acid load delivered to the duodenum is the product of the acid secreted and the amount emptied. Patients with duodenal ulcer secrete more acid and empty their stomachs more rapidly than do normal persons, and the acid load in the duodenum is greater. However, in duodenal ulcer patients and normal subjects, most of the chyme is delivered to the duodenum at pH 3.0 or greater.

It is not the pH of duodenal contents that determines the rate of gastric emptying, but the neutralizing ability of the duodenum. Acid entering the duodenum both slows gastric emptying and evokes duodenal, pancreatic, and biliary secretion to neutralize itself. A steady state is reached in which the rate of emptying of acid from the stomach is matched by the rate at which acid is neutralized in the duodenum.

Gastric Emptying of a Meal

The volume in the stomach and the rate of emptying in the course of digestion of an ordinary meal are shown in Figures 4–11 and 4–12. After the subject had swallowed tubes that permitted sampling necessary for the determination of volume and rate of emptying, he ate a meal of 90 gm of tenderloin steak, ground, cooked, and seasoned with 0.1 gm of NaCl

and a pinch of polyethylene glycol marked with ^{14}C. The steak was accompanied by 25 gm of white bread covered with 8 gm of butter, and it was followed by 60 gm of vanilla ice cream topped with 33 gm of chocolate syrup. The meal was washed down with 240 ml of water. The volume of an identical meal after homogenization was 400 ml; its pH was 6.0, and its osmotic pressure 540 mOsm.

The subject responded to the meal by secreting 800 ml of gastric juice, and consequently the volume in the stomach remained high for almost 2 hours despite early rapid emptying.

Gastric Emptying of Solids

During digestion of a meal, solids must be reduced to less than 1 mm in diameter before they pass through the pyloric canal. Consequently, large lumps of meat remain in the stomach as long as 9 hours, whereas an equal amount of ground beef leaves quickly. Because solids are slowly emptied, they continue to stimulate gastric secretion; there is a large volume in the stomach for a long time, and emptying of this volume evokes prolonged pancreatic and biliary secretion.

Lumps are pushed toward the pylorus by each peristaltic wave, and the systolic contraction of the terminal antral segment grinds them. Resulting fragments less than 0.5 mm are emptied with the liquid phase, and remaining lumps are thrown backward by each retropulsive, shearing movement. Eventually, digestion and trituration make them small enough to enter the duodenum. Because the rate of emptying of lumps of meat depends upon their rate of digestion in the stomach, emptying of their constituents is approximately linear, about 28% per hour. Whereas pyloroplasty or distal antrectomy has little effect upon emptying of liquids, those operations, which remove the barrier to lumps, allow lumps to leave the stomach quickly. Presence of solids in the stomach has no effect upon the concurrent emptying of liquids.

Undigested residues, or undigestible objects, such as a diamond ring swallowed by a thief, are expelled from the stomach by the strong antral contractions occurring during the gastric phase of the interdigestive myoelectric complex.

Mechanism of Regulation of Gastric Motility*

In the regulation of gastric motility, three pathways are involved: (1) reflex mechanisms whose afferent and efferent fibers are in sympathetic and parasympathetic nerves, (2) reflexes not involving higher centers but operating through intrinsic plexuses or the celiac plexus, and (3) blood-borne hormones liberated from the intestinal mucosa in response to chemical properties of the chyme.

Ninety percent of the 30,000 fibers in the vagus nerves are afferent, and an unknown proportion of these have their receptors in the stomach. Records of activity of the afferent fibers are obtained by dissecting the cervical vagus so that only a few fibers are in contact with recording electrodes. Then the stomach is stimulated in various ways and places until a mode and locus of effective stimulation of the receptor organs connected with the fibers are found. Some slowly adapting fibers respond to increased tension evoked either by passive stretch or by active contraction. Other slowly adapting fibers respond to distention of the stomach not accompanied by a rise in intragastric pressure, thus signaling the size of the stomach. Still other fibers fire when the pH of gastric contents is 3.0 or lower; others when it is above 8.0. In addition, numerous afferent fibers in the splanchnic nerves enter the cord and ascend in the same region as homologous somatic afferents. Impulses carried by the afferent fibers affect feeding behavior. Distention of the stomach increases electrical activity in the satiety center of the hypothalamus and decreases firing in the feeding center. The vagal fibers carry activity that is projected to the orbital cortex and to the amygdaloid region; the activity carried by the splanchnic afferents is projected to the somatosensory cortex and, by way of connections with the reticular formation, to the central excitatory system of the brain. These fibers are the afferent limbs of diverse reflexes affecting blood pressure and contraction of somatic musculature whose central pathways and significance are poorly understood. Activity in the vagal fibers appears to affect the vagal nucleus, which in turn affects gastric motility. In sheep and goats, coordinated reticuloruminal movements are entirely dependent on such reflexes. Afferent stimulation of the vagus elicits rhythmic contractions at 10-sec intervals, whereas similar efferent stimulation merely produces uncoordinated movements. In the cat, stimulation of an afferent branch of the abdominal vagus nerve near the lower part of the esophagus is usually followed by relaxation of gastric muscle, increased secretion of gastric juice, and secretion of enzymes by the pancreas.

Afferent nerves mediating pain have their receptors in the gastric muscle wall and perhaps in the mucosa. The major cause of deep gastric pain is tension in the wall of the stomach. Afferents mediating this sensation travel with sympathetic nerves, for the sensation is completely abolished by bilateral preganglionic sympathectomy. No pain or other sensation is caused by acid in the normal stomach. Insensitivity of the mucosa to stimuli causing pain is relative, not absolute. The mucosa may be cut or actually eroded without pain, but crushing of the mucosa (or the muscle) is painful. The pain of gastric or duodenal ulcer, an intense burning sensation referred to the epigastrium, appears when the ulcerated surface is exposed to acid, but its exact cause is unknown. It has been attributed to action of acid on exposed nerves, to spasm of gastric muscle, and to inflammation. Although there are

---

*This section summarizes the obviously very elementary knowledge of the subject contained in Western literature. Russian physiologists have extensively studied interoception and have described many autonomic and somatic reflexes whose afferent limbs arise in the digestive tract. For example, if water is placed in the stomach of a dog through a fistula, diuresis follows as the water is absorbed from the intestine. But diuresis also follows sham administration of water after a conditioned reflex to distention of the stomach is established. A glimpse of this literature, which I am unable to evaluate, can be seen through Bykov K.M.: *The Cerebral Cortex and the Internal Organs* (trans. Gantt W.H.). New York, Chemical Publishing Co., Inc., 1957. More recent literature can be read in the English translations of the *Sechenov Physiological Journal of the USSR* and the *Bulletin of Experimental Biology and Medicine, USSR*.

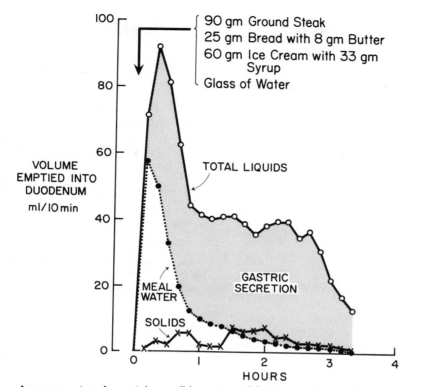

Fig 4–12.—Average rate of emptying solids and liquids of a meal and the accompanying secretion. The data are the average found in seven normal human subjects. (Adapted from Malagelada J-R.: *Gastroenterology* 72:1264, 1977.)

abundant nerve endings at the edge of the ulcer crater, they are deep in the tissue and no more exposed to acid than other tissue components. There are few or no nerve endings in the crater itself, and those present appear to be insulated by layers of tissue. Typical pain can occur when there is no acid in the stomach. The pangs of ulcer pain are associated with contractions of the stomach (Fig 4–13), but there is no evidence that the contractions are different from those in the normally painless stomach. Tissue around the ulcer is inflamed and indurated; pressure on such areas is painful. Ulcer pangs may be caused or exacerbated by movements that distort the area. Contraction of the gastric muscle momentarily occludes venous outflow, and increased venous pressure within the ulcer area may also contribute. Anticholinergic drugs reduce pain, probably more by diminishing motility than by reducing acid secretion.

Both sympathetic and parasympathetic efferent fi-bers influence the stomach. In general, sympathetic stimulation reduces the frequency and rate of conduction of the basic electrical rhythm of the stomach, and it decreases the strength of contraction of the circular muscle. The major effect of sympathetic stimulation is probably upon the activity of the neurons of the intramural plexuses. Central stimulation of the hypothalamic defense area or of the surrounding pressor area causes, in addition to widespread vasoconstriction, prompt inhibition of vagally induced gastric motility. In the absence of ongoing vagal stimulation, sympathetic discharge produces only small and sluggish inhibition of gastric muscle, perhaps as the result of overflow of adrenergic mediator released near blood vessels. Distention of the duodenum or jejunum also causes slowing of the antral rhythm and weakening of antral contractions. The efferent pathway is in part by way of adrenergic sympathetic fibers. Under usual circumstances, sympathetic nerves have lit-

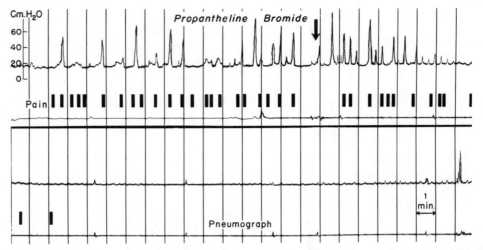

Fig 4–13.—Relation between contractions of the stomach and pain in a patient with peptic ulcer. Pressure was recorded by means of a small balloon within the stomach, and the patient signaled when he felt a pang of ulcer pain. At the arrow he was given the anticholinergic drug propantheline bromide. (Courtesy of C. F. Code.)

tle or no influence on gastric motility, for most of the numerous persons who have surgical or chemical sympathectomies for hypertension have normal gastric function.

Parasympathetic nerves have both inhibitory and excitatory influences on the stomach. Receptive relaxation of the fundus, which is part of the swallowing reflex, has been described above. Vagal inhibitory fibers also go to the body and antrum. Slight distention of the esophagus causes rapid and deep relaxation of the stomach. Afferent impulses activating vagally mediated reflex relaxation of the stomach also arise from nerve endings in many abdominal structures; among them are receptors in the duodenum sensitive to the chemical composition of the chyme. Instillation of cottonseed oil into the duodenum of a dog decreases the frequency of the basic electrical rhythm (normally 5.4 cycles per min) by about 1 cycle per min; it reduces the number of action potentials, and therefore the strength of contraction, of antral circular muscle. This inhibitory effect is abolished by transthoracic vagotomy. The mediator is neither acetylcholine nor catecholamines.

Both gastric secretion and gastric motility are reduced by acid, fat, and protein digestion products and solutions of high osmotic pressure in the duodenum

and jejunum. At least part of this effect is mediated by hormones released from the intestinal mucosa. However, the latent period for inhibition of motility is as short as 20–40 sec, and at least part of the inhibitory effect is mediated through nervous connections between small intestine and stomach, and, consequently, it is called the enterogastric reflex. The inhibition produced by acid in the first 5 cm of duodenum is in part effected through the intrinsic plexuses connecting duodenum and stomach. There is also a reflex arc, described in Figure 1–1, by which stimuli such as 100 mN HC1 in the duodenum or jejunum inhibits gastric motility; acid stimulates receptors of afferent fibers, which travel to the celiac plexus and there synapse with efferent sympathetic fibers.

Many forms of central nervous activity having emotional concomitants (e.g., pain and anxiety, sadness, withdrawal, hostility) also have expression in changes in gastric motility and secretion, but the particular effect is unpredictable. Gastric emptying may be completely suppressed for 24 hours following a painful injury; but when an experimental subject experiences severe pain by plunging one hand in ice water or contracting the forearm muscles under ischemic conditions, gastric motility may not be reduced.

REFERENCES

Abrahamsson H.: Studies on the inhibitory nervous control of gastric motility. *Acta Physiol. Scand.* [*suppl. 88*] 390:1, 1973.

Brooks F.P., Evers P.W. (eds.): *Nerves and the Gut.* Thorofare, N.J., Charles B. Slack, Inc., 1977.

Bykov K.M.: *The Cerebral Cortex and the Internal Organs* (trans. Gantt W.H.). New York: Chemical Publishing Co., Inc., 1957.

Code C.F.: The mystique of the gastroduodenal junction. *Rendic. R. Gastroenterol.* 2:20, 1970.

Code C.F., Carlson H.C.: Motor activity of the stomach, in Code C.F. (ed.): *Handbook of Physiology:* Sec. 6. *Alimentary Canal,* vol. 4. Washington, D.C.: American Physiological Society, 1968, pp. 1903–1915.

Code C.F., Carlson H.C., Szurszewski J.H., et al.: A concept of control of gastrointestinal motility, in Code C.F. (ed.): *Handbook of Physiology:* Sec. 6. *Alimentary Canal,* vol. 5. Washington, D.C.: American Physiological Society, 1968, pp. 2881–2896.

De Groot J.: Organization of hypothalamic feeding mechanisms, in Code C.F. (ed.): *Handbook of Physiology:* Sec. 6. *Alimentary Canal,* vol. 1. Washington, D.C.: American Physiological Society, 1967, pp. 239–248.

Duthie H.L. (ed.): *Gastrointestinal Motility in Health and Disease.* Lancaster: MIT Press, 1978.

Edwards D.A.W., Rowlands E.N.: Physiology of the gastroduodenal junction, in Code C.F. (ed.): *Handbook of Physiology:* Sec. 6. *Alimentary Canal,* vol. 4. Washington D.C.: American Physiological Society, 1968, pp. 1985–2000.

Fisher R.S., Boden G.: Gastrin inhibition of the pyloric sphincter. *Am. J. Dig. Dis.* 21:468, 1976.

Hunt J.N.: A possible relation between the regulation of gastric emptying and food intake. *Am. J. Physiol.* 239:G1, 1980.

Hunt J.N., Knox M.T.: Regulation of gastric emptying, in Code C.F. (ed.): *Handbook of Physiology:* Sec. 6. *Alimentary Canal,* vol. 4. Washington, D.C.: American Physiological Society, 1968, pp. 1917–1936.

Hunt J.N., Stubbs D.F.: The volume and energy content of meals as determinants of gastric emptying. *J. Physiol.* 245:209, 1975.

Ingram W.R.: Central autonomic mechanisms, in Field J.

(ed.): *Handbook of Neurophysiology,* vol. 2. Washington, D.C.: American Physiological Society, 1960, pp. 951–978.

Jansson G.: Extrinsic control of gastric motility. *Acta Physiol. Scand.* [*suppl.*] 326:1, 1969.

Kelly K.A.: Gastric emptying of liquids and solids: Roles of proximal and distal stomach. *Am. J. Physiol.* 238:G71, 1980.

MacGregor I.L., Martin P., Meyer J.H.: Gastric emptying of solid food in normal man and after subtotal gastrectomy and truncal vagotomy with pyloroplasty. *Gastroenterology* 72:206, 1977.

McShane A.J., O'Morain C., Lennon J.R., et al.: Atrumatic non-distorting pyloric sphincter pressure studies. *Gut* 21:826, 1980.

Malagelada J-R.: Quantification of gastric solid-liquid discrimination during digestion of ordinary meal. *Gastroenterology* 72:1264, 1977.

Meyer J.H.: Gastric emptying of ordinary food: effect of antrum on particle size. *Am. J. Physiol.* 239:G113, 1980.

Moilan J-P., Roman, C.: Discharge of efferent vagal fibers supplying gastric antrum: indirect study by nerve suture technique. *Am. J. Physiol.* 235:E366, 1978.

Morgan K.G., Muir T.C., Szurszewski J.H.: The electrical basis for contraction and relaxation in canine fundal smooth muscle. *J. Physiol.* 311:475, 1981.

Sharma K.N.: Receptor mechanisms in the alimentary tract: Their excitation and functions, in Code C.F. (ed.): *Handbook of Physiology:* Sec. 6. *Alimentary Canal,* vol. 1. Washington, D.C.: American Physiological Society, 1967, pp. 225–238.

Stoddard C.J., Smallwood R., Brown B.H., et al.: The immediate and delayed effects of different types of vagotomy on human gastric myoelectric activity. *Gut* 16:165, 1975.

Stubbs D.F.: Models of gastric emptying. *Gut* 18:202, 1977.

Thomas J.E., Baldwain M.V.: Pathways and mechanisms of regulation of gastric motility, in Code C.F. (ed.): *Handbook of Physiology:* Sec. 6. *Alimentary Canal,* vol. 4. Washington, D.C.: American Physiological Society, 1968, pp. 1937–1968.

Wolf S.: *The Stomach.* New York, Oxford University Press, 1965.

5

Motility of the Small Intestine

When chyme is delivered by the stomach to the duodenum, a large volume of fluid is added as pancreatic and biliary secretions. In addition, water moves in both directions across the duodenal mucosa as nutrients are digested and absorbed, and as a result there is a large net increase in the volume of fluid handled by the duodenum (Fig 5–1 and 5–2). The major part of digestion and absorption occurs in the duodenum and jejunum; undigested and unabsorbed residues are delivered from the ileum to the colon. The duodenum is short, about 20 cm on the average, and the jejunum and ileum together are 7–8 m long. Chyme moves slowly through this tube, at a rate such that the residue of one meal leaves the ileum as another enters the stomach. As befits its digestive and absorptive functions, the most important motion of the small intestine is segmentation, controlled by a basic electrical rhythm originating in the longitudinal muscle near the entrance of the bile duct. This mixes chyme with digestive juices and repeatedly exposes the mixture to the absorptive surface of the mucosa. Although segmentation accomplishes some downward propulsion, chyme is also moved by small peristaltic contractions, which, during normal digestion, move only a few centimeters before dying out. Peristaltic rushes rapidly traversing the whole length of the small intestine are abnormal, as are spasm, reverse peristalsis, and complete absence of movement. During the interdigestive period a band of intense segmentation followed by peristalsis sweeps from the stomach through the duodenum to the terminal ileum, and another, almost identical, band of segmentation follows in about an hour. Smooth muscle and intrinsic nerves of the small intestine are alone sufficient to perform segmentation and peristalsis, but extrinsic innervation affects the occurrence and modifies the strength of these movements and mediates reflexes by which stimuli acting in one part of the intestine govern activity in the rest of the bowel.

Segmentation

When electrodes are placed in longitudinal array in the duodenal muscle of an unanesthetized dog, the electrical control activity, or basal electrical rhythm, is found to originate near the duodenal bulb and to move down the duodenum at the rate of 19–20 cm/sec (Fig 5–3). The record obtained when the electrodes are arranged radially shows that all points in cross section are in phase. The electrical cycle is 3.4 sec long, and the rhythm is 17–18/min. Spike potentials signaling muscular contraction occur intermittently at a definite phase of the control rhythm, and the interval between contractions is some multiple of 3.4 sec. Spike potentials appear sequentially in the longitudinal direction, and the wave of contraction moves in the aboral direction. However, 80% of the contractions travel less than 3 cm.

The following is the original description of segmentation seen by fluoroscopic examination in the cat (Fig 5–4):

Rhythmic segmentation is by far the most common . . . mechanical process to be seen in the small bowel. . . . A small mass of food is seen lying quietly in one of the intestinal loops. Suddenly an undefined activity appears in the mass, and a moment later constrictions at regular intervals along its length cut it into little ovoid pieces. . . . A moment later each of these segments is divided into two particles, and immediately after the division neighboring particles rush together, often with the rapidity of flying shut-

70

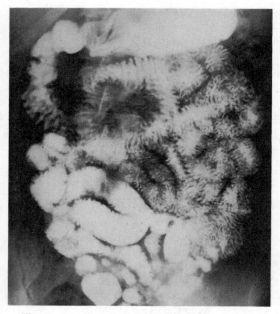

Fig 5–1.—X-ray photograph of the small intestine of a normal man. The shadow thrown by the barium sulfate shows characteristic patterns of the mucosa. (Courtesy of F.J. Hodges and J.N. Correa.)

tles, and merge to form new segments. The next moment these new segments are divided, and neighboring particles unite to make a third series, and so on.[*]

The process described by Cannon has also been observed in the dog, in the rat, and in man. In the human duodenum, segmental contractions are of two types. The first are eccentric contractions, confined to a segment less than 2 cm long; these do not empty the segment completely and do not cause any change in intraluminal pressure. The second are concentric contractions, which usually succeed in emptying a segment greater than 2 cm in length and which generate luminal pressure as high as 20 mm Hg. The empty segment refills as the annular contraction disappears.

In 12 normal human subjects the frequency of seg-

---

[*]Cannon W.B.: *The Mechanical Factors of Digestion.* New York: Longmans Green & Co., 1911, pp. 131–133. Cannon was a first-year medical student at Harvard in 1896 when he made the basic discoveries of modern gastroenterological physiology.

mentation was found to be $11.80 \pm$ SD 0.32/min at the end of the duodenum and $9.39 \pm$ SD 0.18 in the ileum. In some subjects the descent of frequency was stepwise; in others it was linear. The rate of segmentation is little, if at all, affected by extrinsic nerves, by feeding or fasting or by castor oil. However, vigor and amplitude of contractions vary greatly. Segmenting movements may be absent or barely visible in fasting, and they become strong immediately after feeding. Activity may wax and wane in cycles; and influences such as fright, which stimulate the thoracolumbar nervous system, depress segmenting movements. Stimulation of vagal nerves to the stomach and upper small intestine augments antral peristalsis and intestinal segmentation. It is probable that increased frequency of firing in vagal fibers during the interdigestive period promotes segmentation as well as antral peristalsis. In the opposum, but not in the cat, cholecystokinin stimulates phasic contraction of duodenal circular muscle. Whether in man cholecystokinin released during digestion of a meal stimulates segmentation is unknown.

Although segmenting movements are only slightly progressive, they move chyme in the aboral direction (Fig 5–5). A single segmental contraction squirts chyme lying beneath it in both directions. Because the frequency of contraction is greater in a higher part of the intestine than in a lower part, on the average more chyme is propelled in a downward than in an upward direction. Flow of chyme is governed by resistance as well as by pressure generated by contraction. A segmental contraction offers resistance; and because segmental contractions are less frequent aborally than orally, the average resistance is less in a lower part of the small intestine than in a higher part. Hence, chyme flows more readily downward than upward.

Nature of the Intestinal Gradient

The gradient of the frequency of segmental contraction is determined by the basic electrical rhythm generated in each segment of the small intestine. The source of the basic electrical rhythm in the small intestine is not exactly known. There is poor electrotonic coupling between the longitudinal and circular layers, and therefore the older idea that the rhythm originates in the longitudinal layer and is conducted electronically to the circular layer may not be correct. Each layer may be capable of generating pacemaker

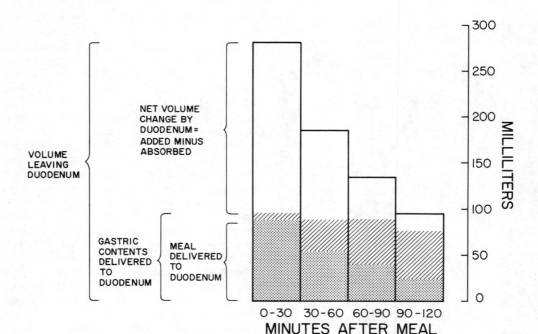

Fig 5–2.—Volumes of fluid delivered to and leaving the duodenum during digestion of a 400-ml mixed meal whose composition is given in Fig 4–13. Values are the averages of 16 studies in 11 normal human subjects. (Adapted from Miller L.J., et al.: *Gut* 13:699, 1978.)

potentials, or there may be a pacemaker in the myenteric plexus. The following description of the gradient is based on observations made on the intestine of the anesthetized cat. Essentially similar results have been obtained on the dog and the monkey.

Recordings from closely spaced electrodes placed on the serosal surface of the intestine, beginning high in the duodenum, show that the frequency of the basic electrical rhythm is constant for some centimeters. In the example given in Figure 5–6 the frequency from mid-duodenum to upper jejunum is 18/min. Below this, a length of jejunum has a frequency of 17/min. Between the two frequency plateaus is a short length of intestine whose frequency waxes and wanes between the two limits of 18 and 17/min. The entire length of intestine exhibits frequency plateaus, each of a lower frequency than the one above it and each connected to its neighbors by a short segment whose frequency waxes and wanes. At the terminal ileum the last plateau has a frequency of only 12.5/min.

If the intestine is cut into small bits, each piece exhibits its own intrinsic rhythm. The frequency of

the intrinsic rhythm diminishes in a linear, not a stepwise, manner from duodenum to terminal ileum. Each segment of muscle is, in effect, an oscillator having a characteristic frequency.

In the intact intestine, an orally located oscillator having a higher intrinsic frequency is coupled with its aborally located neighbor, which has a lower intrinsic frequency. If the coupling is strong enough, the oscillator having the higher frequency drives that having the lower frequency, and the second in turn drives a third. Consequently, a series of points along the intestine has a basic electrical rhythm of uniform frequency, that set by the frequency of the most proximal segment. This accounts for the frequency plateaus. At some distal point along the plateau, the driving frequency is so much higher than the intrinsic frequency of the driven cells that the cells having the lower intrinsic frequency have difficulty following the driving frequency. Consequently, there is a zone of waxing and waning frequency between the two plateaus. In this zone the basic electrical rhythm propagates with diminishing velocity, ending in propaga-

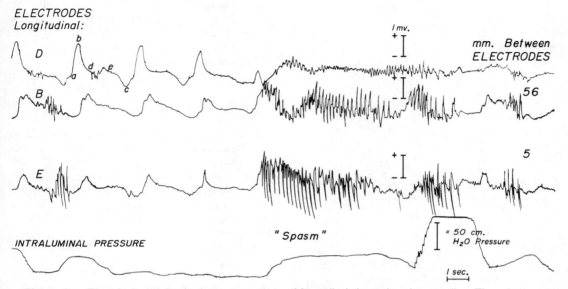

ELECTRODES
Longitudinal:

D

B

56

E

5

INTRALUMINAL PRESSURE

"Spasm"

= 50 cm.
H₂O Pressure

I mv.

mm. Between
ELECTRODES

I sec.

Fig 5–3.—Electrical activity and pressure recorded from the duodenum of an unanesthetized dog. At an earlier operation, the duodenum was brought out through the abdominal wall and covered by a skin flap so that it formed a permanent handle-like loop. Electrical recording was made from needle electrodes inserted into the muscle through the skin; pressure was recorded by means of an open-tip catheter inserted through a thin-walled hypodermic needle. The dog was given morphine by slow intravenous injection. The left half of the record shows the basic electric rhythm with some spiking. Pressure increases following spiking. At the middle of the record, spiking becomes prolonged, and there is a corresponding increase in intraluminal pressure. (From Bass P., et al.: *Am. J. Physiol.* 201:287, 1961.)

tion failure. Therefore, the frequency at the zone and beyond is determined by a local generator whose frequency is one cycle less. Eventually, the more orad pacemaker takes over again, and the frequency in the zone temporarily increases.

If the intestine is transected, the frequency of the basic electrical rhythm and of segmental contraction of the intestine below the cut falls, for coupling between driving and driven parts has been broken at the cut. The frequency followed by the section below the transection is the intrinsic frequency of the muscle at the point of transection. When the segment below the cut is warmed a few degrees, the frequency at the point of warming rises and so does the frequency of as much as 75 cm of immediately distal intestine.

The frequency of the basic electrical rhythm and of segmental contraction of the entire small intestine is higher than the intrinsic frequency of its component cells. In the duodenum there is some process whose nature is unknown that raises the frequency of that part above its intrinsic rate.

Short Propulsive Movements

Chyme is moved down the intestine by short, weak propulsive movements; it would be disadvantageous if long, vigorous ones drove chyme rapidly downward without allowing time for digestion and absorption. In human subjects observed by cineradiography an annular contraction originating in the upper duodenum may be seen to move as a stripping wave into the upper jejunum. Such a wave pushes chyme before it; and when the chyme is propelled into the jejunum, it rarely returns in any significant quantity into the duodenum. The wave travels with a mean velocity of 1.2 cm/sec for an average distance of 15 cm.

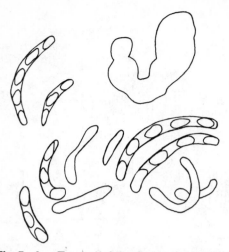

Fig 5–4.—Tracing of the fluoroscopic shadow of a cat's intestine during digestion of a meal of starch mixed with contrast medium. The parts marked with ovals were undergoing segmentation. (From original tracing of W.B. Cannon.)

On fluoroscopic examination of subjects whose intestinal contents are mixed with barium, the shadow is often seen to swing back and forth like a pendulum. The to-and-fro movement is caused by sequential contractions, first above and then below. There is no specific "pendular movement" of the intestinal musculature.

Although motility of the small intestine is increased following feeding, its propulsive activity is actually reduced. In a dog, propulsive activity can be measured by allowing isotonic fluid at body temperature to flow into the upper small intestine at a pressure of about 1 mm Hg. The quantity of fluid transported downward in unit time reflects the ability of the intestine to move its contents. The rate of transport is reduced about 80% when the animal is fed. The reduction in flow is caused by increased contraction of the intestine, which narrows its lumen and increases resistance to flow. On the other hand, progress of chyme may be abnormally rapid in the atonic bowel, for resistance to flow is small. This relation between motor activity and transit is exemplified by the action of drugs. Morphine causes spasm of the duodenum by suppressing activity of inhibitory neurons (see Fig 5–3) and increases phasic and tonic activity of the smooth muscle of the gastrointestinal tract. Reduced

rate of transit permits desiccation of gut contents, thus contributing to constipation. Castor oil and its active ingredient, ricinoleic acid, are not, contrary to popular opinion, irritants. A few milliliters of either introduced into the duodenum causes a brief initial increase in activity followed by prolonged inhibition.

Peristalsis

Peristalsis is a progressive wave of contraction of successive rings of circular muscle moving a shorter or longer distance down the small intestine. The peristaltic wave moves down the intestine at the velocity of the electrical control activity, and the frequency with which peristaltic waves follow one another is some submultiple of the frequency of the electrical control activity.

In the normally innervated small intestine, the excitability of the circular muscle is strongly influenced by the intrinsic plexuses, which, in turn, are influenced by their extrinsic innervation.

Distention of the intestine elicits the peristaltic reflex, which consists of a compound wave of contraction above and relaxation below. First, the longitudinal muscle contracts, and then the circular muscle contracts. The two contractions have a phase difference of 90 degrees, for the circular muscle begins to contract when the contraction of the longitudinal muscle is half completed.

Although the reflex is unimpaired by extrinsic innervation, it is abolished by asphyxia and application of cocaine to the mucosa. Treatment with hexamethonium blocks conduction in the ganglia and abolishes contractions of the circular coat without affecting preceding contractions of the longitudinal one. However, if the reflex function of the intrinsic plexuses is completely blocked by hexamethonium, a peristaltic wave can be made to pass down the intestine by infusing a parasympathomimetic drug such as bethanechol. This increases the excitability of the circular layer so that it responds with action potentials to the electrical control activity.

The myenteric plexus contains a neural organization activated by distention. A neurone traveling downward from the integrating center at the point of distention liberates a nonadrenergic inhibitor which relaxes the circular smooth muscle below. At the same time, a cholinergic neurone excited by distention causes contraction at and above the point of dis-

Dog's jejunum segmenting

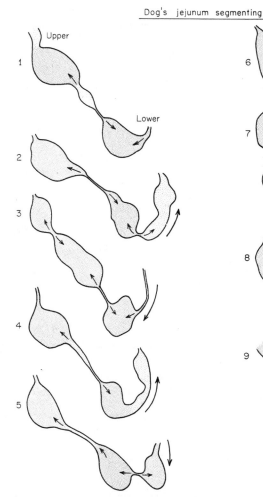

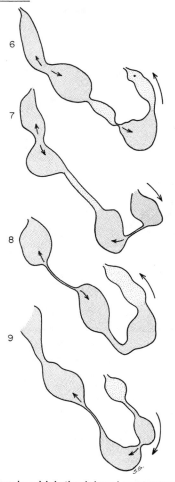

Fig 5-5.—Movement of chyme by segmentation. A dog had been fed a meal mixed with the x-ray contrast medium, barium sulfate, and continuous cinefluorographic pictures were taken as the contrast medium entered the upper jejunum. The tracings in this figure were made from the film to show the locus of the medium at approximately 1-sec intervals. The *arrows* show the direction in which the jejunal contents were moving at the instant represented by the tracing. Progress of the contents into the two distal segments shows that the intestinal contents can be moved downward from above by segmentation without peristalsis. (From Davenport H.W.: *A Digest of Digestion*. Chicago: Year Book Medical Publishers, 1975. Adapted from a film by H. C. Carlson.)

tention. Finally, a descending excitatory pathway having a long latency stimulates the relaxed circular muscle to contract. If the bolus has been pushed into the relaxed portion of the intestine, the distention it causes and the increased tension in the wall resulting from the delayed contraction causes a repetition of the cycle a little farther down the intestine. Thus, a peristaltic wave consisting of relaxation below the point of stimulation and contraction above moves downward.

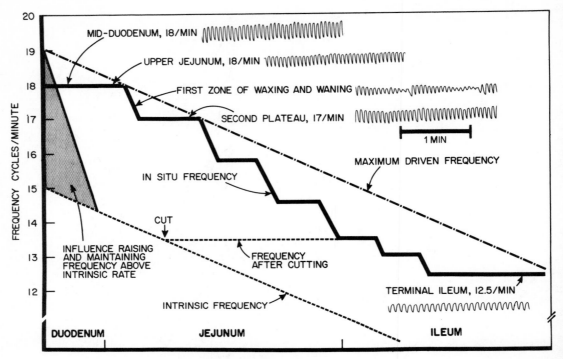

FREQUENCY CYCLES/MINUTE

MID-DUODENUM, 18/MIN

UPPER JEJUNUM, 18/MIN

FIRST ZONE OF WAXING AND WANING

SECOND PLATEAU, 17/MIN

1 MIN

MAXIMUM DRIVEN FREQUENCY

IN SITU FREQUENCY

CUT

INFLUENCE RAISING AND MAINTAINING FREQUENCY ABOVE INTRINSIC RATE

FREQUENCY AFTER CUTTING

INTRINSIC FREQUENCY

TERMINAL ILEUM, 12.5/MIN

DUODENUM   JEJUNUM   ILEUM

DISTANCE FROM PYLORUS

Fig 5–6.—Frequency of the basic electrical rhythm of the small intestine of the anesthetized cat. The stepwise *heavy line* shows the observed frequency, 18/min in the duodenum and upper jejunum and descending to 12.5/min in the terminal ileum. Segments of the electrical record from which the frequency was measured are shown. The *lower slanting dotted line* shows the intrinsic frequency displayed by short, isolated segments of the intestine. If the intestine is transected at the point labeled CUT, the frequency distal to the cut falls to that of the intrinsic frequency at the point of transection. (Adapted from Diamant N. E., Bortoff A.: *Am. J. Physiol.* 216:734, 1969. Electrical records supplied by A. Bortoff.)

The Interdigestive Myoelectric Complex

Overnight a man's digestive tract undergoes six or eight characteristic sequences of motion called the *interdigestive myoelectric complex*. Similar sequences occur in the fasting dog, and experimental studies using healthy, unanesthetized dogs have given a full description of the complex and have revealed some of the means by which the complex is controlled.

Under surgical anesthesia, electrodes are sewn in a row on the serosal surface of the gastric antrum and along the entire length of the small intestine. Leads from the electrodes are brought to the surface of the body. A dog with implanted electrodes may remain in good health for many months, and records of the electrical activity of the muscle of his stomach and small intestine can be made for hours or days. Strain gauges may be sewn to the muscle to permit recording of movements, or intraluminal catheters may be inserted to measure pressures. A small bolus of contrast medium may be infused through a catheter so that cinefluorography can be used to demonstrate how the complex moves the contents of the bowel.

When a dog has been fasted about 12 hours, the complex occurs with almost clockwork regularity (Fig 5–7). The four phases of the complex can be understood by referring to the figure.

I. The pacesetter potential moves along the intes-

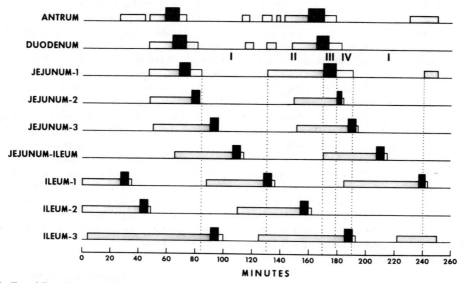

Fig 5–7.—Migrating complexes recorded from the antrum, duodenum, jejunum, and ileum of a fasted dog. See the text for identification and description of the complexes. (Adapted from Code C.F., Marlett J.A.: *J. Physiol.* 246:289, 1975.)

tine, but there are few action potentials. Consequently, there is little or no contraction of the bowel, and its contents lie in stagnant pools.

The record made at the site labeled *Jejunum 1* shows that Phase I lasted between the 83rd and the 130th minute of observation.

II. Persistent but random action potential activity signals contraction in step with the pacesetter potential. In the intestine, this is segmentation; the contents of the bowel move back and forth, but they do not move downward.

At the site of *Jejunum 1,* Phase II occurred between the 130th and the 170th minute of recording.

III. There is a sudden onset of continuously occurring action potentials at the peak of every pacesetter potential, and there are, consequently, strong and progressive waves of peristaltic contraction which rapidly sweep the contents of the bowel downward. This phase ends as abruptly as it began, and between start and finish 60–90 peristaltic waves are generated. The peristaltic waves stop before they travel the whole length of the intestine.

The peristaltic waves of Phase III passed *Jejunum 1* between the 170th and the 180th minute of recording. The fact that Phase III did not begin at *Jejunum*

2 until after it had stopped at *Jejunum 1* shows that the waves did not travel the entire distance between the two parts of the jejunum.

IV. There is a sudden decrease in the frequency of action potentials, and motor activity rapidly dies out at the former site of Phase III which is then occupied by another Phase I.

At *Jejunum 1,* Phase IV lasted between the 180th and 190th minutes, and it was succeeded by a new Phase I between the 190th and 240th minutes.

The period, propagation time, and velocity are characteristic of the particular dog studied. The length of each cycle is 90–114 min, and a complex migrates from stomach to terminal ileum in 105–134 min. The velocity of propagation is 6–12 cm/min in the upper part of the bowel, and it diminishes to 1–3 cm/min at the midpoint of the small intestine, remaining constant for the rest of the intestine's length. In man, the peristaltic rush occurring in Phase III moves with a mean velocity of 2 cm sec^{-1}. It has a length of 10 cm, and therefore it passes one point in 5 sec. However, the frequency and cycle duration of the interdigestive myoelectric complex in man are not so regular as they are in the dog.

In man, several of the six or eight complexes occurring overnight originate in the lower esophageal sphincter and pass over the whole stomach. The rest appear to start in the gastric antrum. The influence setting off the complexes may be a rise in plasma concentration of the hormone motilin. Its concentration is lowest during Phase I, and it rises to about four times that level during Phase III, after which it falls again. What causes plasma motilin to rise is unknown. Acid in the duodenum releases motilin, and it is possible that acid delivered at relatively high concentration from the nearly empty stomach releases enough motilin to start a complex. This, however, hardly accounts for the regularity of occurrence of the complexes. Administration of atropine, of course, abolishes the complexes, because it blocks cholinergic stimulation of gastrointestinal muscle. However, atropine also blocks the rise in plasma motilin, and it is possible that some regularly repeating central nervous event during fasting is expressed periodically through vagal impulses releasing motilin. In man and the dog, feeding abruptly terminates any ongoing complex. The rise in plasma gastrin concentration which follows ingestion of a meal is in part responsible for the termination. Migrating complexes occur in the rabbit and sheep, but, in those animals whose stomachs are seldom empty, feeding has no effect.

Migrating complexes thoroughly empty the stomach and small intestine of any residual contents, and for this reason they have been ascribed a "housekeeping" function. Gastric emptying is, in fact, so complete that a solid object, such as a swallowed coin, which is not emptied during the normal course of digestion, is swept into the small intestine and onward to the colon by an interdigestive complex. Migrating complexes are feeble or absent in some patients with bacterial overgrowth of the small intestine, but whether their absence is responsible for overgrowth has not been determined.

The pattern of the interdigestive myoelectric complex is set by the neuromuscular apparatus of the gut itself. The pattern is not determined by the vagus nerve, for it persists unchanged after proximal vagotomy. After truncal vagotomy the pattern is somewhat disturbed, probably because gastric emptying is delayed, and, hence, the stimulus of acid delivered to the duodenum is distorted. Once the gut is aroused to activity in Phase II, the sequence of segmentation and peristalsis continues until it has traversed the whole intestine. A pouch of the fundus of the stomach of a dog, totally extrinsically denervated by transplanta-tion, has large bursts of activity coincident with bursts of action potentials in the stomach and duodenum. A piece of small intestine with its nerve and blood supply intact may be cut from the middle of the bowel. Each end is brought to the surface of the body so that the isolated loop of intestine may be perfused. The continuity of the rest of the intestine is reestablished by end-to-end anastomosis. Some investigators find that in a dog prepared in this way, the complex migrates to the point of anastomosis, begins at the end of the loop which had been uppermost, travels to the lower end and then begins again at the point of anastomosis. Others have not found such a regular pattern of progress, only increased activity in the loop randomly associated with activity in the rest of the intestine. If the loop is perfused with glucose, migrating complexes do not occur in it although the rest of the intestine is active.

The Duodenal Bulb

The motility of the first part of the duodenum just distal to the pylorus, the duodenal bulb, differs from that of the stomach and of the lower duodenum. During gastric emptying the bulb may be filled during terminal antral contraction, and chyme spills over into the duodenum. In normal persons, there is little or no regurgitation of contents of the bulb into the stomach, for pressure in the pyloric canal is greater than that in the bulb. Most contractions of the bulb occur when the pyloric canal is firmly closed by terminal antral contraction. In patients with gastric ulcers, regurgitation is frequent.

The longitudinal muscle layer of the duodenal bulb has its own relatively rapid basic electrical rhythm. The antral electrical rhythm whose frequency is much less than that of the duodenum is conducted into the duodenal bulb by longitudinal muscle fibers along the lesser curvature. The two electrical rhythms interact (see Fig 4–7) with the result that every fifth or so duodenal depolarization is augmented by depolarization conducted from the antrum; such augmentation can be detected, in the monkey, as far as 17 mm beyond the pyloric junction. Augmented duodenal depolarization is often followed by contraction. This coupling of antral and duodenal rhythms accounts for the fact that most duodenal contractions follow antral contractions.

The duodenal bulb is a sensitive receptor area, responding to qualities of the chyme entering it from the stomach; and osmotic equilibration, neutralization, and intestinal digestion begin there and in the immediately succeeding few centimeters of duodenum.

Movements of the Mucosa and Villi

The mucosal surface of the small intestine, stomach, and colon is thrown into a complex pattern of folds. An important characteristic is the rapidity with which the pattern shifts. In the jejunum, for example, the mucosa may at one moment exhibit a pattern of high longitudinal folds closely connected by transverse creases and, at another, a series of simple circular folds, which cut the lumen into a series of disk-shaped chambers. This movement is independent of contraction of the circular and longitudinal muscle of the gut wall, and folding of the mucosa is not a passive result of larger movements of the intestine. It is the result of contraction of the muscularis mucosae, occurring spontaneously or in response to local stimuli.

The muscularis mucosae of the small intestine of the dog contracts in an irregular rhythm of 3 or 4 times per minute, and occasionally there is a slower rhythm on which the faster is imposed. Mechanical stimulation by gentle stroking or touching causes contraction lasting 30 sec to 1 min, which throws the mucosa into folds, ridges and pits. Solid matter in the intestine, lumps of food, or tapeworms are stimuli for contraction. The chyme itself can be a sufficient stimulus, and stimulation by chyme accounts for the rapid shifts in pattern of folds seen roentgenoscopically in the barium-filled intestine.

Stimulation of the sympathetic nerves causes contraction of the muscularis mucosae and mucosal blanching. The nerves are adrenergic, and epinephrine given topically or IV also causes contraction; it does not inhibit the muscularis mucosae as it inhibits the major muscle of the gut wall. The muscularis mucosae also responds to parasympathomimetic agents (e.g., acetylcholine) with contraction. Vagal stimulation has little effect, although sometimes it may cause small contractions; this may mean that parasympathetic postganglionic fibers do not generally reach the muscularis mucosae. However, afferent fibers from the mucosa pass centrally in the vagus, and these carry a train of impulses whose rhythm corresponds to contractions of the muscularis mucosae. Since the muscularis mucosae contracts during digestion, the

afferent impulses signal activity of the intestine, but their reflex significance is unknown. Afferent fibers in the splanchnic nerves also fire when the mucosa is irrigated with isotonic solutions of glucose or amino acids.

Villi of the intestinal mucosa move either by swaying to and fro or by abrupt contraction. Some of their movement results from contraction of the underlying muscularis mucosae. Partially purified extracts of human and dog intestinal mucosa cause villi to contract. The agent, assumed to be a hormone, is called villikinin. Individual villi contract at intervals, with no relation to the activity of their neighbors. The rhythm is irregular, being faster in the duodenum and jejunum and higher in fed, than in fasted, animals. Individual villi of the ileum seldom contract. The central lacteal of the villus is not completely emptied by its contraction, but lymph flow from the intestine is elevated during periods of increased activity of the villi.

Secretion Accompanying Motility

Movements of the small intestine, segmentation or peristalsis, are accompanied or immediately followed by increased secretion from the intestinal mucosa. Secretion is stimulated by a cholinergic mechanism, for it is suppressed by an anticholinergic drug. The function of secretion is to dilute and lubricate the chyme as it is being mixed. In the case of the interdigestive myoelectric complex, acid and pepsin secretion in the stomach is elevated in the 30 minutes before an activity front passes over the gastric muscle, and as the activity front leaves the duodenum there is a rise in the bicarbonate and amylase concentrations in duodenal contents.

Retrograde Movement and Effect of Reversal of Intestine

Segmentation and peristaltic movements carry most of the chyme downward, but with each contraction some fraction is transported upward. This accounts for the fact that some of any soot or other easily identifiable substance placed in the ileum can eventually be recovered from the stomach. Reverse peristalsis, which is rare or entirely absent from the human small intestine, need not be invoked to explain retrograde transport. During the spastic contraction of the whole duodenum accompanying nausea, its contents are squirted upward into the stomach, with the result that the vomitus is bile-stained, although reverse peristalsis may or may not have occurred. Surgical reversal of an intestinal segment is not necessarily fatal. Reversal changes the direction of peristalsis to one opposing downward movement; but if the reversed segment is short enough (under 20 cm) and if the diet is sufficiently bland, chyme can flow through the reversed segment. In fact, transplantation of the dog's pyloric antrum, with its direction reversed, to midjejunum does not seriously impede normal progression of intestinal contents. If the reversed segment is long or if the diet contains undigestible bits of solid, the reversed segment blocks movement, and fatal dilatation of the bowel at the upper end of the segment follows.

The normal aboral progress of the basic electrical rhythm in the muscle of the gastric antrum or the intestine can be reversed by retrograde pacing through a series of electrodes sewn on the serosal surface. In that case gastric emptying or progress of chyme through the intestine is delayed.

The Ileocecal Sphincter

The ileocecal sphincter, separating the terminal ileum from the colon, is normally closed. When an open-tip catheter is drawn from the dog's ileum into the colon, a zone of elevated pressure, about 1 cm long, is found in the cone-shaped segment composing the ileocolic junction. In man, the zone is 4 cm long, and its resting pressure is about 20 mm Hg greater than colonic pressure. Action of the sphincter is partially governed by extrinsic innervation. Stimulation of splanchnic or lumbar colonic nerves causes the sphincter to contract. The nerves act through adrenergic alpha receptors. Vagal stimulation causes the sphincter to contract through a cholinergic mechanism. The sphincter is also controlled by intrinsic mechanisms, for extrinsic denervation does not destroy its function. One factor in keeping it closed is a myenteric reflex from the cecum. Mechanical stimulation of the cecal mucosa or distention of the cecum causes pressure in the sphincter to rise and delays passage of chyme through it. On the other hand, the ileocecal sphincter relaxes each time a propulsive wave passes along the last few centimeters of the ileum, and on such occasions a small amount of chyme may squirt through the sphincter. Distention of the terminal ileum in man or dog causes a fall in

sphincter pressure (Fig 5–8). In man, the hormone gastrin in physiological doses causes increased segmentation in the terminal ileum at a frequency of 6–9/min, and at the same time it causes sphincter pressure to fall. Heightened activity of the ileum accompanies gastric secretion and emptying, and therefore emptying of the ileum through the sphincter is associated with eating. This relation is called the *gastroileal reflex*. The ileocecal sphincter has a valvelike action, which normally prevents regurgitation from cecum to ileum. It is not essential for this purpose; in carnivorous animals, such as the raccoon, the bear, the mink, and the skunk, which have no ileocecal sphincter, regurgitation is kept at a minimum by the generally downward direction of intestinal movement. The same gradient is adequate even in animals that have a sphincter, for its surgical removal in man and experimental animals is not followed by any significant regurgitation.

Long Intestinal Reflexes

The gastroileal reflex is one example of the influence of the activity of one part of the digestive tract on that of another, and some additional examples have been dignified with the name of reflex. Inhibition of gastric motility when the ileum is distended is called the *ileogastric reflex*. Immediate cessation of all intestinal movement (adynamic ileus) follows distention of an intestinal segment, rough handling of the intestine during abdominal surgery, or peritoneal irritation. Adynamic ileus operates through three pathways: general sympathetic discharge, the peripheral reflex pathway through the celiac and mesenteric plexuses (see Fig 1–1,*D*), and the intramural plexuses. Gaseous distention follows paralysis, exacerbating and perpetuating the condition. Cessation of intestinal movements as the result of distention is called the *intestino-intestinal reflex*. Receptors are in the longitudinal muscle, and they are activated by stretch. Inhibition ceases as soon as distention is relieved, as by draining the obstructed segment through a tube. The intestino-intestinal reflex is both facilitated and inhibited by higher nervous centers. Many cardiovascular reflexes are affected by impulses arising in small-intestinal receptors.

Hormones also influence intestinal motility. Serotonin (5-HT) is probably a neurotransmitter between sensory and motor neurons in the peristaltic reflex. It is released into intestinal venous blood during peri-

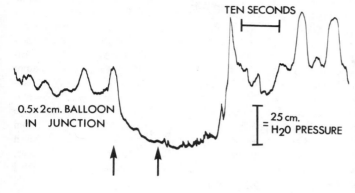

TEN SECONDS

0.5×2cm. BALLOON
IN JUNCTION

$= 25$ cm.
H_2O PRESSURE

Fig 5–8.—Record of pressure measured by means of a 0.5 × 2 cm balloon having a volume of 0.35 ml placed in the cone-shaped junctional zone between the ileum and the colon of an unanesthetized dog. The zone, or ileocolic sphincter, displays rhythmic fluctuations of pressure. At the first *arrow* a balloon in the ileum was distended by injection of 1 ml of water; at the second *arrow* it was emptied. (From Kelley M.L., Jr., Gordon E.A., DeWeese J.A.: *Am. J. Physiol.* 211:614, 1966.)

stalsis and when intraluminal pressure rises. Intravenous infusion of 5-HT into human subjects in a dose of 0.5 mg/min for 8 min stimulates upper and lower intestinal motility. Gastrin administered intravenously in a dose of 25–50 μg to human subjects also increases upper intestinal contractions.

Intestinal Blood Flow

Mesenteric, renal and coronary blood flow, heart rate, and arterial pressure have been measured by telemetry in unrestrained baboons at rest, during eating and after eating. The animals weighed between 22 and 26 kg, and the following figures are the means of the variables cited. At rest, mesenteric blood flow was 212 ml/min compared with renal blood flow of 150 and coronary blood flow of 25 ml/min. When the animals were given a meal of apples, oranges, and bananas, their heart rate rose at once from 72 to 142 beats per minute, and their mean arterial pressure went from 96 to 120 mm Hg. At the same time, mesenteric blood flow fell, and mesenteric resistance rose. These immediate changes probably reflected the excitement of fasted animals upon receiving food. Within 15 min, heart rate and mean arterial pressure returned nearly to normal, and mesenteric flow had risen to 373 ml/min chiefly as the result of diminished sympathetic vasoconstriction. Mesenteric flow remained high for an hour and slowly declined in 4 hours to control level.

Qualitatively similar results have been obtained in conscious dogs and in a chimpanzee who washed down three peanut butter sandwiches with a quart of milk.

No method for accurately measuring intestinal blood flow in normal human beings has been devised. Measurements made in anesthetized patients at laparotomy have found intestinal blood flow to be between 29 and 70 ml min^{-1} 100 gm of intestine^{-1}. Between 50% and 90% of the flow went to the mucosa and submucosa.

There are probably no parasympathetic vasodilator nerves to the stomach, small intestine, and colon. Vasodilation following parasympathetic stimulation is secondary to increased secretory and motor activity. There is abundant adrenergic sympathetic innervation of intestinal vascular smooth muscle, and sympathetic stimulation or infusion of norepinephrine increases vascular resistance by constricting precapillary sphincters. However, flow tends to return to the baseline value during prolonged stimulation or infusion as the result of autoregulatory escape. Reflexes within the intramural plexuses also control blood flow; gentle stroking of the jejunal mucosa of a cat causes a transient increase in flow through vessels of the villi. This reflex is blocked by local anesthetics and by antagonists of 5-HT but not by atropine.

Blood flow through the intestinal mucosa is locally controlled by metabolism, intrinsic reflexes, and by response of vascular smooth muscle to blood pressure. Blood flow increases when oxygen demand increases, as it does in other tissues. It also rises when the partial pressure of oxygen in arterial blood is reduced; in that case oxygen consumption is maintained by greater oxygen extraction. Presence of chyme in the lumen increases blood flow through the mucosa, but not through the muscle, chiefly through a local reflex. Duodenal blood flow increases when the pH of chyme is 3 or lower. Of the major constituents of chyme in the process of digestion, only long-chain fatty acids and bile salts cause significant mucosal vasodilation. Bile salts themselves have little effect, but they greatly enhance the hyperemic effect of digested food. During absorption of glucose with sodium or of amino acids with sodium, previously closed mucosal capillaries open; because total blood flow through the tissue is unchanged, oxygen extraction increases. Cholecystokinin, secretin, and gastrin all cause vasodilation when infused, but their role in postprandial hyperemia has not been determined. When blood pressure falls, vascular resistance decreases as the result of relaxation of vascular smooth muscle in response to diminished stretch. Villous vessels have a greater capacity to relax than do vessels in the deeper part of the mucosa; this is important in maintaining oxygenation of the tips of the villi. However, during hemorrhagic shock, circulation through the villi is inadequate, and their tips become necrotic. When arterial pressure is low, venules draining the intestine contract as the result of a local reflex; capillary pressure and capillary filtration are maintained. Arterial resistance vessels usually contract when arterial pressure is high. When portal pressure rises, the common result is contraction of precapillary resistance vessels. In both these instances an untoward increase in capillary pressure and filtration is prevented.

Gastrointestinal capillaries are more permeable to plasma proteins than are most other capillaries with the result that gastrointestinal interstitial fluid has a higher concentration of plasma proteins than does, for example, muscle interstitial fluid. Consequently, the colloid osmotic pressure difference across intestinal capillaries is small, and a rise in capillary hydrostatic pressure causes substantial net filtration and an expansion of interstitial volume. Mature intestinal epithelial cells are extruded from the tips of the villi, leaving a gap between cells. When interstitial volume and pressure are high, interstitial fluid containing plasma proteins is forced through the gap into the lumen.

Blood vessels supplying intestinal villi form hairpin loops, and consequently they are the site of countercurrent exchanges. Each villus has a central arteriole, which arises from the submucosal network and which runs in the core of the villus without branching. This vessel has no muscular coat. At the tip of the villus the central vessel arborizes into a dense, subepithelial capillary network, which descends the villus and returns blood to venules at the base of the villus. The mean transit time of plasma in a villus is 4–8 sec at rest, about 1 sec during vasodilatation, and 20–30 sec during low perfusion occurring in hemorrhagic shock. Plasma skimming occurs at the origin of the central arteriolar vessel, and the hematocrit of blood flowing through the villi is only 50%–60% of that in major arterial vessels.

This anatomical arrangement permits countercurrent exchange of substances between blood flowing in opposite directions in the two limbs of the hairpin loop. The effect depends upon whether the substance enters the loop at the tip or the base. If an ion such as sodium is delivered to the blood at the tip of the loop by active absorption, it is in a higher concentration in the venous blood descending the loop than in the ascending arterial blood. The osmotic pressure of blood in the descending loop is also higher than in the ascending arterial blood. Consequently, sodium tends to diffuse from the descending vessel into the ascending one, and its concentration in the blood flowing toward the tip of the villus increases (Fig 5–9). At the same time, water flows along its osmotic gradient from the ascending blood to the descending blood, and this also increases the sodium concentration and osmotic pressure of the blood reaching the tip of the villus. As long as active absorption of sodium continues, countercurrent multiplication keeps the concentration of sodium and the osmotic pressure high at the tip of the villus. The osmotic pressure in a cat's intestinal villus during active absorption of sodium may be as high as 1,200 mOsm at the tip.

Any substance entering the veins at the tip of the villus that can diffuse through the walls of the blood vessels undergoes the same process of multiplication. Oleic acid absorbed during fat digestion and carbon dioxide produced by metabolism in the epithelial cells are concentrated at the tips of the villi by countercurrent multiplication.

On the other hand, oxygen arriving at the arteriolar end of the vessels is short-circuited by diffusion into venous blood. If a slug of ^{51}Cr-labeled erythrocytes whose hemoglobin carries isotopically labeled oxygen is injected into a small artery serving a length of intestine, the isotopic oxygen appears in the venous blood draining the intestine 1–2 sec before the erythrocytes. Because oxygen can pass from arterial to venous blood without traversing the end-loops of the capillaries, the partial pressure of oxygen in the end-loops is low.

The effectiveness of countercurrent multiplication

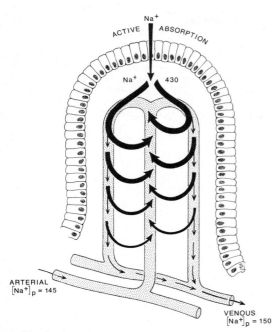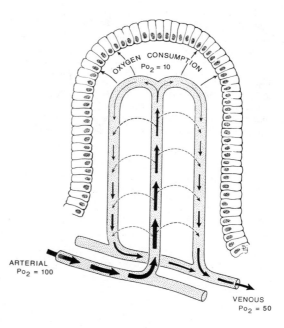

Fig 5–9.—Countercurrent effects in intestinal villi. **Left,** the multiplication of the concentration of sodium at the tip of a villus during active absorption of the ion. **Right,** production of a low partial pressure of oxygen at the tip of a villus as the result of short-circuit transfer of oxygen from the ascending arteriole to the descending venules.

or dilution is a function of the rate of blood flow through the hairpin loops. An increase in the rate of flow tends to wash out the concentration gradients, and a decrease in flow exaggerates the gradients. In the resting state, enough oxygen reaches the end-loops to meet the needs of the epithelial cells, and when blood flow increases as the result of active hyperemia, the partial pressure of oxygen in blood reaching the end-loops increases. When blood flow is substantially slowed, as in hemorrhagic shock, the partial pressure of oxygen at the tips is so low that epithelial cells die and slough off. This kind of necrosis of the tips of the villi can be prevented by perfusing the intestinal lumen with oxygenated fluid.

REFERENCES

Cheung D.W., Daniel E.E.: Comparative study of the smooth muscle layers of the rabbit duodenum. *J. Physiol.* 309:13, 1980.

Code C.F.: The interdigestive housekeeper of the gastrointestinal tract. *Perspect. Biol. Med.* 22:49, 1979.

Code C.F., Marlett J.A.: The interdigestive myo-electric complex of the stomach and small bowel of dogs. *J. Physiol.* 246:289, 1975.

Code C.F., Marlett J.A., Szurszewski J.H., et al.: A concept of control of gastrointestinal motility, in Code C.F. (ed.): *Handbook of Physiology:* Sec. 6. *Alimentary Canal,* vol. 5. Washington, D.C.: American Physiological Society, 1968, pp. 2881–2896.

Daniel E.E., Gilbert J.A.L., Schofield B., et al.: (eds.): *Proceedings of the Fourth International Symposium on Gastrointestinal Motility.* Vancouver, Mitchell Press, 1974.

Edwards D.A.W., Rowlands E.N.: Physiology of the gastroduodenal junction, in Code C.F. (ed.): *Handbook of Physiology:* Sec. 6. *Alimentary Canal,* vol. 4. Washington, D.C.: American Physiological Society, 1968, pp. 1985–2000.

Faulk D.L., Anuras S., Christensen J.: Chronic intestinal pseudoobstruction. *Gastroenterology* 74:922, 1978.

Fleckenstein P.: Migrating electrical spike activity in the fasting human small intestine. *Am. J. Dig. Dis.* 23:769, 1978.

Furness J.B., Costa M.: Adynamic ileus, its pathogenesis and treatment. *Med. Biol.* 52:82, 1974.

Granger D.N., Richardson P.D.I., Kvietys P.R., et al.: Intestinal blood flow. *Gastroenterology* 78:837, 1980.

Haglund U.: The small intestine in hypotension and hemorrhage. An experimental cardiovascular study in the cat. *Acta Physiol. Scand.* [suppl.] 387:1, 1973.

Lewis T.D., Collins S.M., Fox J-A.E., et al.: Initiation of

duodenal acid-induced motor complexes. *Gastroenterology* 77:217, 1979.

Lundgren O.: The circulation of the small bowel mucosa. *Gut* 15:1005, 1974.

Melville J., Macagno E., Christensen J.: Longitudinal contractions in the duodenum: Their fluid-mechanical function. *Am. J. Physiol.* 228:1887, 1975.

Miller L.J., Malagelada J-R., Go V.L.W.: Postprandial duodenal function in man. *Gut* 19:699, 1978.

Stewart J.J., Bass P.: Effects of ricinoleic and oleic acids on the digestive contractile activity of the canine small and large bowel. *Gastroenterology* 70:371, 1976.

Thomas P.A., Kelly K.A.: Hormonal control of interdigestive motor cycles of canine proximal stomach. *Am. J. Physiol.* 237:E192, 1979.

Thompson D.G., Wingate D.L., Archer L., et al.: Normal patterns of human upper small bowel motor activity recorded by prolonged radiotelemetry. *Gut* 21:500, 1980.

Vatner S.F., Patrick T.A., Higgins C.B., et al.: Regional circulatory adjustments to eating and digestion in conscious primates. *J. Appl. Physiol.* 36:524, 1974.

6

Movements of the Colon

THE COLON OF an adult man receives between 500 and 2,500 ml of chyme each day. This contains undigested and unabsorbed residues of food in addition to water and electrolytes. Man's usual diet contains some material, chiefly walls of plant cells, which is not easily digested higher in the tract. This residue accumulates and is to some extent digested in the cecum. The contents of the cecum and of the ascending colon are kneaded by nonpropulsive churning movements, and they are slowly dried by absorption of water and electrolytes until they are the consistency of mush. Haustral movements, multihaustral contractions, peristalsis, and mass movements slowly move the contents of the colon toward the rectum. Net transport is increased during digestion of a meal, and the sigmoid colon and rectum are frequently filled at the end of a meal. Distention of the rectum arouses the defecation reflex whose lowest nervous center is in the sacral segment of the spinal cord. If the reflex is facilitated, or not inhibited, by higher centers, the act of defecation, involving coordinated movement of both voluntary and involuntary muscles, follows. If defecation does not occur, the rectum relaxes, and its contents are often returned to the sigmoid colon. Tension in the wall of the rectum is reduced, and defecation is postponed. Feces remaining in the rectum or sigmoid colon may be dried to rocklike masses.

Structure of the Colon

In man, the longitudinal muscle of the colon is concentrated in three bands, the taenia coli; and a thin layer of longitudinal muscle is present in the intervening spaces (Fig 6–1). The walls of the cecum and ascending colon are usually folded into sacs, or haustra, by contraction of the circular muscle. Haustra are not fixed structures. The proximal end of the colon is closed by the ileocecal sphincter, and the distal end by a thickening of the circular muscle, which forms

the powerful internal anal sphincter. The taenia coli broaden at the sigmoid colon, and a thick layer of longitudinal muscle completely encircles the anal sphincter. The longitudinal muscle ends by fanning into muscle fascia and skin in the perineal region. Its contraction elevates and shortens the anal canal. This canal, distal to the internal sphincter, is surrounded by striated muscle fibers arranged in two groups, the superficial and the deep external anal sphincters, which overlap to a greater or lesser degree with the smooth muscle of the internal sphincter. The puborectalis muscle arches around the uppermost part of the anal canal; it has no posterior attachment. It has the same innervation as does the external anal sphincter, and it is contracted except during defecation. It pulls the upper part of the anal canal forward so that the axis of the anal canal forms a right angle with the axis of the rectum. This angle is obliterated when the puborectalis muscle relaxes during defecation. The angle is also reduced when the hips are flexed more than 90 degrees, the usual position for defecation. The anus passes through and is supported by the pelvic diaphragm, striated muscle, which forms the pelvic floor.

Measurement of blood flow through the human colon made on anesthetized patients during abdominal operations found flow to average 18 ml/min 100 gm of tissue, with a range of 8–35. These values are similar to those obtained in anesthetized cats. In the cat, stroking the mucosa of the colon causes an increase in mucosal flow up to 50% of control values. In the proximal colon this hyperemia is the result of a local reflex involving release of 5-HT; in the distal colon the reflex acts through the pelvic nerves.

Innervation of the Colon

Both myenteric and submucous plexuses are present throughout the colon. Their organization and function

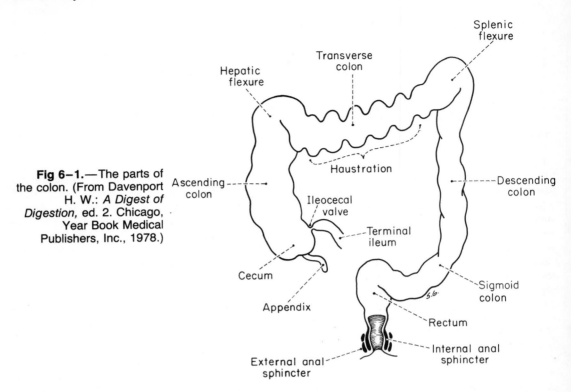

Fig 6–1.—The parts of the colon. (From Davenport H. W.: *A Digest of Digestion,* ed. 2. Chicago, Year Book Medical Publishers, Inc., 1978.)

are similar to those in the rest of the bowel; they form a competent nervous system capable of mediating most of the secretory, motor, and vascular reflexes largely independent of extrinsic innervation.

Extrinsic parasympathetic innervation comes to the proximal colon by way of preganglionic fibers of the vagus. Anatomists dispute how far distally vagal influence goes, but it perhaps reaches the first third of the transverse colon. Parasympathetic preganglionic fibers to the colon reach it by the pelvic nerves, which arise chiefly from the sacral segments of the cord; these overlap vagal innervation of the proximal colon. The preganglionic fibers are cholinergic, and they synapse with cells in the intrinsic plexuses, whose postganglionic fibers are likewise cholinergic. Therefore, parasympathomimetic drugs, anticholinesterases, and acetylcholine congeners have the same effect on the colon as stimulation of these nerves. In the cat, which has a colon similar to man's, stimulation of vagal fibers to the colon increases rhythmic segmental mixing movements in the proximal colon only; the distal colon is unaffected. Stimulation of the pelvic nerves facilitates expulsive movements of the

entire colon; contraction is not rhythmic but sustained. An exception is that the internal anal sphincter is relaxed. The pelvic nerves, but not the vagus, carry vasodilator fibers to the colonic mucosa; the mediator is probably kallikrein.

Preganglionic cholinergic sympathetic fibers arise from lumbar roots and pass through the sympathetic chain to end in the inferior mesenteric ganglia, two to four in number, contained in the inferior mesenteric plexus. Postganglionic sympathetic fibers to the distal colon travel in the inferior mesenteric and hypogastric nerves. Postganglionic fibers also reach the colon from the splanchnic nerve.

Many adrenergic postganglionic sympathetic fibers end on blood vessels where they are vasoconstrictor. Many other similar fibers end within the ganglia of the intrinsic plexuses, modulating ongoing reflex activity. For example, splanchnic fibers to the colon mediate the intestinocolonic reflex by which the colon is inhibited when the intestine is distended. In addition, postganglionic sympathetic fibers end on or near smooth muscle cells in some parts of the colon where they affect local function. The ileocecal sphincter is under local control, being relaxed by distention of the terminal ileum and contracted by pressure on the cecal mucosa, but impulses reaching the sphincter through the splanchnic nerves cause pressure in the sphincter to rise. Numerous adrenergic sympathetic fibers innervate muscle of the sigmoid colon and internal anal sphincter where they participate in the defecation reflex.

Many afferent nerves traveling centrally with efferent sympathetic fibers carry impulses which arouse the sensations of fullness and pain when the colon is distended, and they are the afferent limb of many somatic, cardiovascular, and gastrointestinal reflexes.

The prevertebral ganglia are not merely way-stations in efferent pathways; they are integrating centers for the colon and, probably,* for the rest of the gut as well. For example, postganglionic neurones in the inferior mesenteric ganglion receive excitatory impulses carried in afferent fibers from the colon, most of whose cell bodies are in the wall of the colon itself. The receptors in the colonic wall are mechanoreceptors whose frequency of firing increases with intraluminal pressure and with propulsive contractions. The same postganglionic neurones also receive input from preganglionic fibers of central origin. The output of the postganglionic fibers reduces or abolishes motility of the distal colon, and, consequently, colonic motility is in part governed by synaptic integration within the inferior mesenteric ganglion of inputs originating both peripherally and centrally. A large fraction, from one third to one half, of the postganglionic cell bodies in the celiac and superior mesenteric ganglia also receive excitatory synaptic input from the colon.

Electrical Control Activity of the Colon

Electrical activity of the colon operates on the same principles as in the stomach and small intestine: basic rhythms in the longitudinal muscle influence the occurrence of action potentials in both longitudinal and circular layers and therefore determine mixing and propulsion. However, as compared with gastric and intestinal electrical control activity, that in the colon

---

*Other examples: Distention of the ileum inhibits bile flow in preparations in which the celiac ganglion has been decentralized. Inhibition of gastric contractions by hydrochloric acid in the duodenum is eliminated by celiac ganglionectomy.

is highly irregular, particularly in the ascending and sigmoid parts.

Although many records have been obtained from human subjects by means of electrodes passed through the anus and clipped to the mucosa, the most comprehensive observations have been made on human subjects undergoing cholecystectomy. In such a patient electrodes are sewn to the taenia coli at selected parts of the colon. The leads are brought out along with the T-tube draining the biliary tract and are connected to instruments for recording and analysis. The following description is drawn from one such set of observations on 15 patients begun 36–48 hours postoperatively and continued until the electrodes were removed along with the T-tube.

Frequency of electrical control activity varies from cycle to cycle and from minute to minute at the same electrode. At one site in the ascending colon, dominant frequencies of 2.0, 3.2, 5.5, and 10.2 cycles min^{-1} may be found, but the frequency varies at any one peak by 1–2 cycles min^{-1}. Multiple frequency peaks are found throughout the colon. In general, there is a region beginning in the proximal transverse colon and ending, usually, in the distal transverse colon where high frequencies occur. There the mean frequency is, 10.9 cycle min^{-1}. In the ascending colon and in the descending colon, both high and low frequencies are found, but the dominant frequency is 4.2 cycles min^{-1}.

If the electrical wave moves progressively from one point to another, the signals are phase locked; the signal at the distal point has a definite temporal relation to the signal at the proximal point. If activity at one point is independent of that nearby, the signals are not phase locked. Phase locking is found in the middle part of the colon which is dominated by higher frequencies, and when action potentials accompany the control activity, contractions are progressive. In the more proximal and more distal parts of the colon, the electrical control activity is not phase locked; this is consistent with random contractions mixing but not propelling the contents.

Multihaustral propulsion, long peristaltic waves, and mass movements are rare in the colon, and their electrical signs have not been fortuitously recorded.

Movements of Cecum and Appendix

The cecum contains only gas or the remains of previous meals when it is being slowly filled by the gastroileal reflex, which empties the terminal ileum. In the cat and the dog, distention of the wall of the cecum during filling initiates antiperistalsis. These waves begin at a tonic ring of contraction high in the ascending colon and pass slowly backward into the cecum. They are not preceded by a wave of relaxation. Six or so successive waves may be present at one time, and periods of antiperistaltic activity occur in cycles lasting 2–8 min. Their effect is to drive the contents of the colon into the cecum and to mix them thoroughly, but they never cause regurgitation through the ileocecal sphincter. Antiperistaltic waves are not prominent in man, and some experienced radiologists say they have never seen them. Others report that antiperistaltic waves do occur in man, that they begin in the transverse colon close to the hepatic flexure, and that they are weak and shallow. The vermiform appendix of man is filled from the cecum. When the cecal content contains contrast medium, the appendix throws a shadow, which goes through wormlike writhing movements. At other times, the shadow is divided into five or more beads. Then the appendix empties itself, presumably by peristalsis, only to be refilled a few moments later.

Types of Colonic Movement

When the human colon is studied by time-lapse cineradiography, the chief types of movement seen are haustral shuttling, segmental propulsion, multihaustral propulsion, and peristalsis. The last two may produce mass propulsion of colonic contents. The frequency with which these forms occur in human subjects (each observed for a full hour) is given in Table 6–1.

Annular contractions in the transverse and descending colon fold the mucosa into sacs called haustra. The contractions form and reform at different sites, and the commonest form of movement detected in a subject fasting and at rest is a shuttling one in which liquid or semiliquid haustral contents are displaced short distances in both directions by apparently random segmental contractions of the circular muscle. The contractions are not progressive. There is little or no net movement of the fecal mass; it is slowly kneaded while water is absorbed. Similar contractions occur in the sigmoid colon, and they are responsible for the ovular shape of well-formed feces.

Segmental propulsion occurs when the contents of a single haustrum are displaced into the next segment, and thence into the one beyond, without being returned to the first. Subsequent contraction of the muscle surrounding the third and more distal haustra may expell their contents in both directions, but net movement of feces is in the aboral direction. In one instance, such propulsive activity involved an 18-cm section of colon over a period of 5 min. Retropulsion by means of an identical but adoral displacement of contents through two or more haustra is about two thirds as common as orthograde propulsion.

Systolic multihaustral propulsion occurs when a number of adjacent segments contract more or less simultaneously. Part or all of their contents is displaced into a neighboring length of bowel whose interhaustral folds are obliterated. The recipient segments may in turn contract in the same manner. A train of three such systolic multihaustral contractions occurring in about 10 min is shown in Figure 6–2. Infrequently, the recipient segment, instead of contracting as a unit, may undergo segmental haustration in which one after another of its original segments, beginning at the proximal end, recovers its original shape and size. The surplus contents slowly advance, distending still more distal parts of the colon.

Peristalsis in the colon consists of a progressive wave of contraction, which advances steadily, pushing the fecal mass ahead of it, at the rate of 1–2 cm/min. Muscular relaxation precedes the wave of contraction, and the relaxed segment is frequently filled with gas. The wave is followed by sustained contraction, which keeps the contracting segment of bowel closed and empty from 5 min to an hour. Reverse peristalsis does occur, but it is extremely rare.

TABLE 6–1.—PERCENTAGE OF SUBJECTS SHOWING VARIOUS TYPES OF COLONIC MOVEMENTS OBSERVED BY TIME-LAPSE CINERADIOGRAPHY FOR A FULL HOUR'S FASTING, AFTER A MEAL, OR AFTER INTRAMUSCULAR INJECTION OF 0.25 MG CARBACHOL*

	FASTING	AFTER A MEAL	AFTER CARBACHOL
No. of subjects	101	63	29
Haustral shuttling	38%	13%	13%
Segmental propulsion	36%	57%	52%
Multihaustral propulsion	9%	17%	41%
Peristalsis	6%	8%	28%

*From Ritchie J.A.: *Gut* 9:442, 1968.

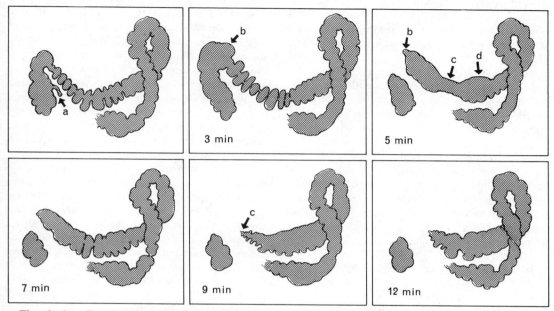

Fig 6–2.—Drawings made from cineradiographic studies of multihaustral propulsion in the normal human colon. At the time of the first observation *(upper left)*, barium-impregnated ileal contents *(arrow a)* were entering the cecum. Three minutes later, contraction of the cecum propelled a large part of its contents upward to distend the region of the hepatic flexure *(arrow b)*. After another 2 min, this section also contracted, and its contents were distributed over the proximal half of the transverse colon. Haustral markings disappeared over most of this length, and there was some narrowing of the mass in 2–3 in. of the middle section *(arrow c)*. Material forced out of the narrow section was accommodated by distention of the next four haustra *(arrow d)*. Four minutes later, as haustration was returning, most of the proximal half of the transverse colon between *b* and *c* also contracted. All the contents expelled from this section lodged in the distal half of the transverse colon. The conical outline at *c* in the fifth picture is typical of multihaustral contraction. (Adapted from Ritchie J.A.: *Gut* 9:442, 1968.)

Haustral shuttling is the most frequent type of movement in the fasting subject, but its occurrence diminishes after eating or when the colon is stimulated by a parasympathomimetic drug. Then segmental and multihaustral propulsion becomes more frequent. Persons having more than one bowel movement a day show 3 times as many propulsive movements as do those who habitually have only one daily evacuation.

Propulsive movements of a normal subject force the contents of the colon forward at a mean rate of 8 cm/hr. However, retropulsion forces the contents backward at 3 cm/hr, so that net forward movement is only 5 cm/hr. Net movements in a constipated subject, one having fewer than four bowel movements a week, is 1 cm/hr. After feeding, the forward rate of propulsion in a normal subject is 14 cm/hr. Retropulsion does not increase proportionately, and net propulsion becomes 10 cm/hr. After intramuscular injection of carbachol, net propulsion rises to 20 cm/hr.

Rate of Passage Through the Intestine

High-residue diets pass more rapidly through the entire gut more rapidly than do low-residue diets. The effect is probably entirely upon the colon, for addition of bran to the diet actually decreases the rate of flow of digesta through the jejunum. Individual particles of food residues go back and forth under the influence of intestinal movements, and several particles ingested together may be evacuated at widely different times. The residue of one meal catches up and mixes with that from previous meals, and a particular stool is a blend of residues of food eaten over several previous days.

Two thirds of normal subjects eliminate all but traces of a small amount of barium sulfate taken with a meal within 3 or 4 days, but the rest require 5 or more days to do so. Since chromate is not absorbed from the intestine, rate of passage may be measured by giving a small amount of sodium chromate labeled with ^{51}Cr by mouth; individual stools are subsequently collected, and their radioactivity is measured. In one such study on ten normal subjects, the time before the initial appearance of ^{51}Cr in the stool ranged from 4 to 10 hours, and the final appearance of ^{51}Cr was between 68 and 165 hours. There were distinct differences among subjects; some excreted ^{51}Cr relatively early; others late.

In another study, each of 59 normal adults on a daily diet containing 14.4 gm of crude fiber swallowed 20 radiopaque pellets shortly after breakfast. On the average, 16 of these pellets were eliminated in the stool on the first day, and all had been passed by the sixth.

In another study, normal persons were fed a different kind of pellet on each of three successive days. The results demonstrated that there is a pool, probably in the cecum and right colon, where residues are mixed, for in 83% of the instances, one or more stools contained all three kinds of pellets.

Pressures in the Colon

Pressures within the lumen of the colon are measured by open-tip catheters or by a catheter whose opening is covered by a small balloon. The open-tip system accurately records pressure, but it gives no information about muscular contraction. The small balloon records pressure, but it also responds to mechanical deformation produced by contraction, making no distinction between the two. Since contraction can occur without change in pressure, and change in pressure can be produced by distant contraction, a record obtained from a balloon alone may be misleading. Combined pressure recording (preferably by means of an open-tip catheter) and cineradiography give more readily intelligible information. The following paragraphs are based on such studies.

Extensive shifts of colonic contents often occur

with very little change in intraluminal pressure, and large variations in pressure may not be accompanied by transport of feces. Pressure within the lumen of the colon is generated by a combination of contraction and resistance. If adjacent segments of bowel are patent, their resistance is low, and contraction in one segment pushes its semiliquid contents into the area of low resistance with a pressure gradient of only a few mm Hg. If interhaustral contractions completely or nearly completely occlude the lumen, contraction can raise intraluminal pressure 70–80 mm Hg without displacing the contents of the haustrum.

Very slow changes in pressure, amounting to 2–3 mm Hg, occur over a period of 2–5 min in the human descending and sigmoid colon. Local contractions are isometric—i.e., without change in volume—if the colonic segment is closed at both ends. Although a high pressure in a single segment may be attained, the subject experiences no discomfort. Concurrent contractions of many segments are often painful. Contractions may be isotonic—i.e., without change in pressure—if the contents are free to move. These local contractions are often irregularly rhythmic, one frequency being about 12/min and another 2/min. In a normal subject at rest, 14 pressure peaks, averaging 27 mm Hg, appear each hour and pressure is elevated above baseline 10%–19% of the time. If the subject is engaged in an emotionally charged conversation about any topic likely to arouse anxiety, the frequency of contractions rises to an average of 34/hr, and pressure is elevated 22%–45% of the time.

Propulsive movements, consisting of waves of contraction moving from one segment to another in either direction, are preceded by a zone of raised pressure as gas is forced ahead of them. The pressure reached depends more on resistance than on contractile force. On the other hand, peristaltic contractions are preceded by a zone of low pressure as the wall of the colon relaxes ahead of the advancing wave of contraction. Solid feces, scybala, appear to be propelled by peristalsis only when resistance is drastically reduced.

Propulsive movement usually distends the colon ahead of it, and if the segment distended has been quiet, distention evokes a secondary contraction. After a latent period of 3–4 sec, the longitudinal muscle contracts, and within the next 4 sec, contraction of the circular muscle follows. If contraction is strong, the feces are pushed back.

The slowness of secondary contraction explains

why gas moves through the colon so much faster than feces. Liquids or semisolids fill each haustral pocket before moving on to the next, and they are stopped by secondary contraction. Gas, however, rapidly passes through the interhaustral constrictions; although it distends each segment as it passes, it has blown by before a secondary contraction can stop it.

Intraluminal pressure changes in children or adults with uncomplicated diarrhea are actually subnormal, but intraluminal pressure waves in persons with constipation are more frequent and stronger than those in normal subjects (Fig 6–3). The explanation of this apparant paradox is that pressure is generated by contraction plus resistance. In diarrhea, no resistance is offered by local segmental contractions, and feces trickle along until they reach the rectosigmoid area, where the defecation reflex is aroused. In constipa-

tion, strong segmental contractions offer resistance, and they impede movement of feces.

This explanation is confirmed by observations on the effects of drugs or dietary fiber. A bolus infusion of ricinoleic acid, the active ingredient of castor oil, causes a brief spasm followed by relaxation prolonged for as much as 45 minutes. Thus, ricinoleic acid, which causes diarrhea, is not a stimulant. On the other hand, morphine, which causes constipation, stimulates the colon to contract. A person on a high-fiber diet has more frequent, larger, and softer stools than when he is on a low-fiber diet. If we may extrapolate from observations made on the monkey, the average pressure in his descending colon during the high-fiber diet is about half that prevailing during the low-fiber diet.

The Gastrocolic Reflex and Mass Propulsion

Stimulation of the colon during or immediately after a meal is the result of the *gastrocolic reflex*. The reflex is, in part, stimulated by inflow from the ileum which itself increases after a meal. In one series of cineradiographic observations on human subjects, ileal inflow to the colon substantially increased after lunch in 8 of 37 subjects; in the whole group, net colonic propulsion rose on the average from 3 to 10 cm/hr. In those subjects with the greatest ileal inflow, net propulsion was 34 cm/hr. However, the gastrocolic reflex also occurs in a patient whose intestinal contents cannot enter the colon. Consequently, it must be mediated as well by nerves or hormones. Within 10 minutes after a meal eaten quickly, distal colonic activity is at its peak, which suggests that the reflex is neurally mediated.

In a subject given an anticholinergic drug, this early response is prevented, but his colon experiences a later increase in motility at the time hormones released by a meal should be reaching their peak. Fat in the diet, which releases cholecystokinin, is the predominant stimulus of colonic motility after eating, but it has not been demonstrated that cholecystokinin is the responsible hormone. Whole proteins and amino acids, but not carbohydrates, inhibit the response to fat, and thus the effect of a whole meal is a balance of stimulation and inhibition.

The increased activity in the right colon that follows eating, or sometimes discussion of food or listening to a lecture on defecation, stimulates mass pro-

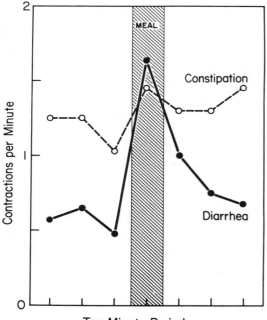

Fig 6–3.—Average number of contractions in successive 10-min periods observed by means of a 7 × 10 mm balloon 20 cm from the anal margin in 12 patients with diarrhea and in 14 patients with constipation. The meal consisted of two eggs, bread, butter, fruit, and two cups of tea. (Adapted from Waller S.L., Misiewicz J.J., Kiley N.: *Gut* 13:805, 1972.)

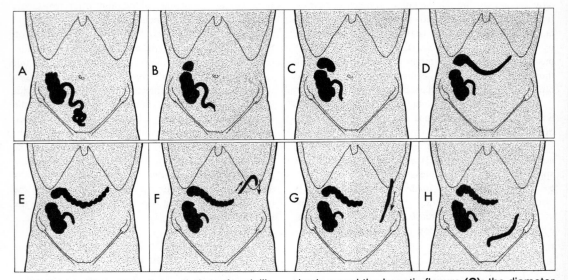

Fig 6–4.—The original description of a deliberately evoked mass movement reads: "A normal man had an ordinary breakfast with two ounces of barium sulfate at 7 A.M. At 12 noon the shadow of the end of the ileum, the cecum and the ascending colon was visible **(A)**. He then had an ordinary luncheon consisting of meat, vegetables and pudding. During the meal the cecum and ascending colon became more filled owing to the rapid emptying of the end of the ileum, and toward the end of the meal a large, round mass at the hepatic flexure became cut off from the rest of the ascending colon **(B)**. Immediately after the meal was finished some of this was seen to move slowly round the hepatic flexure **(C)**; the diameter of the separated portion then became suddenly much smaller, the large round shadow being replaced by a long narrow one, which extended from the hepatic flexure almost to the splenic flexure **(D)**. The shadow was at first uniform, but in a few seconds haustral segmentation developed **(E)**. About 5 minutes later the shadow suddenly became still more prolonged and passed round the splenic flexure **(F)**, down the descending colon **(G)** to the beginning of the pelvic colon **(H)**." (Adapted from Hertz A. F., Newton A.: *J. Physiol.* 47:57, 1913.)

pulsion. When primed by movement of contents from the right colon, mass propulsion often begins in the middle of the transverse colon as a series of multihaustral movements or as peristalsis (Fig 6–4). Contents advance into a narrow, tubular section of bowel at least 20 cm long. Bowel contents are transported at various speeds, commonly 2–5 cm/min, as far as the pelvic colon. On some occasions, mass propulsion may originate in the cecum when ileal outflow is increased by feeding; then rapid transfer of some cecal contents fills the sigmoid colon and rectum. This extensive process, called mass movement or mass peristalsis, is described in Figure 6–4. Mass peristalsis is stimulated when the cathartic drug, oxyphenisatin, comes into contact with the colonic mucosal surface.

In one study of 27 ambulant patients, mass movements followed a meal only when the patients moved around, but propulsive activity of the colon rarely occurred if the patients lay still after eating. This agrees with the well-known clinical finding that continued bed rest tends to produce constipation.

Internal and External Anal Sphincters

The thick terminal portion of the circular smooth muscle of the rectum, 2.5–3 cm long, forms the internal anal sphincter surrounding the anal canal. It is overlapped distally by the striated muscle of the external anal sphincter.

The external anal sphincter and the puborectalis muscle are composed of striated muscle, and they contract only if their motoneurones are intact. The external anal sphincter is normally in a state of tonic contraction resulting from reflex activation through dorsal roots in the sacral segments, and the sphincter remains contracted even in sleep. However, the usual degree of contraction is minimal; stronger contraction occurs during distention of the rectum or when the urge to defecate is voluntarily resisted. Maximal contraction can be sustained for only 50 sec. The puborectalis muscle, which has the same innervation as the external anal sphincter, is also tonically contracted. The receptors from which many of the afferent impulses arise are in the sphincter itself. They operate like muscle spindle systems of other skeletal muscles, and proprioceptive feedback from them maintains discharge of motoneurons whose discharge over ventral root fibers excites contraction of the sphincter. When the dorsal root fibers are destroyed, as in tabes dorsalis, resting contraction of the sphincter, but not voluntary contraction, disappears. Stimulation of the receptors by stretch accounts for the strong and sometimes painful contraction of the external anal sphincter that follows sudden distention of the anus.

Such sudden distention is followed by a gasp and stimulation of respiration, a fact useful to those dealing with respiratory depression. Whereas the internal anal sphincter is not under voluntary control, the external sphincter is; reflex contraction can be modified by voluntary effort.

Because it is voluntary muscle, the external anal sphincter is paralyzed after destruction of the lower spinal cord, but the internal anal sphincter, being smooth muscle, is not.

The internal anal sphincter is usually in a state of nearly maximal contraction; its major reflex response is to relax. Activity of the internal anal sphincter is controlled in three ways: (1) Because the sphincter's muscle is continuous with the circular muscle of the rectum, the intrinsic plexuses are its most important innervation; extrinsic innervation reinforces action of the intrinsic plexuses. In the complete absence of extrinsic innervation, the internal anal sphincter is contracted, and it reflexly relaxes when the rectum is distended (Fig 6–5). Neurones mediating this relaxation are purinergic. (2) The internal anal sphincter receives

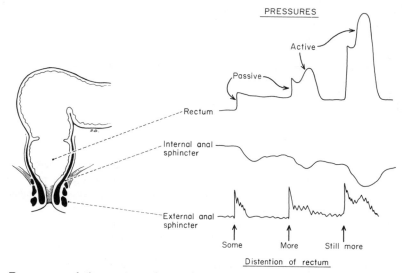

Fig 6–5.—Response of the rectum, internal anal sphincter, and external anal sphincter to distention of the rectum. Mild distention of the rectum stretches its wall and causes a passive increase in pressure. More distention is followed by further increase in pressure resulting from active contraction. Still more distention is followed by greater active contraction, and subsequent contractions may occur rhythmically at 20-sec intervals. Each increase in pressure in the rectum is accompanied by a decrease in pressure in the internal anal sphincter and an increase in pressure in the external anal sphincter. (From Davenport H.W.: *A Digest of Digestion*. Chicago: Year Book Medical Publishers 1975. Adapted from Denny-Brown D., Robertson E.G.: *Brain* 58:256, 1935; and from Schuster M.M., et al.: *Bull. Johns Hopkins Hosp.* 116:79, 1965.)

extrinsic innervation by way of the hypogastric nerves. Some are postganglionic sympathetic fibers, and the sphincter has a dense adrenergic innervation. In the normally innervated sphincter, these nerves, acting through alpha adrenergic receptors, contribute to the maintenance of sphincter pressure. (3) The sphincter also receives extrinsic parasympathetic purinergic inhibitory innervation, and this participates in relaxation of the sphincter upon distention of the rectum.

Responses of the sphincters have been measured by placing small balloons within them. The average peak pressure found 2 cm from the anal verge is between 25 and 120 mm Hg in a normal adult at rest, and the internal anal sphincter is responsible for 85% of this pressure. When the rectosigmoid portion of the colon or the rectum is distended, the internal anal sphincter relaxes and the external anal sphincter contracts (see Fig 6–5). The greater the degree of distention, the greater is the magnitude of the response. Relaxation of the internal anal sphincter occurs whether or not the colon and rectum themselves contract following distention. If distention is transitory, relaxation is likewise brief; but during prolonged distention of the rectum, the internal anal sphincter usually returns to its contracted state. This reflex relaxation does not require extrinsic innervation, for it is found in patients whose sacral nerve roots have been destroyed. Contraction of the external anal sphincter does require integrity of extrinsic innervation. Reflex contraction occurring when the rectum is moderately distended, as with 50 cc of air, is a major factor in maintaining continence. In incontinent patients, the reflex is diminished or absent. However, the external anal sphincter relaxes when the rectum is grossly distended with 200 cc at a pressure of 45–55 mm Hg. Any activity (except defecation and micturition) that increases intra-abdominal pressure increases external anal sphincter contraction. Micturition is accompanied by sudden and complete relaxation of the external anal sphincter, and gas is frequently passed.

Distention of the rectum in a patient with anal fissures causes relaxation of the internal anal sphincter as in a normal person, but relaxation is followed by reflex contraction, which elevates pressure in the anal canal by 20–30 mm Hg above resting pressure.

Defecation

The rectum is usually empty and collapsed; the angulation caused by the contraction of the puborectalis

muscle and helical folds of the mucosa resist progress of feces into it. Because the rectum contracts more than does the sigmoid colon, there is a reversed gradient of motility; this gradient empties contents of the rectum, including suppositories, upward.

When contractions of the upper parts of the colon force feces into the rectum, the rectum contracts and the internal anal sphincter relaxes. In paraplegics, whose external anal sphincter is paralyzed, this is enough to produce automatic defecation. In normal persons, contraction of the external anal sphincter prevents expulsion of feces and permits build up of volume in the rectum. As volume increases, pressure increases; and when intrarectal volume is 150–200 cc and pressure 55 mm Hg, there is complete relaxation of both internal and external anal sphincters, and the contents of the rectum are expelled.

Distention of the rectum brings the urge to defecate. However, as anyone who has had diarrhea knows, fullness of the rectum is not an essential part of urgency. The massive and strong contractions of the hyperexcitable colon are enough; and in persons who have strong rhythmic contractions, a sense of fullness of the rectum and a desire to defecate accompany each contraction, despite emptiness of the rectum.

When the rectum is filled to about 25% of its capacity by 100 ml of feces, and when its pressure has increased by about 18 mm Hg, the urge to defecate is experienced. Volumes and pressures in the rectum can be trebled before the need becomes imperative. During each increment of filling there is a transitory relaxation of the internal anal sphincter and contraction of the external anal sphincter. However, as the urge to defecate increases, the external anal sphincter and the puborectalis muscle relax, and relaxation is facilitated by centers above the spinal cord. The complex act of defecation follows. Contraction of the longitudinal muscles of the rectum and distal colon shortens the rectum, and the angle between the distal colon and rectum is straightened. Pressure rises within the rectum. These changes are enough in themselves to expel feces at rates ranging from stately slowness to explosive abruptness. They frequently empty the distal colon as high as the splenic flexure.

As the rectum empties, there is rebound contraction of both the internal and external anal sphincters, but voluntary facilitation of defecation can maintain relaxation of the sphincters until the rectum is completely empty.

In a normal person, evacuation is assisted by descent of the diaphragm to its extreme inspiratory position; at the same time the glottis is closed and contraction of the chest muscles on the full lungs raises both intra-abdominal and intrathoracic pressures. The hemodynamic consequences of this maneuver are an abrupt rise in arterial pressure as increased intrathoracic pressure is transmitted across the wall of the heart, stoppage of venous return with a rise in peripheral venous pressure, a fall in stroke output, and a subsequent fall in arterial pressure. Death while straining at stool can result from cerebral vascular accidents due to the rise in intracranial pressure, from ventricular fibrillation or coronary occlusion, possibly related to the decreased stroke output, from dissecting aneurysm, and from pulmonary emboli caused by mural clots that have been dislodged by changes in cardiac transmural pressure. Intra-abdominal pressure is further increased to 100–200 mm Hg by powerful contractions of the abdominal muscles. Greatest pressures with easiest evacuation occur in the squatting position. Persons sitting on high modern water closets bend forward to assist abdominal contraction to raise intra-abdominal pressure. The greatest straining, measured both by duration and by intra-abdominal pressure reached, occurs with the smallest stool; the greater the fecal mass, the less the effort to evacuate it. During defecation, the abdominal contents are supported by the pelvic diaphragm, which draws the anus up over the fecal mass and resists prolapse of the anus and rectum.

The defecation reflex is further augmented by tactile stimuli of the skin and anus so that actual passage of the feces reinforces the reflex. Arousal of the micturition reflex by a full bladder also facilitates defecation. Although micturition may precede defecation, the stream of urine is usually interrupted during defecation while the stool is actually passing; then, after the bulk of the feces is evacuated, micturition resumes and the bladder is emptied. The final emptying of the bladder may be accompanied by contractions of the empty rectum.

The efferent pathways of defecation reflex are parasympathetic and cholinergic; therefore the reflex is made more vigorous by parasympathomimetic drugs. Prolonged administration of the destructive anticholinesterase diisopropylfluorophosphate results in diarrhea, which eventually becomes bloody as the mucosa is rubbed off by repeated vigorous contractions of the muscle of the colon. Defecation is entirely normal in the complete absence of sympathetic innervation. In man, transection of the cord above the lumbosacral region results in early incontinence of feces, but the reflex soon returns, so that autonomous evacuation follows a mass movement in the proximal colon.

The fact that voluntary facilitation or inhibition of defecation exists shows that nervous activity of the highest centers projects on the medullary and sacral centers. Voluntary inhibition of defecation is expressed in strong contraction of the striated muscles of the pelvic diaphragm and external anal sphincter. Depression of motility of the whole colon also occurs, for the movements of colonic segments viewed through a colostomy have been seen to cease when the subject was attempting to restrain defecation. When voluntary closure of the external sphincter is effective, feces may remain in the rectum, and subsequent relaxation of the rectum, reducing tension in its wall, removes the stimulus for defecation. Likewise, when the urge to defecate following inflation of a balloon in the rectum is successfully defeated, the pressure in the rectum and tension in its wall drop, and the urge passes. Then, if the balloon is further inflated, pressure and tension rise and the urge to defecate returns. Consequently, the rectum may be filled with feces that fail to stimulate defecation until a subsequent mass movement increases their volume. The defecation reflex is also inhibited by pain or fear of pain, especially the pain aroused, as with hemorrhoids, by defecation itself.

When the call to stool is refused, contents of the rectum are often returned to the descending colon for safekeeping. This is accomplished by a series of retroperistaltic progressive movements traveling at 0.5–1 cm/min.

Megacolon

The underlying physiological disturbance in the disease of congenital megacolon is similar to that in achalasia of the esophagus. In fact, megacolon, megaureter, and esophageal achalasia have occurred in the same person. In megacolon, the colon is enormously enlarged and hypertrophied above a narrow segment, which is usually near the rectosigmoid junction. Dilatation and hypertrophy are entirely the result of failure of the feces to pass through the constriction, for they disappear following colostomy or excision of the segment. Failure of feces to pass is the result of lack or coordination of movement in the segment with that of the rest of the colon. Although the constricted part may contract rhythmically, its contractions are not propulsive.

Ganglion cells are completely absent from the affected segment, and there is a greatly reduced tissue concentration of vasoactive intestinal peptide. On the other hand, the segment is richly supplied with adrenergic nerve endings. Perhaps a peptidergic inhibitory mechanism is missing, and contractile influences are unopposed. Excision of the aganglionic segment usually allows establishment of essentially normal defecation, followed by regression of colonic enlargement.

Megacolon can be produced experimentally in the dog. Ganglion cells of the myenteric plexus are killed by ischemia when a segment corresponding to the rectosigmoid junction is perfused for 4 hours with physiological salt solution. After blood flow is reestablished, the perfused segment remains persistently constricted, often with dilatation of the colon proximal to the site.

Psychosomatic Factors

Alterations in function of the colon frequently accompany emotional disturbances. The acute reaction to extreme fright is known to everyone and is described in familiar phrases of the common speech. Medical students facing a serious examination frequently have diarrhea. Chronic dysfunction may be expressed either as depression or as enhancement of activity. In general, reactions to pain, fear, and anxiety produce pallor of the mucosa, reduced secretion of mucus, and inhibition of motility.

If, in response to his troubles, the subject is "grimly hanging on" not only to his life but to his fecal mass as well, he is likely to be constipated. On the other hand, anger, resentment, and hostility, overt or subconscious, are generally associated with hyperemia, engorgement, and increased motility. When extreme, these responses produce the *irritable* or *spastic colon* characterized by uncoordinated, abnormal motor function, abdominal pain, flatulence, and either constipation or diarrhea. As the result of irregular spasms, the fecal masses may be retained for a long time while water is absorbed; they are eventually evacuated as hard, dry balls or, as the result of spasm of the anus, rectum, and sigmoid colon, as pencil-thin ribbons. This state may alternate with diarrhea. However, it is difficult to attribute these bowel symptoms to any abnormal colonic myoelectrical activity asso-

ciated with psychological disturbance. Other patients who are equally disturbed may have similar colonic myoelectrical activity yet be free of bowel symptoms.

A patient with irritable colon has hyperalgesia; a degree of tension in the wall of his colon that would not bother a normal person causes him to feel pain in the hypogastrium, in the iliac fossa and in the anorectal region. During irregular spasms of the sigmoid colon and rectum, the internal anal sphincter relaxes; this, together with the hyperalgesia, produces a sense of urgency without the ability to defecate.

When the behavior of normal subjects or of patients with irritable colon is observed in the laboratory, functional changes are often found to coincide with mood swings. Depression of motility may accompany self-reproach and a feeling of helplessness; hypermotility may occur during expression of hostility. Hyperemia without motility changes occurs during embarrassment; the colon of a male medical student being examined through a sigmoidoscope blushed when he learned that a young woman was peering through the instrument. However, it is impossible to predict what emotional states will have an objective correlative in colonic activity; there may be no relation between the two; or the relation may be a quixotic one. The colon of one woman thoroughly studied on four occasions had a very similar pattern of motility when she was calm and when she was weeping freely during discussion of distressing life experiences. Pressures in the colon of a 58-year-old patriotic Briton were measured when he was recovering from a slight stroke. There were no changes when his physician probed the patient's responses to his hemiplegia, the high probability of future disablement, and the likelihood of his being thrown on the parish, but his colon rose to a perfect storm of activity when talk was turned to the Royal Family.

REFERENCES

Bovier M., Conella J.: Nervous control of the internal anal sphincter of the cat. *J. Physiol.* 310:457, 1981.

Cerulli M.A., Nikoomanesh P., Schuster M.M.: Progress in biofeedback conditioning for fecal incontinence. *Gastroenterology* 76:742, 1979.

Christensen J.: Myoelectric control of the colon. *Gastroenterology* 68:601, 1975.

Connell A.M.: Motor action of the large bowel, in Code C.F. (ed.): *Handbook of Physiology:* Sec. 6. *Alimentary Canal,* vol. 4. Washington, D.C.: American Physiological Society, 1968, pp. 2075–2092.

Daniel E.E., Bennett A., Misiewicz J.J., et al.: Symposium on colon function. *Gut* 16:298, 1975.

Dickinson V.A.: Maintenance of anal continence: A review of pelvic floor physiology. *Gut* 19:1163, 1978.

Hulten L.: Extrinsic nervous control of colonic motility and blood flow: An experimental study in the cat. *Acta Physiol. Scand. [suppl.]* 335:1, 1969.

Ihre T.: Studies on anal function in continent and incontinent patients. *Scand. J. Gastroenterol. [suppl.]* 9:1, 1974.

Kreulen D.L., Szurszewski J.H.: Nerve pathways in celiac plexus of the guinea pig. *Am. J. Physiol.* 237:E90, 1979.

Latimer P., Sarna S., Campbell D., et al.: Colonic motor and myoelectrical activity: A comparative study of normal subjects, psychoneurotic patients, and patients with irritable bowel syndrome. *Gastroenterology* 80:893, 1981.

Martelli H., Devroede G., Arhan P., et al.: Some parameters of large bowel motility in normal man. *Gastroenterology* 75:612, 1978.

Mendeloff A.I.: Defecation, in Code C.F. (ed.): *Handbook of Physiology:* Sec. 6. *Alimentary Canal,* vol. 4. Washington, D.C.: American Physiological Society, 1968, pp. 2140–2146.

Patel P.D., Picologlou B.F., Lykoudis P.S.: Biorheological aspects of colonic activity. II. Experimental investigation of the rheological behavior of human feces. *Biorheology* 10:441, 1973.

Sarna S.K., Bardakjian B.L., Waterfall W.E., et al.: Human colonic electrical control activity (ECA). *Gastroenterology* 78:1526, 1980.

Schuster M.M.: Motor action of rectum and anal sphincters in continence and defecation, in Code C.F. (ed.): *Handbook of Physiology:* Sec. 6. *Alimentary Canal,* vol. 4. Washington, D.C.: American Physiological Society, 1968, pp. 2121–2139.

Schuster M.M.: The riddle of the sphincters. *Gastroenterology* 69:249, 1975.

Snape W.J., Carlson G.M., Matarazzo S.A., et al.: Evidence that abnormal myoelectrical activity produces colonic motor dysfunction in the irritable bowel syndrome. *Gastroenterology* 72:383, 1977.

Stoddard C.J., Duthie H.L., Smallwood R.H., et al.: Colonic myoelectrical activity in man: Comparison of recording techniques and methods of analysis. *Gut* 20:476, 1979.

Thompson W.G., Heaton K.W.: Functional bowel disorders in apparently healthy people. *Gastroenterology* 79:283, 1980.

Truelove S.C.: Movements of the large intestine. *Physiol. Rev.* 46:457, 1966.

Weems W.A., Szurszewski J.H.: Modulation of colonic motility by peripheral neural input to neurones of the inferior mesenteric ganglion. *Gastroenterology* 73:273, 1977.

Young S.J., Alpers D.H., Norland C.C., et al.: Psychiatric illness and the irritable bowel syndrome. *Gastroenterology* 70:162, 1976.

7

Vomiting

VOMITING, OR EMESIS, is the forceful expulsion of gastric and intestinal contents through the mouth; being forceful, it is distinguished from the gentle regurgitation occurring in infants, whose lower esophageal sphincter is inadequately closed. Preceding or accompanying vomiting are tachypnea, copious salivation, dilatation of the pupils, sweating and pallor, and rapid or irregular heartbeat—all signs of widespread autonomic discharge. In man, these are usually associated with the psychic experience of nausea. It is impossible to tell whether an animal showing the same signs is nauseated; and because all the signs can occur in a decerebrate man or animal, they cannot be equated with the experience of nausea. Labored breathing with the mouth partly or entirely closed, and retching, the involuntary movements of vomiting without actual expulsion of vomitus, may precede vomiting. The involuntary muscles of the esophagus, stomach, and small intestine participate with movements that assist in expelling vomitus. These five components may be mixed in any proportion: (1) the sensation of nausea, (2) associated autonomic discharge, (3) spasm or reverse peristalsis in the duodenum, (4) retching, and (5) actual expulsion.

Mechanics of Vomiting

Vomiting is usually preceded by retching, which defeats the normal antireflux mechanisms.

In the dog and cat, each episode of retching is preceded by dampening of the electrical control activity of the small intestine. Then an intense burst of spikes signalling concentric contractions of the muscle of the small intestine usually begins in the last quarter of the intestine and migrates toward the stomach at the rate of 2–3 cm sec^{-1}. Retching begins as this wave of reverse peristalsis passes over the duodenum. Such reverse peristalsis has not been observed in man, simply because adequate recording technique has not

been applied to the problem. Nevertheless, either reverse peristalsis or duodenal spasm pushes bile-stained intestinal contents into the stomach to be expelled with the vomitus. Contents of the lower intestine may also be regurgitated during prolonged vomiting. The colon becomes active, and defecation follows.

The first somatic act of retching is a deep inspiration. The glottis is closed, and intrathoracic pressure falls far below atmospheric pressure. At the same time, strong contractions of the abdominal muscles raise intra-abdominal pressure, and the pressure gradient from abdomen to thorax may be as great as 200 mm Hg. A sliding hiatal hernia occurs at each retch; the abdominal portion of the esophagus and a large cone of stomach are forced through the hiatus of the diaphragm. That portion of the stomach which is in the thorax is expanded by the large pressure gradient across its wall. A tear may begin in the gastric mucosa and cross the gastroesophageal junction; such a Mallory-Weiss tear is a frequent cause of upper gastrointestinal bleeding. Alternatively, the lower end of the esophagus may rupture. During retching, a strong annular contraction at the angular notch nearly divides the body of the stomach from the antrum. While the body of the stomach is flaccid, the antrum contracts so that gastric contents are shifted into the upper part of the stomach (Fig 7–1,B), and then gastric contents are emptied, as through a funnel, into the relaxed esophagus (see Fig 7–1,C). Because the pharyngoesophageal sphincter is closed, none of the gastric contents enters the pharynx or mouth. As abdominal contraction slows and stops, the abdominal wall moves outward; at the same time the chest wall moves inward and the diaphragm downward. The stomach refills from the esophagus as rapidly as it empties. The esophagus remains dilated, and there is no sign of esophageal peristalsis. With return of most of the esophageal contents into the stomach, the

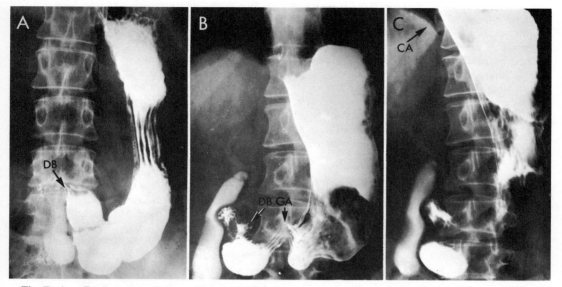

Fig 7–1.—Radiographs taken of an erect human subject. **A,** before retching began. The duodenal bulb *(DB)* is distended with barium, and the stomach has not yet begun to contract. **B,** taken during retching. The duodenal bulb *(DB)* has emptied, and its contents appear to have passed back into the stomach. The proximal part of the gastric antrum *(GA)* is tightly contracted, and there has been a shift of barium from the lower to the upper part of the stomach. **C,** taken during emesis. The cardia *(CA)* is elevated and opened widely. Barium is being forced out of the stomach and up the esophagus. There is a long contracted segment in the lower part of the stomach. (From Lumsden K., Holden W.S.: *Gut* 10:173, 1969.)

esophagus quickly collapses symmetrically along its entire length; but the lower esophageal sphincter remains open. The retching reaction does not hesitate at this point, but proceeds with immediate repetition of the cycle.

During retching, secretion of mucus by the stomach increases, but secretion of acid falls.

The act of vomiting builds up on a developed retch. While the esophagus is full of gastric contents, there is a sudden rise in intrathoracic pressure. Strong contractions of abdominal muscles increase intra-abdominal pressure and force the diaphragm high into the thorax. The larynx and hyoid bone are drawn forward, and high intrathoracic pressure, which may be 100 mm Hg above atmospheric pressure, forces the contents of the esophagus through the pharyngoesophageal sphincter and out of the mouth. Because abdominal contraction is maintained, there is limited reflux from the esophagus into the stomach, and at the end of expulsion the esophagus is still full. After several seconds, peristalsis begins in the upper esoph-

agus; the esophagus empties into the stomach, and the lower esophageal sphincter closes. If a significant volume remains in the stomach after several seconds, a second episode of cyclic filling and emptying of the esophagus begins, culminating in further expulsion of vomitus. Finally, the voluntary muscles relax, and normal respiration resumes.

The Vomiting Center

Straightforward vomiting unaccompanied by retching, called *projectile* vomiting, is governed by a vomiting center in the dorsolateral border of the lateral reticular formation of the medulla, lying ventral to the solitary tract and its nucleus. Projectile vomiting follows electrical stimulation of this region in experimental animals with a latent period no longer than is required for the initial inspiration, and it ceases abruptly at the end of stimulation. The center in the lateral reticular formation is closely associated anatomically and functionally with the respiratory center, vasomotor nuclei, and a locus whose stimulation produces retching without vomiting. The vomiting center coordinates activities of these neighboring structures to produce the complex response. The heavy, rhythmic respiratory movements of retching may precede or follow expulsion of vomitus, but retching is distinct from vomiting and may occur for many hours without actual vomiting. In the cat, electrical stimulation of the vestibular nucleus has occasionally produced retching that could not be converted to vomiting even by filling the animal's stomach with milk.

Afferent Pathways

The vomiting center is activated only by afferent impulses, which arise from many parts of the body; it is not directly stimulated by emetic substances carried to it by the blood. Among the effective stimuli exciting afferent impulses are: tactile stimulation to the back of the throat; distention of the stomach or duodenum to a pressure of about 20 mm Hg; distention or injury of the uterus, renal pelvis, or bladder; a rise in intracranial pressure; rotation or unequal stimulation of the labyrinths; acceleration of the head in any direction; and many kinds of pain, such as that attending injury to the testicles. Many young children can vomit voluntarily to express disapproval of parental actions, and some adults may do so, often by thinking about nauseating experiences. Persistent psychogenic vomiting in adults occurs in persons locked in an inescapable hostile relationship within the family. Such persons may become severely hypokalemic.

There are two general pathways by which emetic substances or chemical changes in body fluids affect the vomiting center. The first is by way of the chemoreceptor trigger zone consisting of a restricted group of receptor cells in the area postrema, on the floor of the fourth ventricle. Stimulation of the chemoreceptor trigger zone by emetics in the blood or cerebrospinal fluid elicits vomiting. Destruction of the trigger zone eliminates the response to centrally ap-

plied emetics and also the vomiting accompanying uremia, radiation sickness, and motion sickness. Other pathways lie in many afferent nerves, especially from the digestive tract, which are activated by drugs or noxious substances. The most sensitive receptors are in the first part of the duodenum. Ipecac placed in the human stomach through a fistula has no effect until it passes through the pylorus; then it immediately induces vomiting. Copper sulfate stimulates receptors whose fibers run centrally in both vagus and sympathetic nerves; but after complete denervation of the gut, absorbed copper sulfate still causes vomiting by stimulating the chemoreceptor trigger zone. After ablation of the trigger zone, even a lethal dose of copper sulfate will not cause vomiting.

The emetic drug apomorphine acts directly upon the chemoreceptor trigger zone, and it is dopaminergic. Drugs commonly used to ward off motion sickness are antihistaminics (H_1 blockers) with significant anticholinergic activity and the tendency to produce drowsiness. Marijuana and its major active ingredients suppress vomiting in patients undergoing cancer chemotherapy by action on forebrain centers as well as on the vomiting center.

REFERENCES

Borison H.L., Wang S.C.: Physiology and pharmacology of vomiting. *Pharmacol. Rev.* 5:193, 1953.
Hill O.W.: Psychogenic vomiting. *Gut* 9:348, 1968.
Lumsden K., Holden W.S.: The act of vomiting in man. *Gut* 10:173, 1969.
McCarthy L.E., Borison H.L., Spiegel P.K., et al.: Vomiting: Radiographic and oscillographic correlates in the decerebrate cat. *Gastroenterology* 67:1126, 1974.
Smith C.C., Brizee K.R.: Cineradiographic analysis of vomiting in the cat: I. Lower esophagus, stomach and small intestine. *Gastroenterology* 40:654, 1961.
Stewart J.J., Burks T.F., Weisbrodt N.W.: Intestinal myoelectric activity after activation of central emetic mechanism. *Am. J. Physiol.* 233:E131, 1977.
Watts H.D.: Lesions brought on by vomiting: The effect of hiatus hernia on the site of injury. *Gastroenterology* 71:683, 1976.

PART III

Secretion

8

Salivary Secretion

THE SALIVARY GLANDS of man secrete 0.5–1 L of saliva a day, the rate of secretion ranging from barely perceptible to as high as 4 ml/min on maximal stimulation. In keeping the mouth wet, saliva facilitates speech and lubricates food for swallowing. It is essential for dental health, since caries occurs in the absence of secretion. Decreased secretion in dehydration makes the mouth dry, and this contributes to the sensation of thirst. Saliva dissolves sapid substances, making them available for taste. Human saliva contains the α-amylase ptyalin, which, when mixed with food by chewing, begins the digestion of starch. Saliva secreted in response to noxious or unpleasant substances dilutes them and helps to cleanse the mouth, and its bicarbonate content neutralizes acids.

Structure and Innervation of Salivary Glands

The mouth contains numerous small buccal glands, together with some purely mucous glands on the inferior surface and margin of the tongue. These glands do not secrete enough saliva to keep the mouth comfortably moist; and so, a man having no functioning major salivary glands must frequently rinse his mouth with small drinks of water. Most saliva is secreted by three pairs of large glands: the parotid, the submaxillary, and the sublingual. The parotid glands are "serous" glands, for their acini contain only one type of cell, which secretes fluid devoid of mucin. Their secretion is therefore watery compared with the mucin-containing saliva from the mixed submaxillary and sublingual glands, whose acini contain cells capable of secreting mucoproteins. The acini lead to short intercalated tubules, which empty into interlobular tubules and then into the excretory ducts. The cells of the acini and the two kinds of tubules have ultrastructure similar to that of other secretory cells, and the physiological evidence shows that all three structures participate in formation of saliva. There is a sharp transition at the point where the tubules become excretory ducts.

All three pairs of glands receive sympathetic innervation from the superior cervical ganglion. These fibers are all adrenergic, liberating norepinephrine, and they are distributed to blood vessels and secretory cells. Stimulation of sympathetic fibers to all glands causes vasoconstriction; in man, stimulation of the sympathetic trunk in the neck or injection of epinephrine causes secretion of fluid by the submaxillary but not by the parotid glands. Sympathetic stimulation of the parotid glands causes them to secrete amylase, which is washed out by the fluid produced under parasympathetic stimulation. Preganglionic parasympathetic innervation reaches the glands from the cranial outflow by way of the glossopharyngeal nerve (IX), the chorda tympani branch of the facial nerves (VII), and the hypoglossal nerve (XII). These synapse in, or close to, the glands; their postganglionic fibers are distributed to all secretory cells. Postanglionic fibers innervated by 5–10 preganglionic fibers converge on each secretory cell, and their stimulation causes copious secretion. The mediator is acetylcholine, and injection of parasympathomimetic drugs produces equal but no greater response. Atropine administration causes dry mouth by blocking the effect of acetylcholine on secretory cells. All salivary secretion in man, with the exception of paralytic secretion, is a response to nerve impulses; there is no humoral control.

At rest, acinar cells are electrically coupled; ions and organic compounds can pass through cytoplasmic bridges from one cell to another throughout an acinus and occasionally to adjacent acini. Coupling permits coordination of activity despite irregularities of innervation. Calcium released intracellularly upon stimulation tends to uncouple the cells.

Salivary glands atrophy when they are not used,

and a decrease in the flow of nerve impulses to the glands in both the sympathetic and parasympathetic fibers is responsible for the atrophy. The rat is a nocturnal animal; when it is fed solid chow, there is a diurnal rhythm of protein synthesis in its salivary glands. Synthesis is decreased by day and increased by night. If the solid food is replaced by a liquid diet, the glands atrophy, their store of secretory products falls to a low level, and there is no diurnal pattern of protein synthesis. Increased stimulation causes hypertrophy of the salivary glands.

Composition of Saliva

Figure 8–1 shows representative values for the composition of human parotid saliva secreted in response to a parasympathomimetic drug. The saliva used for analysis was being secreted at a steady rate in response to a graded dose of the drug, and the graph was constructed from many observations made at rates of secretion covering the range of 0.1–4 ml/min. At high secretory rates, the sodium and chloride concentrations are greater than at low rates, but they never equal their concentrations in plasma. The total osmotic pressure is always hypotonic in man, being, at maximal secretory rates, about two thirds of the plasma value. Dehydration and repletion lead to corresponding changes in saliva osmolarity. Phosphate concentration is constant at 6 mEq/L at rates above 1 ml/min. In the same range of secretion rates, the concentration of magnesium is only 0.06 mEq/L.

Calcium bound to amylase is secreted into saliva under β-adrenergic stimulation, but when the gland is under parasympathetic cholinergic stimulation, calcium is secreted independently of amylase. Then its concentration in saliva may become as high as 11 mEq/L.

The pH of saliva secreted by the unstimulated human parotid gland ranges from 5.45 to 6.06. On stimulation, the pH of parotid saliva rises by two pH units to a maximum of 7.8. After secretion, saliva becomes more alkaline in the mouth as the result of loss of dissolved carbon dioxide. The reason for the pH rise is that the partial pressure of carbon dioxide remains approximately constant at all secretion rates, but the bicarbonate concentration rises as rate of secretion increases.

Parotid glands on opposite sides may secrete at different rates in response to the same stimulus, but the same relation between rate and electrolyte secretion obtains for each. There are detectable differences between individuals, and a small day-to-day variation occurs in any one person. Human submaxillary and sublingual salivas have the same general relation between rate of secretion and sodium and potassium concentrations.

Reasons for Variations in Composition of Saliva

Systematic variation in the composition of juice according to rate of secretion is characteristic of all digestive glands. Three general types of explanation are given:

1. The composition of the juice as it is extruded from the secreting cells may itself be variable. This explanation is usually intuitively rejected, apparently

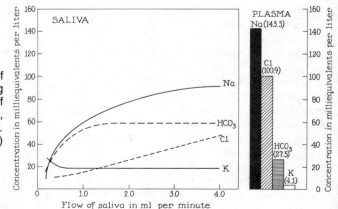

Fig 8–1.—Average composition of the parotid saliva of three young women as a function of the rate of secretion. (From Thaysen J.H., Thorn N.A., Schwartz I.L.: *Am. J. Physiol.* 178:155, 1954.)

on the ground that it violates some presumed constancy of nature.

2. The juice as collected may be a mixture of two or more juices, each secreted at constant composition but at different rates by distinct cells. This idea explains why, in the case of mixed glands such as the submaxillary, samples of saliva may contain greatly differing amounts of mucin, for mucin is secreted by one type of cell, while its aqueous menstruum is secreted by another. This principle accounts for much of the variability of gastric secretion, and it will be more fully discussed when the stomach is considered.

3. A secretion extruded from cells at constant (or perhaps variable) composition may subsequently be acted upon by other cells, such as those of the gland's ducts. This theory of salivary secretion, which asserts that a primary secretion from acinar cells undergoes secondary changes while passing down the ducts, is the most coherent, if still fragmentary, explanation of salivary secretion. The theory is shown diagrammatically in Figure 8–2.

Primary Secretion by Acinar Cells

Almost nothing is known about the mechanism of secretion of saliva in man. Most experimental evidence has been obtained using glands, chiefly the parotid gland, of rats, cats, dogs, and rabbits. There are large differences in function between different glands in the same species and between similar glands in different species. Consequently, the generality of the following description is suspect.

Fluid has been obtained by micropuncture from the acini and intercalated ducts of salivary glands of several species. The concentrations of sodium, chloride, and bicarbonate in the fluid are approximately equal to their concentrations in plasma water. The concentration of potassium in fluid obtained from nonsecreting glands is higher than the plasma concentration, but it falls as the glands secrete. The fluid is usually slightly hypertonic. Thus, the fluid that begins its journey down the ducts is similar to, if not identical with, an ultrafiltrate of plasma.

As a microelectrode passes through the basal membrane of a resting acinar cell of the cat's sublingual gland, it detects a potential difference of about −31 mV, the inside of the cell being negative to the outside. A value of −95 mV would be predicted on the basis of the potassium concentration gradient, and the measured value obviously does not agree with the predicted one. Neither does it agree with the calculated sodium potential (+ 29 mV) or the chloride potential (− 12 mV). None of these ions appears to be in electrochemical equilibrium across the membrane. Since the concentration of potassium is high in the resting cell, potassium may be continuously accumulated, and perhaps sodium is extruded. According to this view, the resting potential is established by the processes continuously transporting the ions.

A similar and almost identical potential difference exists across the apical membrane of the cell. The membrane is electrically inexcitable, and it does not show a propagated action potential when it is stimulated. Two seconds after the cell is stimulated through its parasympathetic nerve supply, the basal potential difference rises to −56 mV, and later the apical potential may rise to −43 mV. These secretory potentials may also be produced by the injection of acetylcholine into the gland's arterial supply. They are not all-or-none potential changes, such as the action potentials of nerve and muscle, for they may be graded by varying the strength of stimulation. They are likewise independent of the resting potential of the membrane. The resting potential may be set at any value by passing a current through the membrane. When this is done, secretory potentials of the usual magnitude still follow stimulation.

Most or all of the water contained in saliva is secreted by the acinar cells; although modification of the composition of acinar fluid occurs in the ducts, the ducts have little effect upon the volume secreted. Water flow from interstitial fluid through the cells into acinar fluid is probably secondary to secretion of electrolytes.

Tubular Exchanges

Fluid leaving the acini and entering the tubules and ducts is similar to an ultrafiltrate of plasma, but saliva leaving the gland has a very different composition. Therefore, major modifications of acinar fluid must occur in the ducts. Such exchanges have been identified and measured by micropuncture and microperfusion, and some of the results are summarized in Figure 8–2.

The relative positions of some of the exchanges along the tubules have been determined by the method described in Figure 8–3. If small amounts of two tracer substances, *A* and *B*, are injected into the arterial supply of a gland secreting at a constant rate,

Fig 8–2.—A scheme of electrolyte exchange during secretion of saliva by the parotid gland. The vascular arrangement is that occurring in the rabbit, and the electrolyte exchanges are derived from experiments on rat, rabbit, and dog parotid glands. (From Davenport, H.W.: *A Digest of Digestion,* ed. 2. Chicago, Year Book Medical Publishers, Inc., 1978.)

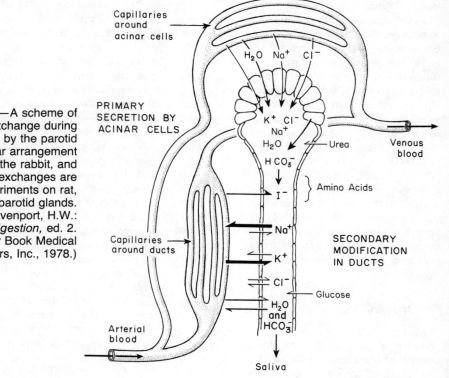

Fig 8–3.—Method of locating the place at which a substance crosses the salivary tubules to enter the saliva. The gland is stimulated to secrete at a constant rate, and the saliva is collected in small, serial samples. A small volume of solution containing two isotopically labeled substances, *A* and *B,* is quickly injected into an artery close to the gland. Substance *A,* because it crosses a more distal portion of the tubule, appears in the saliva and reaches a peak concentration earlier than substance *B,* which enters a more proximal part of the tubule. The relative positions of entry of iodide, sodium, potassium, chloride, water, and bicarbonate, shown in Figure 8–2, were determined in this manner. (Adapted from Burgen A.S.V., Emmelin N.G.: *Physiology of the Salivary Glands.* London, Edward Arnold & Co., 1961.)

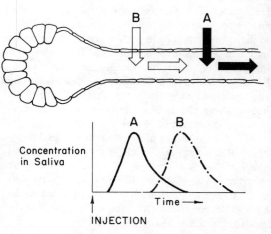

their patterns of efflux into the saliva show the relative positions at which they crossed the tubular wall. Since substance *A* can penetrate at a more distal site, it appears in the saliva first. Such experiments have shown that urea enters the tubular fluid soon after it leaves the acini. Iodide is next and is followed by sodium. Potassium (and rubidium and cesium) enter midway. Chloride (and bromide) cross still farther down, and water and bicarbonate exchange at the most distal region. Amino acids and glucose enter saliva at the position indicated in Figure 8–2.

Electrolyte Exchanges

Samples obtained by micropuncture at the beginning of the excretory duct contains Na^+ at 117 ± 32 mEq/L, whereas samples collected at the duct opening contain Na^+ at 21 ± 15 mEq/L. The potential difference across the ducts is 80–90 mV with the lumen negative with respect to the blood, and therefore Na^+ must be actively transported by the cells of the duct against Na^+'s electrochemical gradient. The ouabain-inhibitable, Na^+-activated ATPase presumably responsible for Na^+ pumping is located on the contraluminal side of the cells of the ducts. Net transport of sodium out of the lumen is actually the resultant of two opposing unidirectional fluxes. If radioactive $^{24}Na^+$ is injected into the carotid artery supplying an actively secreting parotid gland, it appears in saliva within 5 sec. After subtracting circulation time of 1–2 sec and deadspace time of 1–2 sec, the actual transfer time from blood to saliva is less than 3 sec. Considering the dimensions of the salivary glands, this could be achieved in such short time only if some sodium were transferred from blood to saliva without passing through the acinar cells. The simplest explanation is that sodium moves across the walls of the ducts from blood to lumen despite the fact that net flow of sodium occurs in the opposite direction.

Reabsorption of sodium from the ducts accounts for the relation between salivary sodium concentration and rate of flow of saliva shown in Figure 8–1. At low rates of flow, saliva is present in the ducts for a long time, and sodium reabsorption can be nearly complete. At the lowest rates of secretion, salivary sodium content may be as low as 1–2 mEq/L. As flow rate increases, reabsorption of sodium becomes less complete, and salivary sodium concentration rises.

K^+ enters saliva from two sources. Within 0.6 sec after cholinergic stimulation, K^+ is released from acinar cells, and there is a transient rise in K^+ concentration in saliva. Later, K^+ is taken up by acinar cells. In addition, K^+ is actively secreted into the lumen by cells lining the ducts. If fluid perfusing a duct is held stationary, its K^+ concentration rises as high as 135 mEq/L. This transport of K^+ against its electrochemical gradient appears to be coupled with Na^+ transport in the other direction.

Cl^- reabsorption is along Cl^-'s electrochemical gradient, and it may be entirely passive.

Little is known about HCO_3^- secretion in the ducts except that it must be an active process. Because only a small fraction of HCO_3^- secretion is inhibitable by carbonic anhydrase inhibitors, some persons think that HCO_3^- must be transported as such from interstitial fluid to saliva. On the other hand, if HCO_3^- in the fluid perfusing a salivary gland is replaced by acetate, then acetate replaces HCO_3^- in saliva. The same replacement occurs in the pancreas where HCO_3^- secretion is apparently not the result of HCO_3^- transport. (See chap. 10 for a description of HCO_3^- secretion.)

Although measurement of the movement of tritiated water between blood and saliva shows that water molecules can move in both directions across the cells forming the gland's ducts, the ducts are relatively impervious to bulk flow of water along an osmotic gradient. Active reabsorption of sodium and an accompanying anion reduces the osmotic pressure of fluid within the ducts, and the epithelial cells of the ducts can establish and maintain an osmotic gradient of 40–60 mOsm/L. This means that the water-filled channels across the ducts have a very small radius and thus offer high resistance to bulk flow of water. The channels are probably in the tight junctions between cells. The permeability to water increases as the rate of flow of saliva increases. Because sodium reabsorption is more complete at low rates of flow and osmotic permeability is low, saliva secreted at a low rate has a lower osmotic pressure than that secreted at a high rate.

When a salivary gland is stimulated, the tight junctions between the cells of the ducts, but not those between acinar cells, become leaky. The gland's permeability to urea increases about 5 times as the rate of secretion changes from lowest to highest. Stimulation of sympathetic nerves to the salivary glands permits compounds which are normally excluded—glucose, for example—to enter saliva. In fact, a compound having a molecular radius as large as 0.81 mμ can diffuse into saliva after sympathetic stimulation; but inulin, whose molecular radius is 1.47 mμ cannot.

Hormones and Composition of Saliva

The sublingual glands of man and the cat and the parotid glands of the rat and of sheep secrete isotonic or slightly hypertonic saliva. A single 13-gm gland of a sheep secreted 1–4 L/day; this provides a large volume of alkaline juice, which maintains a neutral pH and fluid consistency of the fermenting contents of the animal's rumen. Sheep parotid saliva contains sodium and potassium in a ratio of about 180:10. If a fistula is made so that the juice of a single parotid gland is lost, there is rapid depletion of 400–600 mEq of sodium and a collapse of the extracellular fluid volume. As this occurs, the ratio of sodium to potassium in the saliva reverses, and it may become 10:180. The major part of this reversal is mediated by products of the adrenal cortex, for if these are removed, the animal, despite its severe sodium depletion, continues to secrete saliva with a high ratio of sodium to potassium. In such a depleted, adrenalectomized sheep, the administration of adrenal mineralocorticoids is followed by a fall in salivary sodium and a rise in potassium. Aldosterone acts directly upon the ducts to increase reabsorption of sodium and secretion of potassium.

The generalization that the effect of adrenal mineralocorticoids is to lower salivary sodium and raise potassium concentrations applies to man as well. Administration of aldosterone reduces the salivary Na:K ratio; and excess of endogenous aldosterone, occurring in primary or secondary aldosteronism, likewise reduces the ratio, which rises after the cause of excess hormone secretion has been corrected. On the other hand, patients with adrenal cortical insufficiency secrete saliva with a high Na:K ratio, and the ratio falls during appropriate mineralocorticoid treatment. The concentration of aldosterone in saliva is independent of flow rate, and it is strongly correlated ($r = 0.96$) with the plasma concentration.

Salivary Iodide Secretion

Although only the most minute amounts of iodide are normally present in plasma, the tubules of salivary glands of some species are capable of accumulating and secreting iodide. The ion can be secreted into saliva at a concentration 10 or more times the plasma level. Glands capable of secreting iodide at high ratios of saliva to plasma include the parotid glands in man, the dog, and the guinea pig and the submaxillary glands in the mouse and man. Accumulation by tubular cells has been demonstrated by radioautography. When $^{131}I^-$ is injected into the carotid artery, it appears in saliva within 3 sec, a time too short for secretion by the acinar cells. Iodide secretion shows self-depression, for the ratio of salivary to plasma concentration falls toward unity when the plasma iodide concentration is raised. Furthermore, the ability of the glands to concentrate iodide is reduced by thiocyanate, perchlorate, and nitrate, just as the ability of the thyroid gland to accumulate iodide is reduced. (A similar situation obtains in the stomach.) Thiocyanate itself may be a normally secreted constituent, for saliva gives the color reaction for thiocyanate with ferric ions, and the intensity of color increases when food containing thiocyanate is eaten or tobacco smoke is inhaled.

No organic iodine is present in human saliva, but the dog parotid saliva (and gastric juice and bile) contains iodotyrosine linked in peptide bonds in protein.

Salivary Proteins

Cells of the resting salivary glands contain histologically demonstrable granules whose number diminishes after prolonged secretion. They are called zymogen granules because they are thought to be the locus of storage of enzymes about to be secreted. The major enzyme of saliva is the α-amylase ptyalin, which is secreted by man, apes, pigs, rats, mice, guinea pigs, and squirrels but not by horses, cats, or dogs. It is a hydrolytic enzyme which splits α-1,4-glucosidic linkages, a reaction that is practically irreversible because its equilibrium is far in the direction of hydrolysis. The enzyme's pH optimum is 6.9, and the enzyme is stable between pH 4 and 11. It is inactive without chloride ions, with full activity occurring in 0.01 N Cl^-. Native starch is a mixture of two glucose polymers: (1) unbranched amylase, which is a chain of glucose residues linked through α-1,4-glucosidic bonds, and (2) branched amylopectin, which contains, in addition to straight chains formed by α-1,4-glucosidic bonds, 3%–5% branching linkages formed by α-1,6-glucosidic bonds. Salivary amylase hydrolyzes only α-1,4-linkages within the chain; it does not attack the terminal α-1,4-linkages, nor does it break the α-1,6-branching linkages. Consequently, the products of salivary amylytic digestion of amylopectin are maltose (glucose-glucose; α-1,4-linkage), maltotriose (glucose-glucose-glucose; α-1,4-linkage) and a mixture of dextrins containing α-1,6-branches and averaging six glucose residues per molecule. When the enzyme acts on starch paste, the immediate rupture of only 0.1% of the glucoside linkages produces particles 1/100th the original size, and there is a large drop in the viscosity of the suspension.

The physical properties of saliva depend on the kind of protein it contains, not on protein concentration. Parotid saliva, whose viscosity is only slightly greater than that of water, may contain more protein than sticky, stringy submaxillary saliva. A family of glycoproteins makes up about 8% of parotid salivary protein, and in addition to amylase, parotid saliva of man contains several ATPases, 5'-neucleotidase, esterase, kallikrein, and alkaline phosphatase. The juice of the submaxillary and sublingual glands contains mucin, which is responsible for the lubricating action of the saliva. There are two principal types: those rich in neuraminic acids and N-acetylgalactosamine but devoid of blood-group activity, and others with blood-group activity containing much L-fucose.

Under adrenergic stimulation the submandibular glands of adult male mice secrete both nerve growth factor and epidermal growth factor in high concentrations. The latter peptide is identical, except for two terminal amino acids, with urogastrone, which appears in human urine. It is not known whether urogastrone comes from human salivary glands.

Salivary export proteins are enclosed as secretory granules within a phospholipid membrane. When exocytosis occurs, there is a partial fusion of a granule's membrane with the luminal membrane of the cell, and the contents of the granule are discharged. The luminal membrane is then a patchy mosaic of the two membranes, and eventually that part contributed by the granule is removed and degraded. This process results in turnover of phospholipids. When phosphate containing the radioactive isotope ^{32}P is present, stimulation of secretion is accompanied by a large increase in the specific activity of the phospholipids; there is a 400% increase in that of phosphoinositide and of phosphatidic acid and a much smaller increase

in that of phosphatides containing choline or ethanolamine. At the same time, there is only a small increase in the turnover of the fatty acid and glycerol parts of the phospholipid molecules.

Nervous Control of Salivary Secretion

The neural apparatus governing salivary secretion consists of afferent nerves to secretory centers in the medulla and efferent innervation to the glands by way of both parasympathetic and sympathetic outflows. Afferent innervation comes chiefly from the mouth, pharynx, and olfactory area, with that from the taste buds being most important. Afferent input comes from other parts of the nervous system, and, for example, profuse salivation accompanies the sensation of nausea. Afferent information is processed into appropriate outputs in the lower brain stem without involvement of higher centers, and both sympathetic and parasympathetic outputs are reflexly excited.

In the cat, unilateral stimulation of the medulla results in either ipsilateral or bilateral secretion by parotid and submaxillary glands. When bilateral responses occur, ipsilateral is greater than contralateral secretion. The responsive centers are in the dorsolateral region of the reticular formation, dorsomedial to the spinal trigeminal nucleus, and dorsal to, and at the level of, the facial nucleus. From there, fibers exit in the ventrolateral portion of the medulla by way of the facial nerve to the sublingual and submaxillary glands and by way of the glossopharyngeal nerve to the parotid glands. These parasympathetic fibers are preganglionic, and they synapse with postganglionic fibers in or near the glands themselves. Despite differences in efferent pathways, there is no sharp division between centers. Rostral portions of the center supply the submaxillary, and caudal portions the parotid gland; but there is an intermediate portion supplying both.

Sympathetic innervation supplies all the glands by way of preganglionic fibers in the cervical sympathetic trunk that synapse in the superior cervical ganglion. Centers for the sympathetic outflow have not been clearly delimited. Afferent innervation is included in the intramedullary oral afferent systems, such as the solitary tract and portions of the spinal trigeminal nucleus and tracts; but, as with all medullary centers, additional afferent systems converge on the final common path. For example, salivation is associated with swallowing, licking, and chewing movements and with affective reactions.

Sympathetic stimulation acts through the beta receptors on the secretory cells, and therefore isoproterenol stimulates salivary secretion. Saliva obtained during such stimulation has a very high concentration of potassium and bicarbonate. The reason is that isoproterenol causes only a little secretion of primary fluid by acinar cells, and the cells of the ducts have the opportunity to remove most of the sodium and to secrete much potassium and bicarbonate into the small volume of fluid passing through them.

Stimulation of the parasympathetic efferent fibers causes secretion of a large volume of serous juice by the parotid glands, secretion of a large volume of juice and mucoprotein by the submaxillary gland, and vasodilatation in both glands. Sympathetic stimulation causes contraction of myoepithelial cells on the ducts, and their contraction helps to empty viscous saliva into the mouth.

Paralytic Secretion of Saliva

If one chorda tympani nerve in a dog or a cat is cut, the submaxillary gland on the cut side shows paralytic secretion. This secretion by the denervated gland ranges from scanty to copious, begins 3 days after denervation, and occurs in circumstances when the adrenal medulla is stimulated. Paralytic secretion thus appears to be similar to dilatation of the denervated pupil, another example of Cannon's law that denervated structures become exquisitely sensitive to the chemical mediator that had been their stimulus when the nerves were intact. The denervated gland is, in fact, much more sensitive to injected epinephrine or norepinephrine than is a normally innervated one.

Stimuli for Salivary Secretion in Man

Secretion of saliva is almost abolished during sleep. When saliva is collected by spitting out the contents of the mouth every 2 min, a procedure in itself minimally stimulating, the basal flow occurring in the absence of any obvious stimuli is about 0.5 ml/min. The submaxillary glands contribute 69% of resting flow, the parotid glands 26%, and the sublingual glands 5%. Upon stimulation by acid in the mouth, the rate of secretion by the parotid glands increases proportionately more than that of the other glands until they supply two thirds of the total flow.

Resting flow is reduced during dehydration, anxiety, fear, or severe mental effort. Chewing tasteless wax at the normal rate of 40–80 strokes/min raises secretion to an average of 2.3 ml/min. When chewing is unilateral, the glands on the active side secrete vigorously, while those on the inactive side secrete only a little. Secretion increases with the size of the bolus and with the pressure required to chew it. The most effective stimuli are substances arousing sensations of taste, and acid lemon drops are among the most potent. When the mouth is rinsed with a sapid solution in order to stimulate the whole gustatory receptor field, the secretory response is roughly proportional to the log of the stimulus strength. Maximal secretion averaging 7.4 ml/min follows rinsing of the mouth with 0.5 M citric acid. Smells also stimulate, amyl acetate calling forth secretion at about twice the basal rate.

Our knowledge of conditioned reflexes is built on the observation that a dog's salivary glands secrete in circumstances associated with feeding when food is not actually placed in its mouth. An unconditioned reflex is one that does not depend on previous experience. The secretion of saliva in response to food in the mouth is an unconditioned reflex; the stimulus evoking it is the unconditioned stimulus. If the unconditioned stimulus is presented together with the conditioned stimulus, which in itself would not evoke salivary secretion, a conditioned reflex is eventually established. Then, presentation of the conditioned stimulus alone calls forth salivary secretion. Thus it happens that sound, associated naturally or experimentally with food, comes to be a sufficient stimulus for salivation, and, when presented, the dog's mouth waters.

The universal impression that man's mouth waters at the thought of savory food, the clink of the cocktail shaker, or the sight of a sizzling steak, supports the notion that conditioned reflexes are also important for human salivation. Unfortunately, careful studies do not confirm this belief. Although a lemon, as a child with mumps is told, is believed to be a strong conditioned stimulus, one subject, whose parotid ducts were cannulated so that secretion could be accurately measured, experienced no increased secretion whatever when he saw a lemon and watched it being cut, squeezed, and sucked by a colleague. When he himself drank 2 ml of the juice, his secretory rate increased 12 times. In another experiment, bacon and eggs were cooked within 4 ft of three hungry subjects. In one, there was no increased secretion, and in the other two, saliva flowed at less than twice the resting rate. One subject reported his mouth watering when objectively there was no increase in his rate of secretion. Conditioned reflexes, at least when tested by these homely means, are weak in man, whatever they may be in the dog. We can account for the notion that the human mouth waters by observing that in the waking state the mouth always contains saliva, although usually we are not aware of it; thinking about food or participating in its preparation makes us conscious of the saliva, and we conclude that the saliva has just been secreted.

Salivary Blood Flow

In the rabbit, anatomical studies have clearly demonstrated that in the parotid gland the tubules and ducts are served by one capillary bed and the acini are served by another. The two capillary beds are in parallel, and there are no arteriovenous shunts. The conclusion, based on physiological evidence, that in the dog's parotid gland blood flows first through a capillary network surrounding the ducts and then through a capillary network surrounding the acini has not been controverted by anatomical studies. Nevertheless, on the basis of the principle of the uniformity of nature, it is likely that blood flows through parallel, rather than through series, capillary beds in all mammalian species.

Capillaries of the salivary glands are highly permeable to large molecules such as cyanocobalamin, whose molecular weight is 1,357. The capillaries have an enormous filtration capacity, the value being 40 times that of muscle capillaries. The large filtration capacity, and perhaps a low colloid osmotic pressure difference across the capillary wall, allows fluid to become readily available for secretion.

Normal resting blood flow through the dog's submaxillary gland ranges from 0.1–0.6 ml/gm of gland per min. This is about 20 times the flow through resting muscle. During maximal stimulation, blood flow increases to 4–6 ml/gm of gland per min, or 10 times that through active skeletal muscle.

Vasodilatation follows immediately upon parasympathetic stimulation. Blood flow through capillaries surrounding the ducts increases first and that through capillaries surrounding the acini a little later. This rapid vasodilatation is mediated by acetylcholine and blocked by atropine. If stimulation is continued after atropine blockade, a slow vasodilatation occurs and eventually reaches that produced by acetylcholine.

Two explanations of the slow, atropine-resistant vasodilatation have been proposed.

1. Salivary glands contain the enzyme kallikrein, which acts on a protein substrate carried in plasma to produce a powerful octapeptide vasodilator, bradykinin. Stimulation of the gland, particularly through an α-adrenergic mechanism, releases kallikrein, and kallikrein catalyses the formation of bradykinin. Evidence against this explanation is that kallikrein occurs only in the apical portion of cells lining the striated ducts, that kallikrein is liberated into saliva and not into interstitial fluid where bradykinin would have the opportunity to act on arteriolar smooth muscle, and that atropine-resistant vasodilatation occurs in a gland perfused with fluid containing no bradykinin precursor.

2. Nerves within salivary glands contain vasoactive intestinal peptide (VIP) as well as acetylcholine. Both are released when the nerves are stimulated; acetylcholine causes secretion, and VIP causes vasodilatation. Evidence for this explanation is that VIP is released into effluent blood when the glands are stimulated, that it is a more potent dilator than bradykinin and that intraarterial injection of VIP in the concentration occurring in venous blood during stimulation causes an increase in salivary blood flow.

Other dilating mechanisms: tissue hyperosmolality and release of other dilators such as purines, potassium, and adenosine, have not been ruled out.

REFERENCES

Babkin B.P.: *Secretory Mechanism of the Digestive Glands,* ed 2. New York, Paul B. Hoeber, Inc., 1950.

Blair-West J.R., Coghlan J.P., Denton D.A., et al.: Effects of endocrines on salivary glands, in Code C.F. (ed.): *Handbook of Physiology:* Sec. 6. *Alimentary Canal,* vol. 2. Washington, D.C.: American Physiological Society, 1967, pp. 633–664.

Burgen A.S.V.: Secretory processes in salivary glands, in Code C.F. (ed.): *Handbook of Physiology:* Sec. 6. *Alimentary Canal,* vol. 2. Washington, D.C.: American Physiological Society, 1967, pp. 561–580.

Emmelin N.: Nervous control of salivary glands, in Code C.F. (ed.): *Handbook of Physiology:* Sec. 6. *Alimentary Canal,* vol. 2. Washington, D.C.: American Physiological Society, 1967, pp. 595–632.

Fraser P.A., Smaje L.H.: The organization of the salivary microcirculation. *J. Physiol.* 272:121, 1977.

Gautvik K.: Studies on kinin formation in functional vasodilatation of the submandibular salivary gland in cats. *Acta Physiol. Scand.* 79:174, 188, 204, 1970.

Kreusser W., Heidland A., Hennemann H., et al.: Mono-

and divalent electrolyte patterns, Pco_2 and pH related to flow rate in the normal human parotid saliva. *Eur. J. Clin. Invest.* 2:398, 1972.

Mangos J.A., McSherry N.R., Irwin K., et al.: Handling of water and electrolytes by rabbit parotid and submaxillary glands, *Am. J. Physiol.* 225:450, 1973.

Martinez J.R.: Water and electrolyte secretion by the submaxillary gland, in Botelho S.Y., Brooks F.P., Shelley W.B. (eds.): *Exocrine Glands.* Philadelphia, University of Pennsylvania Press, 1969, pp. 20–29.

Mason D.K., Chisholm D.M.: *Salivary Glands in Health and Disease.* Philadelphia, W.B. Saunders Co., 1975.

Petersen O.H.: Acetylcholine-induced ion transports involved in the formation of saliva. *Acta Physiol. Scand.* [*suppl.*] 381:1, 1972.

Petersen O.H, Poulsen J.H.: Secretory transmembrane potentials in salivary glands, in Botelho S.Y., Brooks F.P., Shelley W.B. (eds.): *Exocrine Glands.* Philadelphia, University of Pennsylvania Press, 1969, pp. 3–19.

Schneyer L.H., Schneyer C.A.: Inorganic composition of saliva, in Code C.F. (ed.): *Handbook of Physiology:* Sec. 6. *Alimentary Canal,* vol. 2. Washington, D.C.: American Physiological Society, 1967, pp. 497–530.

Schneyer L.H., Young J.A., Schneyer C.A.: Salivary secretion of electrolytes. *Physiol. Rev.* 52:720, 1972.

Thorn N.A., Petersen O.H. (eds.): *Secretory Mechanisms of Exocrine Glands.* New York, Academic Press, 1974.

Wotman S., Mandel I.D., in Rankow R.M., Polayes I.M. (eds.): *Diseases of the Salivary Glands.* Philadelphia, W.B. Saunders Co., 1976, pp. 32–53.

9

Gastric Secretion, Digestion, and Absorption

THE STOMACH RECEIVES diverse amounts of food of heterogeneous composition and consistency at irregular intervals. Stimulated by the act of eating, by the presence of food in the stomach, and by chyme in the intestine, the stomach secretes pepsinogens, hydrochloric acid, intrinsic factor, and mucus. The concentration of acid in gastric contents is determined by the rate at which it is secreted, by neutralization and dilution by food, by other digestive secretions, and by back-diffusion of acid into the mucosa itself. There is some digestion of carbohydrate and protein in the stomach, but gastric absorption of the products of digestion is negligibly small. Some fat-soluble substances, including ethanol, are absorbed, and they may damage the mucosa. The cells of the mucosa are in a dynamic state of growth, migration, and desquamation, and most injury to the mucosa is rapidly repaired. The property of the stomach that normally resists penetration by acid may be impaired; then acid diffusing back into the mucosa may cause it to shed large amounts of fluid and to bleed.

Structure of the Gastric Mucosa

The gastric mucosa is divided, approximately at the notch, into the oxyntic glandular mucosa of the body and that of the pyloric gland area. Separation of the two types of mucosa is not precisely at the notch; generally, the pyloric type of mucosa extends above the notch into the body, particularly along the lesser curvature. The mucosa of the body is covered with simple columnar epithelial cells that contain and secrete mucus and an alkaline fluid. The surface is studded with depressions forming gastric pits, into each

of which the lumina of three to seven glands empty. The area occupied by the pits is 50% of the total surface. The distance between centers of the pits averages 0.1 mm, and there are about 100 pits/mm². The cells at the transition between surface epithelial cells and the glands are neck chief cells, whose cytoplasm stains for mucin. The chief cells, containing and secreting pepsinogens, are the most numerous of the several types of cells in the glands. On the borders of the glands, and especially in the neck region, are the cells responsible for acid secretion.

In the mammalian stomach, acid-secreting cells are on the walls of the gastric tubules, and for this reason they are called *parietal* cells (Latin, *paries* = wall). However, the stomachs of many species of animals contain acid-secreting cells that are not on the walls of glands. For this reason, a general, functional designation is better than a morphological one, and the acid-secreting cells will be called *oxyntic* cells (Greek, *oxys* = sharp). That part of the gastric mucosa containing a dense population of oxyntic cells is called the *oxyntic glandular mucosa.* However, the mucosa lining the gastric antrum, the *pyloric glandular mucosa,* also contains a few oxyntic cells.

The surface of the pyloric gland area is also covered with epithelial cells. The glands in this region are composed of cells resembling neck chief cells. Chief cells and oxyntic cells are absent from the cardiac gland area, a ring of mucus-secreting glands 1–4 cm wide at the junction of the esophagus with stomach.

Gastrin, or properly gastrins, are synthesized, stored, and secreted by between 8 and 15 million G cells in the pyloric mucosa. G cells are most dense near the pylorus. They are flask-shaped; their necks project to the luminal surface of the glands, and the exposed surface is covered with microvilli, which, one supposes, receive stimuli from the lumen of the stomach for release of gastrin. Nerve fibrils from the intrinsic plexuses end near the G cells, conveying excitatory and inhibitory stimuli.

Gastrin is contained in basal storage granules from which it is released into interstitial fluid and thence into portal blood. Gastrin is also released into the lumen of the stomach.

Secretion of Surface and Neck Chief Cells

The approximate composition of the alkaline juice secreted by surface cells is given in Table 9–1. The estimated values for the dog in column 2 were calculated from analyses of samples of juice being secreted at different rates in response to graded doses of histamine. It was assumed that these samples consisted of different proportions of acid and nonacid juices. On the further assumption that variations in the rate of secretion were caused only by variation in the proportion of acid juice of fixed composition, an extrapolation to a hypothetical sample containing no acid was made. The values for man were calculated by applying the same assumptions to data obtained from numerous samples of gastric juice stimulated either by histamine or by insulin hypoglycemia. Data for the dog in column 3 were obtained from samples secreted by an exposed flap of gastric mucosa stimulated by high doses of acetylcholine. The ionic composition of the secretion is close to that of an ultrafiltrate of plasma to which has been added mucus containing 3 mEq of sodium and potassium per liter.

Bicarbonate diffuses through the normal pyloric mucosa from interstitial fluid into the lumen, and it leaks through the oxyntic mucosa when that mucosa's permeability has been increased by damaging agents such as ethanol or aspirin. In addition, bicarbonate is actively secreted into the lumen by surface epithelial cells of both the pyloric and the oxyntic mucosa. Bicarbonate secretion is stimulated by cholinergic impulses and by 16,16-dimethyl prostanglandin E_2. It may also be stimulated when the mucosa is damaged. Stimulation of secretion is mediated intracellularly by cyclic-GMP, and it is accelerated by carbonic anhydrase present in the epithelial cells.

Two kinds of mucus are secreted: soluble mucus from the neck chief cells and gel-forming mucus from

TABLE 9–1.—COMPOSITION OF JUICE SECRETED BY SURFACE EPITHELIAL CELLS OF THE STOMACH

	MAN, EST.* RANGE (MM)	DOG, EST.† MEAN (MM)	DOG, BY ANALYSIS‡ MEAN (MM)
Na$^+$	150–160	155	138
K$^+$	10–20	7	4
Ca^{2+}	3–4	4	5
Cl$^-$	125	133	117
HCO$_3^-$	45	33	23
pH	(7.67)§	(7.54)§	7.42
Protein			10–11 gm/L

*Equals nonparietal secretion, calculated by Hunt J.N.: *Physiol. Rev.* 39:491, 1959.
†Gray J.S., Bucher G.R.: *Am. J. Physiol.* 133:542, 1941.
‡Altamirano M.: *J. Physiol.* 168:787, 1963.
§Calculated from bicarbonate concentration and assumed P_{CO_2} of 40 mm Hg.

the surface cells. The first is absent from gastric juice during fasting, but it is secreted in response to vagal excitation by sham feeding or insulin hypoglycemia.

The visible mucus which forms a layer over the surface of the mucosa is synthesized by the surface epithelial cells and stored as granules at their apical tips. Mechanical irritation of the surface, as by rubbing with a glass rod, stimulates secretion; and more severe stimulation by substances that tend to injure the mucosa causes copious secretion. Splanchnic nerve stimulation at low rates within the physiological range causes secretion of mucus, and so does stimulation of the vagus nerve. Mucus secretion is probably under the same reflex control as the other gastric secretions. Secretion is not stimulated by histamine, but it is by parasympathomimetic drugs.

Mucus is discharged to the surface in three ways: (1) A single granule fuses with the apical membrane of the cell; the granule is extruded by pinching off the fused membrane; and mucus flows from the granule onto the surface of the cell. This is exocytosis. (2) Many granules fuse with the apical membrane, and the entire apical membrane dissolves. All granules at the tip of the cell break up, forming a pool of mucus which is then expelled as a package. Subsequently the remainder of the cell degenerates. This is apical dissolution. (3) A surface epithelial cell loses its basal and lateral attachments, and the entire cell is ejected onto the mucosal surface. Dissolution of the shed cell leaves mucus behind. This is exfoliation.

The mucus gel is 95% an aqueous solution of electrolytes and 5% a network of polymers, each having a molecular weight of about 2×10^6. Each polymer is composed of four subunits, and each subunit has a protein core, three quarters of which is covered with 160 closely packed polysaccharide chains having an average of 15 sugar residues per chain. The polysaccharide chains radiate from the protein core like the bristles of a bottle brush. One quarter of the protein core which is not covered with polysaccharide chains is linked by disulfide bridges with the corresponding part of the protein cores of three other subunits. The subunits themselves are not attacked by proteolytic enzymes, but the part of the glycoprotein molecule in which the four subunits are joined is susceptible to proteolysis by pepsin in gastric juice. When the four subunits are liberated, they become soluble mucus within the gastric contents. Consequently, the thickness of the mucous layer at any time is a function of the rate of secretion and the rate of dissolution. Except in extraordinary circumstances, it is less than 1 mm thick.

The major function of the mucous layer is to lubricate the surface of the stomach. Some persons believe that the unstirred layer of fluid within the gel has a protective function. Hydrogen ions diffuse through a layer of mucus at about one fourth the rate they diffuse through an unstirred layer of water, and if the unstirred layer itself were the only barrier, hydrogen ions would quickly reach a high concentration at the bottom of the layer in contact with the apical membrane of the epithelial cells. However, the cells secrete bicarbonate into the gel, and if the rate of bicarbonate secretion is comparable to the rate of acid diffusion, neutralization of acid within the unstirred layer would protect the epithelial cells. This has not yet been experimentally demonstrated to be the case, and much indirect, contrary evidence supports the idea that the only important function of mucus is lubrication.

Cardiac and Pyloric Glands

In man, a small zone of gastric mucosa (0.5–4 cm in radius) surrounding the esophageal opening is lined with cardiac glands. These tubular glands are composed of mucous cells and have few or no chief or oxyntic cells. Oxyntic cells are definitely present in small numbers in the mucosa lining the pyloric gland area of the stomach. The mucosa of this area is thin (1–1.5 mm) and composed of cells resembling mu-

cous neck cells. It produces a scanty, highly viscid secretion which contains some pepsinogen and which lubricates the surface over which a large volume of chyme moves back and forth during gastric digestion.

Pepsinogens and Pepsins

Pepsinogens are zymogens secreted by the gastric mucosa, and they are converted into active, proteolytic enzymes by acid. They can be separated electrophoretically into two groups. Group I contains five (or more) rapidly migrating proteins, and Group II contains two (or more) slowly migrating ones. Group I pepsinogens occur only in the proximal stomach, but Group II are found throughout the stomach and duodenum. Both groups occur in high concentration in the chief cells of the oxyntic glandular mucosa and in lower concentration in mucous neck cells. Group II pepsinogens are abundant in the pyloric glandular mucosa and in Brunner's glands of the duodenum. Different fractions of pepsinogen, when activated to pepsin, have different pH optima and substrate specificity, and they may also be functionally heterogeneous. Nevertheless, all the zymogens and enzymes will be referred to simply as pepsinogen and pepsin. Little or no pepsinogen can be found in the chief cells 3 hours after feeding, although pepsinogen secretion continues. The granules are not obligatory intermediates in the path of synthesis and secretion. The granules begin to reappear after 6 hours, and they are maximally concentrated after 72 hours.

Pepsinogen is converted to proteolytically active pepsin when it mixes with acid secreted by the oxyntic cells. Activation occurs more rapidly the lower the pH and is almost instantaneous at pH 2.0. Pepsin itself activates pepsinogen, and in the process of activation a small polypeptide is cleaved from the zymogen. Pepsin is maximally active as a proteolytic enzyme between pH 1.0 and 3.0. Although pepsinogen is stable in neutral solution, pepsin rapidly and completely loses its catalytic activity when the solution in which it is dissolved is neutralized. Under some experimental conditions, denaturation of pepsin can be reversed; but for all practical purposes, peptic activity of gastric juice is destroyed by neutralization.

Stimuli for Pepsinogen Secretion

There is a small continuous basal secretion of pepsinogen in man. The most powerful stimulus for additional secretion is vagal stimulation; and, therefore, insulin hypoglycemia, acetylcholine and its congeners and the direct vagal phase of stimulation by feeding cause flow of juice with high pepsin concentration. This effect is mediated by acetylcholine liberated in the neighborhood of the chief cells. Pepsinogen secretion is stimulated by the hormone secretin. Gastrin does not directly stimulate pepsinogen secretion by acting on the chief cells. Gastrin stimulates acid secretion, and acid in contact with the surface of the gastric mucosa stimulates afferent nerves of the intramural plexuses which, through a cholinergic reflex, in turn stimulate secretion of pepsinogen. The several modes of stimulation are synergistic. Infusion of secretin increases the rate of pepsinogen secretion 3 times when the contents of the stomach are neutral and 12 times when the contents are acid. Pepsinogen secretion is inhibited by atropine. In man or the dog, administration of histamine causes a transitory burst of pepsinogen secretion, which has been attributed to the washing-out of the precursor from the tubules of the gastric glands. As the dose of histamine is increased, pepsinogen secretion increases, along with acid secretion, until the maximal rate of acid secretion (about 10 times basal) is reached. The rate of pepsinogen secretion is then 3–4 times basal.

There are minor qualifications to the generalization that there is a correlation between acid and pepsin outputs. For example, secretin, which stimulates pepsinogen secretion, inhibits acid secretion. Secretin stimulates cyclic AMP formation in chief cells and causes them to discharge their pepsinogen granules. Unless otherwise qualified, the statement that one or another factor influences acid secretion can be taken as applying to pepsinogen as well.

Plasma Pepsinogen and Uropepsinogen

Pepsinogens are present in lymph collected from small lymphatic vessels on the surface of the stomach, and they are doubtless present in gastric interstitial fluid. Both Group I and Group II pepsinogens are present in plasma, but the concentration of only Group I pepsinogens is strongly correlated with peak acid output. Group I hyperpepsinogenemia is particularly prominent in persons whose peak acid output is greater than 40 mEq/hr.

Group I hyperpepsinogenemia, a concentration greater than 100 ng/ml, is inherited in man as an autosomal dominant trait, and it is associated with the occurrence of peptic ulcer. In two large kindreds,

duodenal ulcer occurred in about 40% of those with Group I hyperpepsinogenemia and in none of those without it. An individual with the trait is about 5 times more likely to have an ulcer than is one without the trait. This, however, is not true of all patients with duodenal ulcer disease, for in another large kindred, only half the patients with duodenal ulcer disease had Group I hyperpepsinogenemia. Thus, Group I hyperpepsinogenemia is a genetic marker for one, but not all, kind of duodenal ulcer disease.

Group I pepsinogens appear in the urine. The amount excreted per day is roughly correlated with chief cell mass and with plasma pepsinogen concentration, but it has no important physiological or diagnostic significance.

Gastric Rennin, Gelatinase, Tributyrase, and Urease

The gastric juice of newborn calves contains a milk-clotting enzyme, rennin; but the enzyme is probably absent from human stomachs. Pepsin itself has powerful milk-clotting activity and performs rennin's function in the human stomach. Crude pepsin preparations contain a distinct and separable enzyme, gelatinase, which liquefies gelatin about 400 times faster than does pepsin. Its close association with pepsin suggests that it, too, comes from chief cells.

Gastric juice also contains a lipase that is stable at pH 2.0 and has a broad pH optimum in the range of pH 4.0–7.0. It catalyzes hydrolysis of ester bonds at the 1 and 1′ positions of triglycerides. It is much more active against a substrate containing short-chain or medium-chain fatty acids than against long-chain triglycerides. Because the only natural source of short-chain triglycerides is milk, the enzyme is more important for infants than for adults. A lipase is also secreted by pharyngeal glands.

Acid hydrolysis of triglycerides in the stomach is probably insignificant.

The enzyme urease occurs in the gastric mucosa of man and many other mammals. Much of it is contained within bacteria contaminating the mucosa, but some urease not of bacterial origin may be contained in the mucosal cells. There is evidence that fetal gastric mucosa, presumed to be uncontaminated with bacteria, contains urease; that sterile samples of the cytoplasm of the surface epithelial cells of mouse stomach have urease activity; and that more enzyme is in the mucosa than can be accounted for by its bacterial population. Bacteria containing urease are also present in the nose and pharynx, and their number, together with the urease activity of the tissues, can be temporarily reduced by heavy antibiotic medication. Urea available for hydrolysis comes in contact with the enzyme as water containing it flows from the plasma through the mucosa during gastric secretion. The result is that, during acid secretion, a small amount of urea is hydrolyzed in the mucosa.

The products of urea hydrolysis are alkaline: 2 M of ammonia to 1 M of carbon dioxide. Because the amount of ammonia formed is only 1/500th the amount of acid secreted, hydrolysis of urea does not contribute to neutralization of acid in normal subjects. When large amounts of urea (15–25 gm) are ingested, its hydrolysis may neutralize up to half the acid secreted, and the gas eructed actually tastes faintly of ammonia. However, a concentrated solution of urea breaks the gastric mucosal barrier, and when so much urea is swallowed acid also disappears from the stomach by back-diffusion and by neutralization by buffers from the interstitial fluid. The histamine-stimulated juice of uremic patients whose blood urea concentration is about 6 times normal contains 60% as much acid as normal controls, and there is ammonia in the juice. In such a patient, ammonia released in the stomach may contribute to hyperammonemia and encephalopathy. A fall in gastric juice ammonia, either spontaneous or induced by antibiotic medication, is accompanied by an equivalent increase in gastric acidity.

Gastric Hydrochloric Acid

When gastric juice is collected either from the human stomach or from a pouch of the oxyntic gland area of the dog stomach, its composition is characteristically found to be a function of the rate of secretion (Fig 9–1). As the rate of secretion increases, hydrogen ion concentration rises, sodium concentration falls and chloride concentration remains nearly constant. The potassium concentration rises slightly. In man the composition of juice secreted at the fastest rate is approximately [H$^+$] 145, [Cl$^-$] 170, [Na$^+$] 7, and [K$^+$] 17 mN. In the dog the upper limit of [H$^+$] is about 160 mN.

Judged by the interest it arouses among physiologists and laymen alike, the most dramatic component of the juice is acid. Acid provides optimally low pH for pepsin activity, but peptic hydrolysis of protein is

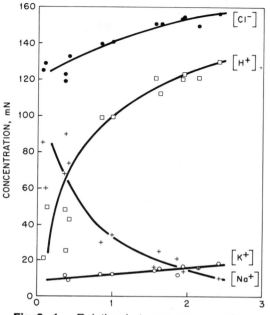

Fig 9–1.—Relation between concentration of electrolytes in the gastric juice of a normal young man and rate of secretion. Secretion was stimulated by intravenous histamine infusion at various constant rates, and the juice to be analyzed was collected after steady rate of secretion had been reached. (Adapted from Nordgren B.: *Acta Physiol. Scand.* 58 *[suppl. 202]:*1, 1963.)

dispensable. The other identifiable function of acid is partially to sterilize chyme before it enters the small intestine. Although the dictum ''No acid, no ulcer'' may be valid, the physiological and pathophysiological roles of acid hardly justify the vast labor expended upon it.

Acid juice has never been caught directly as it issues from the cells secreting it, nor has it been obtained absolutely uncontaminated by at least a small fraction of other gastric juices. Consequently, there is no totally unequivocal answer to the questions: Is the composition of the acid juice constant as it is secreted at various rates, or does its composition vary with the rate of secretion? Is the acid juice pure hydrochloric acid, or does it contain soldium and potassium salts?

The relation between rate of secretion and composition is not invariable; juice high in acid can be obtained at low rates of secretion. When a single subcutaneous dose of histamine is given to a dog, the

acidity of the juice collected from a pouch of the oxyntic gland area of the stomach rises to a maximum
as the rate of secretion rises. However, as the rate of
secretion falls in the fourth quarter-hour after histamine injection, the hydrogen ion concentration of the
juice remains high and continues to be so throughout
the whole period of declining rate of secretion. Other
experiments show that acid concentration can be independent of the rate of secretion; when secretion is
stimulated by continuous administration of histamine,
injection of atropine profoundly reduces the volume,
but not the acidity, of the juice secreted.

Three theories explain variations in the composition
of gastric juice:

1. The acid juice as it issues from the secreting
cells does vary in composition; at low rates of secretion it contains much sodium and little acid, but as
the secretory rate rises, the concentration of sodium
falls and that of hydrogen ion increases until at the
maximum rate of secretion the juice is nearly pure,
isotonic HCl.

2a. Fluid secreted by the oxyntic cells has a constant composition independent of its rate of secretion;
it is nearly pure, isotonic HCl with perhaps a very
small admixture of NaCl and KCl. On the surface of
the mucosa it mixes with one or more components
secreted by other cells of the mucosa. These components likewise have constant composition; they are
similar to an ultrafiltrate of plasma, containing much
sodium and some bicarbonate, and upon mixing with
acid juice the bicarbonate reacts with the acid so that
some hydrogen ions disappear and osmols are lost as
carbon dioxide evolves from the mixture. When acid
secretion is stimulated by histamine or gastrin, the
rate of secretion of the additional components changes
slightly, if at all. Consequently, the composition of
pure oxyntic juice can be deduced by extrapolation of
the analytic data to an infinite rate of secretion.

2b. A variant of this second theory is that the composition of the other components changes with their
rate of secretion.

3. Gastric secretions are modified as they flow over
the surface of the mucosa by exchanges with interstitial fluid. Hydrogen ions diffuse slowly or rapidly
down a very steep concentration gradient from the lumen into the mucosa, and sodium ions diffuse along
an electrochemical gradient from interstitial fluid into
the lumen. The extent to which juice is modified depends on the length of time it is in contact with the
mucosa and on the permeability of the mucosa.

Although gastroenterologists have supported one or
another of these theories with the fervor of Big-enders
attacking Little-enders, there are no experimental observations that establish one as correct and reject the
others as error. On the contrary, a mixture of all three
theories appears to be required to explain the facts.

At very high rates of secretion, gastric juice of the
dog does approach in composition an isotonic solution
of HCl, containing perhaps no more than 3–5 mN Na
and less than 1 mN K; but no amount of manipulation
of the data by mathematical extrapolation can prove
that this is the composition of acid juice secreted at
low rates. A fluid whose composition is similar to that
of an ultrafiltrate of plasma can be collected from the
surface of the gastric mucosa under very special conditions, but there is no direct experimental evidence
that the nonacid juice mixed with acid secretion has
this composition. Finally, both sodium and hydrogen
ions can cross the gastric mucosa; and when the
permeability of the mucosa is high such exchanges
grossly modify acid juice secreted by the mucosa.
However, the part such exchanges play under physiological conditions has not yet been clearly demonstrated.

Sodium and Potassium Outputs in Acid Juice

Sodium in gastric juice comes from two sources:
secretion in acid and nonacid juices from both pyloric
and oxyntic gland areas and from diffusion across the
mucosa. Under normal circumstances the latter contribution is small. As the rate of secretion of acid
juice rises, both concentration (mN) and output (mEq/
min) fall. If, however, the gastric mucosa is abnormally permeable, as after damage by acetylsalicylic
acid in acid solution, sodium output rises with increasing volume output, because a large amount of
sodium diffuses across the mucosa from interstitial
fluid. Sodium output is also large in protein-losing
gastropathy when the mucosa sheds plasma.

Potassium concentration in the basal secretion of
the unstimulated human stomach is higher than in
plasma, being 14–18 mN. In stimulated secretion,
the concentration of potassium is constant when the
rate of secretion is constant. There is a transient increase of potassium concentration of a few milliequivalents per liter, lasting 20 min or more, when
secretion rate rises, and a corresponding transient fall
in potassium concentration when secretion rate de-

creases. In man, the potassium concentration during steady-state secretion stimulated by IV administration of histamine is between 8 and 20 mN. It is lower in the dog, the average being 6 mN, with a range of 3–9. Because potassium concentration varies in a relatively narrow range, the output (concentration times rate of secretion) is positively correlated with the output of hydrogen ions.

Potassium in gastric juice is partly derived from intracellular potassium. If radioactive potassium (^{42}K) is injected IV and acid secretion is stimulated before the isotope has had a chance to equilibrate with intracellular potassium, the specific activity of potassium in the juice is only about half that of plasma potassium.

Loss of gastric juice through vomiting or drainage may deplete the body's stores of sodium and potassium.

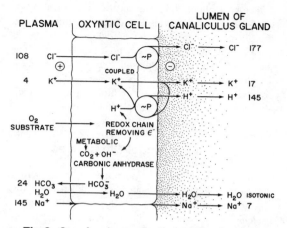

Fig 9–2.—A scheme for the mechanism of secretion of hydrochloric acid.

Secretion of Hydrochloric Acid

At rest, an oxyntic cell contains a few seemingly empty canaliculi and a large number of tubules and vesicles. When the cell is stimulated to secrete, there is an enormous increase in the surface area of the canaliculi as the result of eversion or fusion of tubules and vesicles. The canaliculi then become distended with an acid solution, the product of secretion by the canalicular membrane.

Three intracellular processes result in secretion of hydrochloric acid (Fig 9–2): (1) Acid secretion is absolutely dependent upon oxygen. An electron is removed from a substrate within the cell and passed along the redox system contained in the cytochrome chain. The electron is ultimately transfered to oxygen. This process, in some yet unknown manner, makes H$^+$ available for secretion. (2) An electrogenic pump in the canalicular membrane secretes Cl$^-$ into the lumen of the canaliculi. When there is little or no stimulus for secretion, this Cl$^-$ pump is minimally active, but because it continues to secrete Cl$^-$ into the lumen, the luminal surface of the gastric mucosa is negative with respect to the serosal surface by about 40–60 mV. K$^+$ accompanies Cl$^-$ into the lumen of the canaliculi, probably by simple diffusion down its electrochemical gradient. (3) The canalicular membrane contains a K$^+$-H$^+$ ATPase which effects exchange of H$^+$ from within the cell for K$^+$ in the lumen of the canaliculi. Because the exchange is 1:1, H$^+$ secretion itself is not electrogenic. Secretion of H$^+$ is coupled with secretion of Cl$^-$ so that Cl$^-$ pumping increases

to keep pace with H$^+$ secretion. Pumping of Cl$^-$ and H$^+$ both use energy contained in ATP generated by mitochondria which occupy about 40% of the cell volume.

Transfer of electrons from substrate to oxygen results in generation of OH$^-$ within the oxyntic cell in an amount equal to that of H$^+$ secreted. At rest, the pH of the cell's cytoplasm is 7.1–7.4, but during active secretion the pH rises to greater than 8. The CO$_2$ from the plasma and from cellular metabolism combines with OH$^-$ to form HCO$_3^-$. This reaction is catalyzed by the enzyme carbonic anhydrase which is present in the canalicular membrane. HCO$_3^-$ from within the cell exchanges across the basal membrane for Cl$^-$. Consequently, gastric venous blood has a higher pH than arterial blood when the stomach is secreting acid, and this is reflected in the alkaline tide of the urine.

Some K$^+$ escapes from the canalicular lumen into gastric juice. Some Na$^+$ escapes as well; whether this is through the paracellular route down Na$^+$'s electrochemical gradient as shown in Fig 9–2 or whether Na$^+$ leaks from the oxyntic cell is unknown.

Water moves passively through the oxyntic cell into gastric juice.

An odd aspect of the chloride pump is that, if given a chance, it prefers to transport bromide rather than chloride. If there is bromide in the plasma, as, for example, when bromide salts are ingested, hydrobromic acid is secreted, and the ratio of bromide to chloride in the gastric juice exceeds that in the plasma by a factor of about 1.5. The gastric mucosa also actively secretes iodide ions into the gastric juice. When the plasma of concentration of iodide in the gastric juice to that in plasma is very high (about 15:1), but the ratio falls to unity as plasma iodide rises to 3 mN. The mucosa's ability to concentrate iodide is depressed by thiocyanate and perchlorate, agents which also depress thyroidal accumulation of iodide.

Gastric Potential Differences

The major portion of the potential difference across the oxyntic glandular mucosa is established by the chloride pump in the oxyntic cells. The pump is electrogenic in the sense that it can produce a net transport of electrical charge. It can operate without an exchangeable anion on the mucosal surface of the stomach, and it causes a net flow of chloride ions from the cells to the lumen. When the gastric mucosa is not secreting acid and when its mucosal surface is bathed by a neutral, sodium-containing solution, some mucosal cells pump sodium from the mucosal to the serosal surface. The direction of pumping, movement of a cation from mucosa to serosa, tends to make the mucosal surface negative with respect to the serosal surface. The combined action of the two pumps establishes a potential difference of 40–80 mV across the nonsecreting oxyntic glandular mucosa. Bathing the mucosal surface with acid abolishes sodium pumping and reduces the potential difference.

The potential difference across the pyloric glandular mucosa, which contains few oxyntic cells, is less than 10 mV, the mucosa being negative with respect to the serosa.

The potential difference across the duodenal mucosa is oriented in the opposite direction, since the mucosal surface is positive with respect to the serosal. If a probe bearing an electrode is passed into the duodenum, the potential difference between the mucosal surface and the skin or blood can be measured. When the probe is pulled slowly toward the stomach, an abrupt reversal of potential difference occurs as the probe passes through the pyloric sphincter to come into contact with the pyloric glandular mucosa (Fig 9–3). This reversal of potential difference can be used to identify the location of the pyloric sphincter.

The potential difference may also be used to assess the integrity of the mucosa. Agents that damage the gastric mucosal barrier to acid lower the potential difference. The effect of aspirin is shown in Figure 9–4. Drinking 25 or 50 ml of 80-proof whiskey has the same effect.

Osmotic Pressure of Gastric Juice

The major components contributing to the osmotic pressure of gastric juice are H^+, Na^+, K^+, and Cl^-. Other constituents (Ca^{2+}, Mg^{2+}, PO_4, glucose, urea, and proteins) together make up less than 10 mOsm.

If hydrogen and chloride ions, and perhaps potassium ions, are actively secreted, and if the secreting cells are permeable to water, then, at the same time that the ions are secreted, water should move from blood to gastric juice, merely to maintain osmotic equilibrium. Were this the sole process involved in water movement, gastric acid secretion should always be isotonic with plasma. Nevertheless, in samples of gastric juice having low acidity, there is systematic deviation from isotonicity in the direction of hypotonicity of the juice. The osmotic pressure of human basal secretion collected by stomach tube is 171–276 mOsm. For the dog whose juice can be collected from a pouch of the oxyntic gland area without contamination by other secretions, the maximal deviation from isotonicity is about 40 mOsm of a total of 310 mOsm. Part of this can be explained on the theory that actual samples of gastric juice are a mixture of two or more components. Low acidity of the sample means that it has been reduced by admixture of an originally isotonic acid juice with an alkaline juice that is also isotonic. Since the alkaline juice contains bicarbonate, mixing of it with acid produces volatile carbon dioxide, which disappears from the osmotic balance, leaving a fluid of low osmotic pressure.

When acid secretion is stimulated by histamine, the osmotic pressure of gastric juice rises as the rate of secretion increases. In four of seven human subjects, the ratio of the osmotic pressure of gastric juice to the osmotic pressure of plasma rose to between 0.97 and 1.04; the juice became isotonic. In three other subjects the juice was always hypotonic, the ratio ranging from 0.76 to 0.88. Juice collected from the oxyntic glandular mucosa of the dog stomach has a similar range of osmotic pressure. In two series of experiments the ratio of the osmotic pressure of juice secreted under maximal histamine stimulation to that of plasma was 1.05–1.10; but in a third series of apparently identical experiments, juice was hypotonic to

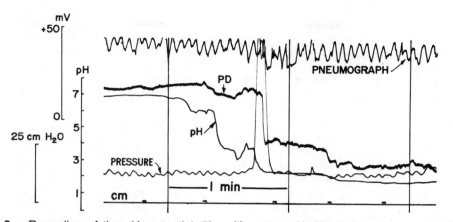

Fig 9–3.—Recording of the pH, potential difference *(PD)*, and pressure from the gastroduodenal junction in man. The tip of the probe by which pH, potential difference, and pressure were measured was withdrawn 1 cm at each signal mark from the left (the duodenum) to the right (the stomach). The pylorus is recognized by the sharp change of potential difference and the simultaneous peak in the pressure recording. The pH started to decrease distal to the pylorous. (From Andersson S., Grossman M.I.: *Gastroenterology* 49:364, 1965.)

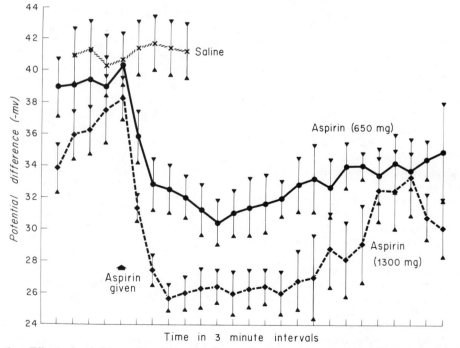

Fig 9–4.—Effect of solutions of aspirin and of isotonic saline upon the potential difference across the human gastric mucosa. Points represent the mean ± SE of observations on ten normal subjects. The mucosal surface is negative with respect to the blood. (From Geall M.G.; Phillips S.F., Summerskill W.H.J.: *Gastroenterology* 58:437, 1970.)

gastric venous plasma in 20 of 30 dogs and isotonic in the rest. If the osmotic pressure of plasma is varied in either direction by infusion of hypo- or hypertonic solutions, there are corresponding changes in the osmotic pressure of acid gastric juice.

When acid juice is secreted, two osmotically active ions, H^+ and Cl^-, are removed from plasma, together with enough water to keep the juice nearly isotonic with plasma. However, the hydrogen ion is manufactured de novo by the mucosa, and the mucosa exchanges a new bicarbonate ion for a chloride ion. The result is that the water of the gastric juice is extracted from plasma, but the total mass of osmotically active particles in plasma is unchanged. In effect, water, not osmols, is removed, and the total osmotic pressure of plasma and other body compartments rises slightly. When acid juice is neutralized and the products are reabsorbed in the intestine, the original osmotic balance is restored. If acid juice is lost by vomiting or gastric drainage, dehydration, consisting of both loss of extracellular fluid and increase in its osmotic pressure, as well as metabolic alkalosis, occurs.

Stimuli at the Cellular Level

Three major stimuli act directly upon the oxyntic cells (Fig 9–5):

1. Histamine is synthesized, stored, and released by enterochromaffin cells different from mast cells within the gastric mucosa. The amount of histamine stored in the mucosa is very large. There is a continuous background concentration of histamine within gastric interstitial fluid, and that concentration increases greatly when the gastric mucosa is injured. Histamine combines with H_2 receptors on the oxyntic cells and thereby stimulates adenylcyclase to catalyze the formation of cyclic AMP within the cells. Cyclic AMP then performs the function of stimulus-secretion coupling with the result that ATP is made available for Cl^- pumping and for K^+-H^+ exchange and HCl secretion. The combination of histamine with H_2 receptors is blocked by cimetidine and other H_2 antagonists, but not by H_1 antagonists. Therefore, cimetidine inhibits histamine-stimulated acid secretion. Histamine is rapidly metabolized within the gastric mucosa, chiefly by methylation. Most methylated derivatives of histamine are inactive, but one N-alpha methyl derivative produced in small amounts is a powerful stimulus for acid secretion.

2. Acetylcholine, and therefore parasympathomimetic drugs such as carbachol, combine with specific receptors on the oxyntic cells and stimulate acid secretion. Acetylcholine is released near the cells from nerve endings, which in turn are stimulated by local and by vagally mediated reflexes. Acetylcholine does not cause formation of cyclic AMP within the oxyntic cells, and it does not influence the formation of cyclic AMP stimulated by histamine. Since cyclic AMP is required for stimulation of secretion, acetylcholine as a stimulant must rely upon the simultaneous action of histamine. The combination of acetylcholine with its receptor is blocked by atropine and similar drugs.

3. Gastrin released from G cells in the pyloric mucosa and intestinal mucosa or from gastrinomas stimulates acid secretion by combining with specific gastrin receptors on the oxyntic cells. Like acetylcholine, gastrin does not stimulate formation of cyclic AMP, nor does it influence the formation of cyclic AMP stimulated by histamine. Combination of gastrin with its receptors is competitively blocked by cholecystokinin, which shares gastrin's active C-terminal amino acid sequence. Since cholecystokinin is a much weaker stimulant of acid secretion than is gastrin, cholecystokinin, by denying receptors to gastrin, inhibits gastrin-stimulated acid secretion. This action of cholecystokinin is not physiologically important. Other hormones may inhibit gastrin-stimulated acid secretion by combining with gastrin receptors.

There is two-way synergism between acetylcholine and histamine. A small amount of histamine increases the effectiveness of a given amount of acetylcholine in stimulating acid secretion, and a small amount of acetylcholine increases the effectiveness of a given amount of histamine. There is likewise a two-way synergism between gastrin and histamine, but there is no synergism between gastrin and acetylcholine. There is a three-way synergism among histamine, acetylcholine, and gastrin. In each of these combinations, histamine has the central role of causing formation of cyclic AMP. These synergisms explain the apparent nonspecific action of drugs in inhibiting acid secretion. Cimetidine, in blocking H_2 receptors, inhibits gastrin-stimulated secretion, and it is therefore able to inhibit hypersecretion produced by the hypergastrinemia of the Zollinger-Ellison syndrome. Likewise, atropine, in blocking the action of acetylcholine, inhibits secretion stimulated by a combination of histamine and acetylcholine or a combination of gas-

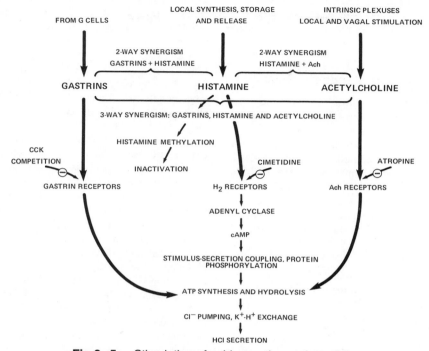

Fig 9–5.—Stimulation of acid secretion at the cellular level by gastrin, histamine, and acetylcholine.

trin, histamine, and acetylcholine. As a result, vagotomy plus cimetidine has been found to be a substitute for gastrectomy in dealing with hypergastrinemia caused by a gastrinoma.

Ammonia and some amino acids, L-phenylalanine being the most effective, stimulate acid secretion by direct effect upon the oxyntic cells. Their effectiveness increases with increasing pH, probably because the compounds can more readily penetrate the cell membrane when their amino groups are less protonated. Such stimulation is not blocked by atropine or cimetidine. Stimulation by L-histidine is blocked by H_2 antagonists, doubtless because the amino acid is first decarboxylated to become histamine. Ethanol stimulates acid secretion by entering into the cells' oxidative metabolism. Enkephalins stimulate acid secretion, but whether they act at the cellular level is unknown.

Somatostatin inhibits acid secretion. Somatostatin is present in the oxyntic mucosa, and its action may

be an example of a local, paracrine effect. Some prostaglandins, particularly 16,16-dimethyl PGE_2, are powerful inhibitors, acting when applied topically to the mucosa in minute amounts.

Two peptides isolated from normal male human urine, when injected in small doses, inhibit histamine- and pentagastrin-stimulated acid secretion. These are called *urogastrone*. One of these peptides contains 53 amino acids, and the other contains the same ones except for arginine at the C-terminus. Urogastrone appears to be identical with a peptide isolated from the submaxillary glands of male mice that promotes epidermal growth. The biological functions of urogastrone and epidermal growth factor are unknown. Bacterial pyrogens, when injected IV, also inhibit acid secretion. Acid secretion is inhibited by many factors whose role in control of gastric secretion is poorly understood. In man, distention of the gastric antrum inhibits basal and pentagastrin-stimulated secretion through a nervous reflex by about 20%. Perfusion of the colon with sodium oleate, liver extract, or hydrochloric acid also inhibits acid secretion, perhaps by releasing an unidentified hormone from the colonic mucosa.

Measurement and Standards of Acidity*

In studying gastric acidity, we must know two quantities: the acidity of a particular sample and the amount of acid secreted in a given time.

The acidity of a solution is the *activity* of hydrogen ions in it, not their concentration. Activity is a dimensionless quantity whose symbol is written a_H+. Although physical chemists are not in complete agreement on the subject, the most commonly accepted definition of pH is the negative logarithm to the base 10 of the activity of hydrogen ions in a solution.

$$pH = -\log_{10} a_H+ \qquad (9.1)$$

In an infinitely dilute solution the activity of hydrogen ions is exactly equal to their concentration $[H^+]$. In more concentrated solutions, activity is usually less than concentration, and this fact is expressed by the equation

$$a_H+ = f[H^+] \qquad (9.2)$$

Here f is the activity coefficient whose numerical value is less than 1 and which must be experimentally determined.

The concentration of hydrogen ions in blood and body fluids except gastric juice is so small that their activity coefficient is probably close to 1, and there is no reason to think that the coefficient varies under most physiological circumstances. Therefore, for these body fluids only, substitution of 1 for f in equation (9.1) gives the familiar equation

$$pH = -\log_{10} [H^+] \qquad (9.3)$$

Gastric juice is not a dilute solution of hydrogen ions, and the activity coefficient of it is less than 1. Its value depends upon both the concentration of hydrogen ions and the sum of the concentrations of sodium and potassium ions. In a solution whose pH,

accurately measured, is found to be 1.00 and in which $[Na^+] + [K^+] = 50$ mEq/L, the activity coefficient is 0.810. Thus, the activity of hydrogen ions is 0.100, but the concentration of hydrogen ions is 0.100/0.810, or 0.124N.†

The amount of acid secreted is estimated by titration with a standard base. A sample of gastric juice properly collected in a given interval may not contain all the acid secreted in the collection period, for acid may be lost by the flow of gastric contents through the pylorus, by back-diffusion of acid into the mucosa, and by neutralization by bicarbonate contained in nonacid secretions. Loss through the pylorus can be minimized by good collection technique or estimated by indicator dilution methods, but acid lost through back-diffusion or by bicarbonate neutralization can be estimated only by calculations of doubtful validity. Acid in gastric contents may also be buffered by proteins and by other buffers contained in nonacid secretions. This moiety of acid can be recovered if the titration with a standard base is carried to the correct end point which lies between pH 7.0 and 8.3.

Acid is secreted by the oxyntic cells as a solution of hydrochloric acid, and in that solution the hydrogen ions are almost completely ionized. Some hydrogen ions combine with buffer groups when secreted acid mixes with buffers present in gastric contents. When a sample of gastric juice is titrated with standard base, the ionized hydrogen ions are first neutralized. This is accomplished when the pH of the solution has reached approximately 3.5, a pH which corresponds to the color change of diaminoazobenzene (Topfer's reagent). The quantity of acid found is commonly called "free acid." When the titration is continued to an end point near neutrality, the buffered hydrogen ions are neutralized, and total titratable acidity is determined. Titratable acidity, expressed as milliequivalents per liter, represents the sum of the hydrogen ion concentration and the un-ionized hydrogen ion concentration expressed in the same units.

The determinants of the acidity of any particular sample of gastric juice are the rate of secretion and

---

*See Baron J.H.: *Clinical Tests of Gastric Secretion; History, Methodology and Interpretation*. New York, Oxford University Press, 1979 for a comprehensive description of the theory, technique, results, and meaning of gastric analysis.

†See Moore E.W., Scarlata R.W.: The determination of gastric acidity by the glass electrode. *Gastroenterology* 49:178, 1965; and Moore E.W.: Determination of pH by the glass electrode: pH meter calibration for gastric analysis. *Gastroenterology* 54:501, 1968. The latter paper contains a table of activity coefficients of hydrogen ions as a function of the concentrations of sodium and potassium in gastric juice.

the rate of neutralization or dilution. Acidity cannot be higher than that of the undiluted, unneutralized acid secretion, in man about 150 mN,‡ and consequently the only sense in which hyperacidity can exist is that the sample of gastric juice is more acid than some supposedly normal sample. A better term is hypersecretion, in which the volume secreted in a given time is above normal. Then, if diluting or neutralizing processes are no more than normal, hypersecretion will result in high acid concentration.

Three data are important in assessing gastric secretion: whether the stomach can secrete acid at all, how much it secretes under standard conditions, and how much it can secrete under the strongest stimulation. Total inability to secrete acid, or achlorhydria (rarely encountered except in subjects with pernicious anemia), is diagnosed when the pH of juice aspirated from the stomach fails to fall below 6.0 after adequate stimulation. The usual stimulus is the "standard dose" of histamine,§ 0.01 mg of histamine acid phosphate (or 0.0036 mg of histamine base) per kg of body weight given subcutaneously. However, this may not be enough. In one large series, 58 subjects were encountered who secreted no acid in response to this dose, but half of them secreted some acid when given 4 times the dose.

Basal Secretion

Basal secretion is determined on subjects who come to the laboratory after fasting overnight. Their stomachs are presumed to be under no extrinsic stimulus. In one group of 615 men, basal secretion of acid, so defined, ranged from 0 to 17 mEq/hr, with a mean of 2.4 ± 2.85. There was a 0.975 probability that no more than 5% of a similar population would secrete more than 6.6 mEq/hr. For 634 women, the range was 0–15 mEq/hr, with a mean and SD of 1.3 ± 2.0.

A naive fasting subject's stomach may not be in a

---

‡An occasionally encountered exception is the acidity of stomach contents of a person who has tried to kill himself by drinking strong acid.

§Histamine is a base, and it is usually dispensed as histamine acid phosphate. One milligram of histamine base is contained in 2.75 mg of histamine acid phosphate, and 1 mg of histamine is contained in 1.7 mg of the more rarely used histamine dihydrochloride. When the dose is reported, it must be made clear how much of which form of the compound was used.

basal state during the test on account of nervous and hormonal stimuli arising out of the situation in which he finds himself. When measurement of basal secretion was repeated almost every week by a highly trained gastroenterologist on himself, basal secretion fell from 3–9 mEq/hr to a low of 0.1 over a period of 41 weeks. He concluded that under truly basal conditions, the nervous and hormonal stimuli for acid secretion are minimal or totally absent.

Maximal Acid Output

The greatest ability of the stomach to secrete acid is measured by collecting gastric secretion in successive 15-min periods after the subject has been given what is thought to be a sufficiently strong stimulus. The sum of the acid outputs in the two highest consecutive 15-min periods is the *peak acid output* (PAO), and the *maximum acid output* (MAO) is the output found in the whole hour after the stimulant is given. Both are expressed as milliequivalents of acid per hour.

In the *augmented histamine test* an antihistaminic drug that blocks all effects of histamine except stimulation of acid secretion is given, and histamine is injected subcutaneously in a dose of 0.04 mg of the acid phosphate per kg. The data in Table 9–2 show that the results are remarkably uniform.

Pentagastrin, which does not require medication to cover side effects, is now substituting for histamine in tests of secretory capacity, and the usual dose is 6 μg/kg given subcutaneously or intramuscularly (IM). In one large study, patients chosen to give a wide range of secretory responses were given pentagastrin, and their peak acid outputs were measured. On another occasion, some of the subjects were given histamine, some were given betazole, a histamine analog, and some were given a repeated dose of pentagastrin. The results given in Table 9–3 show that the peak acid output following pentagastrin matches that obtained by histamine.

Many similar tests have shown that, on the average, peak acid output increases with body weight and with lean body mass, that men secrete more than women, that the rate of secretion falls off after the age of 50, that patients with duodenal ulcers secrete more than normal subjects, and that patients with gastric cancer secrete far less than control subjects. These data define the average, but individuals may not conform to the mean of their class. For one large

TABLE 9–2.—RESULTS OF AUGMENTED HISTAMINE TEST
ON NORMAL ADULTS IN 6 SEPARATE STUDIES*†

	MALES			FEMALES		
STUDY	No. of Subjects	Mean	Range	No. of Subjects	Mean	Range
1	27	22.2				
2	14	22.4	10–35	18	14.6	0.1–31
3	31	23.2		15	15.0	
4	29	22.7	10–42	28	17.2	6–35
5	30	23.3		12	17.7	
6	15	28.8	11–17	21	17.1	3–47

*From Bock O.A.A., et al.: *Gut* 4:112, 1963; references for the 6
studies are contained in this paper.
Acid secreted in first 60 min after subcutaneous injection of 0.04 mg
histamine acid phosphate/kg, expressed as mEq/hr.

TABLE 9–3.—INTRAMUSCULAR PENTAGASTRIN COMPARED WITH OTHER STIMULI AS
TESTS OF GASTRIC SECRETION (PEAK ACID OUTPUT IN MEQ ± SD/HR)*

GROUP	N	PENTAGASTRIN, IM 6 μG/KG	COMPARISON TREATMENT OF SAME SUBJECTS	COMPARISON RESULTS
1	26	21.86 ± 3.32	Histamine, IV, 0.04 μg/kg·hr	24.18 ± 3.43
2	93	29.68 ± 1.53	Histamine, SQ, 0.04 μg/kg	33.12 ± 1.62
3	30	43.61 ± 2.58	Betazole, SQ, 2 mg/kg	46.83 ± 2.53
4	34	21.28 ± 2.70	Pentagastrin, SQ, 6 μg/kg	21.44 ± 2.14

*Multicentre Study: *Lancet* 1:341, 1969.

study of stimulated secretion, the upper limit of normal was set so that there was a 0.975 probability that no more than 5% of the normal population would be above the limit. Nine percent, or one of 11, of those with gastric ulcer, a class secreting on the average less than normal, actually secreted above the upper limit of normal. Although the mean rate of secretion in patients with duodenal ulcer was twice as high as in control subjects, more than half of these patients secreted amounts of acid that fell within the normal range.

The secretory response to stimulation depends in part upon the total number of oxyntic cells in the gastric mucosa. The parietal or oxyntic cell mass has been estimated by counting the cells in stained sections of the mucosa, chosen to give an adequate sample of the total acid secreting volume. In a group of patients scheduled to undergo gastrectomy, the augmented histamine test gave rates of secretion from 0.6 to 82.7 mEq HCl per hour. After a portion of the stomach had been removed, the test was repeated, and the number of oxyntic cells in the resected portion was estimated. The difference between the rates

of acid secretion in the two tests was attributed to the oxyntic cells contained in the part of the stomach removed. The correlation between the decrease in maximal acid output and the number of oxyntic cells removed was almost linear over a very wide range.

Gastrin is a trophic hormone for the secretory cells of the digestive tract (see chap. 2). In rats, increasing or decreasing the circulating concentration of gastrin increases or decreases the maximal secretory capacity. Although there are no comparable data on human subjects, it is probable that the continuously high concentration of gastrin in the plasma of patients with gastrinoma is responsible for their increased oxyntic cell mass and their maximal secretory capacity. It is likely that the exaggerated output of gastrin following a meal in duodenal ulcer patients contributes to their high maximal secretory capacity.

Hypersecretion Caused by a Gastrinoma

Both basal secretion and maximum acid output are enormously increased in a patient with a gastrinoma. One unfortunate man secured the world's record for

MAO by secreting 111 mEq of acid in 60 min.

A gastrinoma occurs as a discrete or diffuse tumor, often in the pancreas but occasionally in other digestive organs. When found, it usually has slowly growing metastases. Because it is not under the negative feedback control exercised by acid, a gastrinoma liberates gastrin at a high rate, and consequently the patient has continuous hypergastrinemia and hypersecretion. Gastrinomas contain more little gastrin than big gastrin, but in the plasma big gastrin is far more abundant than little gastrin. The difference is explained by the longer half-life of big gastrin. Gastrinomas also secrete a form of little gastrin consisting of the N-terminal 13 amino acids. Because this fragment lacks the four amino acids at the C-terminus, it has no biological activity. However, it cross-reacts with antibodies raised to the N-terminus of little gastrin, and it gives falsely high values for the plasma concentration of gastrin in radioimmunoassay. The association between a tumor and hypersecretion was first recognized by Zollinger and Ellison,‖ and one form of the syndrome bears their names.

A patient with a gastrinoma has a very high acid secretory capacity, perhaps on account of the trophic action of gastrin. Because this capacity is continuously used, the basal rate of acid secretion approaches maximal secretory capacity. The consequences of uncontrolled and continuous acid secretion are multiple duodenal and jejunal ulceration and inactivation of digestive enzymes secreted by the pancreas. Steatorrhea, azotorrhea, and diarrhea are frequent results.

Gastrin release by G cells of the antral mucosa is depressed when secretin is injected in pharmacological doses, but secretion of gastrin by gastrinomas is enhanced by secretin. Consequently, a rise in plasma gastrin concentration, rather than a fall, following secretin injection, is evidence that much of the plasma gastrin comes from a gastrinoma. Gastrinomas, as well as G cells in the antral mucosa, are stimulated to release gastrin by a rise in plasma calcium concentration. Gastrinomas are often accompanied by tumors, or perhaps only by hyperplasia, of the parathyroid glands, and the hypercalcemia caused by hyperparathyroidism drives gastrinomas to secrete gastrin.

In very rare instances, hyperplasia of G cells in the

―――――

‖*Ann. Surg.* 142:709, 1955. See Zollinger R.M., Coleman D.W.: *The Influence of Pancreatic Tumors on the Stomach.* Springfield, Ill., Charles C Thomas, Publisher, 1974, for a history of the discovery.

gastric antrum may also be responsible for high plasma concentrations of gastrin. Abnormally large G cells appear not to be inhibited by acid bathing the mucosa, and plasma gastrin concentration is high during fasting and feeding.

Secretion During a Meal:
The Cephalic Phase

Overnight the stomach is emptied of food, and stimuli for secretion, except that occurring intermittently during the interdigestive myoelectric complex, die out. By morning the stomach contains about 20 ml of gastric juice having acidity of 2–40 mN. Then eating a meal stimulates gastric secretion through the cephalic, gastric, and intestinal phases.

Gastric secretion of acid and pepsinogen follows stimulation of afferent nerves in the head; hence, this part of the response is called the cephalic phase. The final common path is the vagus nerve, for the cephalic phase is completely abolished by vagotomy. Vagal impulses reach the stomach by way of the anterior and posterior vagal trunks arising from the esophageal vagus plexus formed by right and left vagus nerves on the surface of the esophagus just above the diaphragm. They are distributed to the anterior and posterior aspects of the stomach and end in synapse with postganglionic cells in the myenteric plexus. This plexus, lying between longitudinal and circular smooth muscle layers, sends connecting branches to the submucous plexus. In man, the submucous plexus consists of a network of nonmedullated fibers in the submucous region. The network contains very few cell bodies. From this network, fibers penetrate the mucosa where they innervate the secretory cells. The mediator is acetylcholine, and it directly stimulates oxyntic and chief cells to secrete. Long-lived parasympathomimetic drugs also stimulate secretion, and the action of these drugs and of acetylcholine is blocked by atropine. Acetylcholine also sensitizes the secretory cells to gastrin, and the magnitude of the response to acetylcholine and gastrin together is far greater than is the sum of the responses to each alone.

Stimulation of gastric secretion through the cephalic phase is demonstrated by sham feeding. A dog is provided with an esophagostomy so that the food it eats falls back into its feeding dish instead of reaching the stomach. Copious gastric secretion occurs during and after sham feeding, and the response is abolished by vagotomy. Sham feeding can be demonstrated in a patient whose esophagus does not empty into his stomach. If such a patient chews food, his stomach secretes (Fig 9–6). The effective stimuli may originate from taste and smell receptors or by conditioned reflexes from other receptors. In man, tasting, smelling, or chewing palatable food, but not indifferent substances such as paraffin wax, is followed by gastric secretion. The results shown in Figure 9–6, obtained during sham feeding, show a strong response to a meal of choice but none to a hospital meal dutifully chewed but not swallowed.

In addition to its effects in releasing acetylcholine near secretory cells, stimulation of the vagus nerve by sham feeding may release gastrin from G cells in the antral mucosa and thereby stimulate acid secretion. There are however, two inhibitory effects which opposes gastrin release. If the pyloric mucosa is bathed with acid, release of gastrin is inhibited. In addition, the vagus nerve contains cholinergic fibers which inhibit gastrin release. Therefore, administration of atropine actually enhances the response to sham feeding, and the rise in plasma gastrin concentration during a meal is greater in vagotomized subjects than in normal persons (see Fig 9–9). The cephalic phase of gastric secretion evokes a secretory response greater than 55% of the maximum acid output. It is not important in man for the reason that the immediate entry of food into the stomach also stimulates even greater secretion. A steak meal, homogenized and fed through a nasogastric tube, stimulates the same amount of secretion as does the steak meal eaten with relish.

In man, reduction of blood glucose to about 45 mg/100 ml, or about half the fasting level, by insulin administration strongly stimulates acid and pepsinogen secretion. Intravenous infusion of insulin at 0.1 units per kg · hr gives a maximum secretory response.

The effect of insulin is prevented by maintaining the blood glucose concentration with glucose infusion. The fall of blood glucose is sensed by cells in the hypothalamus, and excitation is relayed to the stomach along the vagus nerve. Administration of the nonmetabolized sugar 2-deoxy-D-glucose in an IV dose of 50–200 mg/kg of body weight is also a powerful stimulant to gastric secretion by way of the vagus nerve. The sugar probably acts by competing with

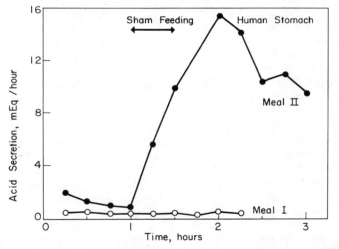

Fig 9–6.—Rate of acid secretion by the stomach of a woman, 24, with complete stenosis of the esophagus. In her usual manner of eating, the food was tasted, chewed, partly swallowed, regurgitated, expectorated, and then placed in the stomach through a gastrostomy. In the two instances recorded here, the food was not placed in the stomach after it had been tasted and chewed. *Meal I* (mean of four experiments) consisted of 8 oz of cereal gruel, eaten dutifully but with obvious distaste. *Meal II* (mean of 12 experiments) was composed of the subject's unrestricted choice: fresh vegetables, salad with dressing, 3 slices of white bread with butter, 2 glasses of milk, potatoes, half a fried chicken or 2 broiled lamb chops or fried ham steak or 2 fried eggs, ice cream, and cake. (Adapted from Janowitz H.D., et al.: *Gastroenterology* 16:104, 1950.)

glucose in the glucose-sensitive cells in the medial forebrain bundle of the hypothalamus.

If the acid secreted by a man in response to insulin hypoglycemia is prevented from reaching the gastric antrum by intragastric neutralization, plasma gastrin rises by about 30% above basal level 45 min after insulin is given. When atropine in a dose of 0.015 mg/kg is given 20 min before insulin is administered, plasma gastrin rises to twice the basal level. This is an example of the cholinergic inhibition of gastrin release.

Vagally mediated secretion does not occur in a vagotomized subject, and therefore the absence of a secretory response to insulin hypoglycemia is used as a test for completeness of vagal denervation. However, epinephrine, which is released during insulin hypoglycemia, in turn releases gastrin, which may stimulate gastric secretion in a patient whose vagotomy is complete.

Secretion During a Meal: The Gastric Phase

The gastric phase of secretion begins when food enters the stomach. The major stimuli are distention and the digestion products of food, chiefly of protein.

It is difficult to measure gastric secretion during a ten-course meal at the Duchess of Guermantes'. Instead, secretory response to a much simpler meal is measured by frequent sampling of gastric contents after a meal tagged with dilution indicators has been eaten or by intragastric titration. The results of frequent sampling have been given in Figures 4–12 and 4–13. The technique of intragastric titration gives somewhat different results.

After a subject has fasted, a double-lumen tube is passed into his stomach, and the residual contents are aspirated. The pH of the stomach is raised to 5.5 by infusion of a standard bicarbonate solution. Then the subject eats a meal. During eating and until the stomach is empty again, the contents of the stomach are frequently sampled through one lumen, and the pH of the sample is measured. The sample is then returned to the stomach. Infusion of the standard bicarbonate solution through the other lumen is adjusted so that the pH of the contents of the stomach remains at pH 5.5. The rate of infusion of bicarbonate measures the rate of acid secretion, and the total amount of bicarbonate infused measures the cumulative secretion of acid.

In the example cited here, the meal was 5 oz of ground sirloin steak cooked and seasoned with salt and pepper, two pieces of toast, and 360 ml of water. The meal contained 39 gm protein, 30 gm fat, 30 gm carbohydrate, and 546 kilocalories. Its total volume was 500 ml and, when homogenized, its pH was 5.5. The subject ate the meal over a period of 30 min, chewing it thoroughly before swallowing it.

The rate of secretion of acid by 6 normal subjects and by 7 patients with duodenal ulcer is given in Figure 9–7, left, and the cumulative acid secreted is given in Figure 9–7, right. Basal secretion, peak rate of secretion during the augmented histamine test and peak rate of secretion in response to the meal are given in Table 9–4. In each group, the natural stimulus of food caused a peak output equal to that achieved with histamine stimulation.

Secretion measured by intragastric titration to pH 5 is greater than that found by frequent sampling and calculation of indicator dilution for the reason that in the intragastric titration method the stomach continues to be distended despite emptying, and the negative feedback inhibition of gastrin release by acid cannot occur.

Distention of the stomach by food stimulates receptors in the wall of the stomach which, through local reflexes in the intrinsic plexuses and through long vagally-mediated reflexes, cause secretion of acid and pepsinogen. In man, these reflexes prolong the cephalic phase of secretion by at least an hour and result in secretion equal to 50% of the maximum acid output. Proximal vagotomy or administration of atropine depress the response to distention.

In the dog, the effect of distention can be shown to result from four distinct reflexes:

1. Distention of the body of the stomach stimulates secretion by the oxyntic mucosa through long and short cholinergic reflexes.

2. Distention of the antrum stimulates secretion by the oxyntic mucosa through cholinergic reflexes.

3. Distention of the body of the stomach stimulates release of gastrin from the pyloric mucosa.

4. Distention of the antrum stimulates release of gastrin from the pyloric mucosa.

The experiments required to sort out these reflexes cannot be done in man, but results obtained in the dog appear to apply to man. Although early experiments failed to show that distention of the human stomach releases gastrin, more recent evidence shows that distention with isotonic NaCl solution does re-

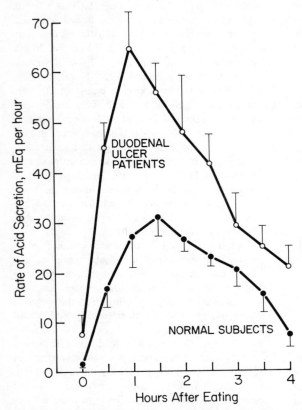

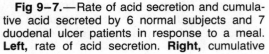

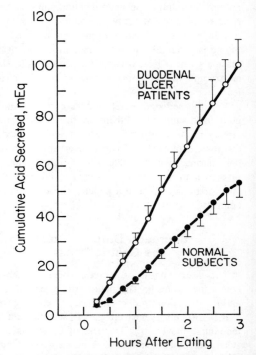

Fig 9–7.—Rate of acid secretion and cumulative acid secreted by 6 normal subjects and 7 duodenal ulcer patients in response to a meal. **Left,** rate of acid secretion. **Right,** cumulative acid secreted. (Adapted from Fordtran J.S., Walsh J.H.: *J. Clin. Invest.* 52:645, 1973, with additional data supplied by J.S. Fordtran.)

lease gastrin. In man, release of gastrin by distention is resistant to antral acidification, and it is enhanced by atropine for the reason that atropine blocks vagal inhibition of gastrin release.

The secretory response to food in the stomach is chiefly determined by the protein content of the food (Fig 9–8) for two reasons: protein digestion products are the most powerful stimulants for release of gastrin, and protein is the chief buffer of the food.

The major stimulus for gastrin release is a neutral solution of L-amino acids, or their polypeptides, bathing the pyloric mucosa. Native proteins are ineffective. Gastrin is released from the G cells into both the plasma and the gastric lumen. Plasma gastrin concentration begins to rise immediately after a meal (Fig 9–9), and the response is essentially the same whether the protein of the meal is derived from eggs

and bacon at breakfast or from steak at dinner. Little gastrin (G-17) is the major component of the early rise, but big gastrin (G-34) rises later and remains elevated longer on account of its slower clearance. G-17 accounts for at least 75% of the secretory response in normal persons.

A meal neutralizes and dilutes acid in the stomach (Fig 9–10). As acid secretion procedes, the pH of gastric contents is quickly reduced to 4.5, and eventually the pH falls sufficiently low to inhibit gastrin release. The time taken to reach inhibition depends upon the bulk and the buffering power of the meal. For meals of equal caloric content, the time is proportional to the buffering power. Because protein is the chief buffer of food, the time, and therefore the total secretory response, is proportional to the protein content of the meal.

TABLE 9–4.—Basal Acid Outputs and Peak Acid Outputs After Histamine Stimulation and During a Meal in 6 Normal Subjects and 7 Patients with Duodenal Ulcer, mEq per hour ± SE*

	BASAL	PEAK HISTAMINE	PEAK MEAL
Normal subjects	1.4 ± 0.7	34.5 ± 2.8	30 ± 4
Duodenal ulcer patients	7.5 ± 2.5	58.2 ± 7.2	64 ± 7

*Adapted from Fordtran J.S., Walsh J.H.: *J. Clin. Invest.* 52:645, 1973, with additional data from J.S. Fordtran.

Addition of sodium bicarbonate or other alkalinizing agents to food increases the secretory response because it neutralizes acid in the stomach but does not in itself stimulate secretion.

Glucose and fat both release small amounts of gastrin. The amount released by glucose is too small to be important. Fat also releases powerful inhibitors of secretion, and its net effect is negative rather than positive. Solutions of calcium salts, including milk, release gastrin and stimulate acid secretion. Milk is effective as an antacid because it dilutes and neutralizes acid. By the time those effects have been overcome, the ulcer patient is ready to drink his next glass of milk with the result that the stimulating effect of calcium in the milk does not become apparent. Aliphatic alcohols, of which ethanol is the most potent, release gastrin in the dog, a confirmed teetotaler, but in man, ethanol, if it releases gastrin at all, has only a trivial effect. Therefore, a solution of ethanol is no better stimulant of acid secretion than the same volume of Adam's Ale. Caffeine does not release gastrin, but decaffeinated coffee does so and stimulates acid secretion, an effect attributable to peptides in the brew.

Protein digestion products in neutral solution (but not intact proteins) stimulate acid secretion when they bathe the oxyntic glandular mucosa. This direct effect upon the oxyntic mucosa is enhanced by concurrent distention of the body of the stomach, and when topical application of protein digestion products is combined with distention, the resulting rate of acid secretion equals that occurring in response to a maximal dose of exogenous gastrin. If the protein digestion products are acidified to pH 2.0, they do not stimulate secretion. The protein digestion products appear to act directly upon the oxyntic cells in a way that is not understood. Their effect is only slightly decreased if

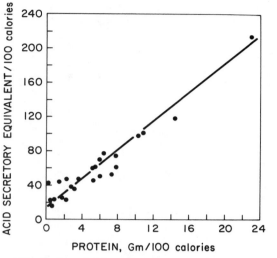

Fig 9–8.—Acid secretory equivalent of 29 common foods in relation to their protein content. The total acid secreted by a dog's vagally innervated pouch in response to a 100-calorie meal of any of the 29 foods divided by the total acid secreted by the same dog's pouch in response to a 100-calorie meal of broiled ground beef, the quotient multiplied by 100, is the acid secretory equivalent per 100 calories. The foods used ranged from canned peaches, containing the least protein per 100 calories and giving the smallest acid secretory equivalent, to haddock, with the most protein per 100 calories and the strongest effect on secretion. (Adapted from Saint-Hilaire S., et al.: *Gastroenterology* 39:1, 1960.)

the mucosa is treated with a local anesthetic or if the responsiveness of the oxyntic cells to acetylcholine or to histamine is abolished by intravenous injection of atropine or an H_2 antagonist. Both atropine and H_2 antagonists suppress the response to a meal in the intact stomach, and consequently it is unlikely that the direct effect of protein digestion products upon the oxyntic cells has an important role in the normal response to a meal.

Intestinal Control of Gastric Secretion

Perfusion of the duodenum or jejunum with chyme, liver extract, or a 10% solution of peptone stimulates gastric secretion. The observed responses range from 5% to 30% of maximal acid output, and they aug-

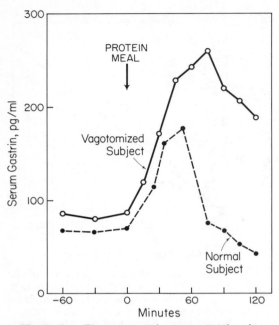

Fig 9–9.—Plasma gastrin concentration in a normal subject and in a subject with truncal va-gotomy in a control period and following ingestion of a protein meal. (Adapted in part from Korman M.G., Hansky J., Scott P.R.: *Gut* 13:39, 1972, and in part from Malagelada J-R., et al.: *Gastroenterology* 70:203, 1976.)

ment the response to concurrently administered histamine or gastrin. Although gastrin can be identified in the intestinal mucosa, release of intestinal gastrin is not responsible for the intestinal phase of stimulation. The concentration of gastrin in plasma may not rise at all during intestinal perfusion with liver extract, or if it does, the rise is small and late. Gastrin responsible for the rise is probably released from the antrum, and its release may be effected by a bombesin-like hormone from the intestinal mucosa.

Another hormone, *entero-oxyntin,* released from the intestinal mucosa and acting directly upon the oxyntic cells, has been postulated to account for the intestinal stimulation of acid secretion. A peptide having the properties of entero-oxyntin has been isolated from intestinal mucosa, and the peptide augments gastric response to histamine and gastrin.

Amino acids directly stimulate the oxyntic cells, and it is probable that a substantial part of the intestinal phase of gastric secretion is the result of stimu-

lation of the oxyntic cells by absorbed amino acids. Intravenous infusion of a mixture of amino acids stimulates acid secretion to the extent of about one third the maximal acid output without affecting plasma gastrin concentration.

Acid secretion declines during the second half of the digestion of a meal, in part because cephalic and gastric stimuli die out and in part because chyme in the duodenum inhibits acid secretion. Acid and fat- and protein-digestion products in chyme are the stimuli for inhibition, and they act through hormones and nerves. Acid in the duodenum inhibits acid secretion as well as gastric motility. Acid in the duodenum lib-

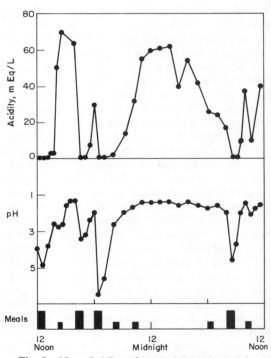

Fig 9–10.—Acidity of material aspirated from stomach of a normal man, age 36, over a period of 24 hours. Results are presented in terms of titratable acidity and pH. Large rectangles represent meals: *lunch*—fish, potato, carrot puree, fruit puree, custard, and tea; *tea*—bread and butter, tea, golden syrup; *supper*—fish, potato, custard, bread and butter; *breakfast*—porridge, boiled egg, bread and butter, tea. Small rectangles represent drinks of milk. (Adapted from James A.H., Pickering G.W.: *Clin. Sci.* 8:181, 1949.)

erates secretin, and secretin is capable of inhibiting acid secretion. (Secretin, however, stimulates pepsinogen secretion.) Early in the digestion of a meal, chyme enters the duodenum with an average pH of 6 (Fig 9–11), and the average pH falls in successive 30–min periods to 5, 4, and 3, respectively. Duodenal neutralization of acid is so rapid and effective that only a short segment of the duodenum is acidified for a brief time (see Fig 4–11). Consequently, there is not enough acid in the duodenum long enough to liberate enough secretin to account for the observed inhibition of acid secretion. It is possible, but unproven, that other hormones liberated from the duodenal mucosa during the intestinal phase of digestion act synergistically with secretin to inhibit acid secretin just as cholecystokinin acts synergistically with secretin to stimulate pancreatic secretion. There may also be an as yet uncharacterized candidate hormone, *bulbogastrone,* which mediates acid inhibition by chyme in the duodenum.

Perfusion of the duodenal bulb of a dog with acid at the rate of 8 mEq/hr, a rate well within the normal range, inhibits pentagastrin-stimulated acid secretion by the dog's stomach. Inhibition cannot be attributed to secretin, for pancreatic secretion is only slightly stimulated. Injection of pure secretin in an amount necessary to produce the same inhibition causes far greater pancreatic secretion. The effect is not me-diated by another hormone, for pentagastrin-stimulated secretion by an extrinsically denervated pouch is unaffected. Separation of the duodenal bulb from the stomach abolishes the inhibition that follows perfusion of the bulb with acid. These results demonstrate that physiological quantities of acid in the duodenum can inhibit acid secretion through a local reflex.

Nervous Influences¶

Nervous influences on secretion, mediated by both branches of the autonomic nervous system, are closely correlated with other nervous activity expressed as behavior. Feeding, which is the behavior most intimately connected with secretion, is controlled at the hypothalamic level. During satiety, activity in the ventromedial region of the hypothalamus inhibits the lateral region; but during feeding, the lateral region dominates. The medial and lateral regions of the hypothalamus also control secretion. Stimulation of the medial region in rats inhibits both food intake and gastric secretion, whereas stimulation of the lateral region increases food intake and gastric secretion. Rats made hyperphagic by destruction of the medial satiety center secrete acid and pepsinogen in the basal state at rates more than double those occurring in control periods. The efferent pathway passes from the hypothalamus through the mesencephalon to the medulla and vagal nucleus.

Activity in higher centers converging on hypothalamic and medullary centers affects secretion. In general, strong physical activity or those aspects of behavior interpreted as expressing rage or the experience of pain inhibit gastric secretion and motility through sympathetic discharge and inhibition of parasympathetic activity. The effect on the stomach may outlast the overt response. In the dog, an innervated pouch of the oxyntic gland area usually secretes when the animal eats. If however, the dog is provoked into a rage, subsequent feeding, even after the dog has calmed, may evoke little or no gastric secretion.

Similar suppression of secretion in rage has been found in man, and literature from the most ancient

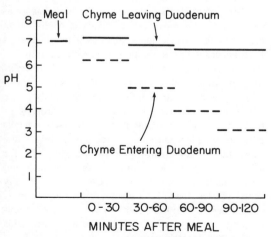

Fig 9–11.—The pH of the meal described in Fig 4–11, and the average pH of the chyme entering and leaving the duodenum in successive 30-min periods. (Adapted from Miller L.J., Malagelada J-R., Go V.L.W.: *Gut* 19:699, 1978.)

¶The problem of the participation of nervous influences in the genesis of peptic ulceration is beyond the scope of this book. See Grossman M.I., Novin D. (eds.): Experimental ulcer produced by behavioral factors. *Brain Res. Bull. (suppl. 1),* 1979.

times contains anecdotes describing cessation of digestion during episodes of sympathetic discharge. In one closely studied human subject whose stomach could be inspected through a fistula, gastric secretion was reduced during depression, and it was subnormal for several months when the dominant psychic state was self-reproach. Increased secretion has been found to accompany aggressive actions. The same fistulous subject, when unjustly reproached, experienced strong feelings of hostility; at the same time, both acid secretion and mucosal blood flow increased about 25% and remained elevated for 2 weeks. The generalization derived from these studies is that, when the affective state is one of fear, sadness or withdrawal, gastric secretion is reduced, but when the dominant element is aggressiveness or the will to fight back, gastric secretion increases. Interpretation of such studies is deeply penetrated by subjective elements, and no one knows how widely these conclusions apply.

Gastric Carbohydrate and Protein Digestion

In the first part of gastric digestion, the contents of the body of the stomach are not mixed with acid. If the starch of the meal has been mixed with salivary α-amylase by chewing, its digestion in the body can continue as long as the pH remains high. In one experiment, two young men who were capable of voluntary regurgitation were fed meals consisting of 150 gm of beefsteak, 150 gm of potato, 50 gm of green peas, 10 gm of butter, and 200 ml of water. The men chewed the food according to their normal habits, and at the end of 30–60 min they vomited at command. The pH of the vomitus was above 3.4. In it, 35%– 48% of the starch had been hydrolyzed to oligosaccharides, but only a trace of the protein had been digested.

Pepsins, acting on the periphery of the mass in the body and in the antrum, reduce the size of lumps of meat and assist in dispersion of fat and carbohydrate by breaking the walls of animal cells. Although exhaustive digestion of representative proteins by pepsins in vitro breaks them into fragments of all sizes, it liberates no more than 15% of their available amino-nitrogen. Consequently, peptic digestion in the stomach cannot be expected to be complete. Peptic digestion is, in fact, unnecessary for adequate digestion and absorption of protein.

When radioiodinated human serum albumin (RISA) is added to a liquid meal taken by normal men, samples recovered by tube from the region of the pylorus contain 85%–90% of the administered ^{131}I in a form precipitable by 5% phosphotungstic acid. This means that the largest part of protein digestion occurs beyond the stomach.

There is no significant gastric absorption of the products of carbohydrate and protein digestion.

Fat digestion products in the duodenum inhibit gastric secretion as well as gastric motility. Perfusion of the duodenum with a 10 mM solution of sodium oleate completely suppresses the gastric response to a meal of liver extract and to injection of gastrin. In such circumstances there is no rise in plasma secretin concentration, and there is only a small, and apparently inconsequential rise in plasma GIP. So far, the hormone or combination of hormones responsible for fat's inhibition of acid secretion has not been identified.

Gastric Digestion of Fat and Absorption of Fat-Soluble Substance

Acid and pepsin tend to break natural or artificial emulsions in the stomach, with the result that fat forms large drops unsuitable for gastric digestion. Liquid or semisolid fats float in gastric contents, and in a man in the upright position the fat in a broken emulsion empties after the rest of the meal has emptied. Acid hydrolysis of ester bonds is negligible in the stomach. Gastric juice contains a lipase distinct from pancratic lipase; the gastric enzyme partially hydrolyzes triglycerides at pH 6.0–8.0. Lipolytic activity is maximal with tributyrin as the substrate, and it falls off as the chain length of the fatty acids in the triglycerides increases, so that gastric hydrolysis of triglycerides of chain length greater than 10 is negligible in the adult and small in the child. Only a few short and medium chain triglycerides are available for gastric hydrolysis. Depot and liver fat of animals and oil of plants contain little or no triglycerides of fatty acids shorter than 10 carbon atoms; and cow's, goat's, and human milk contains only 4%–9% tributyrin and less than 10% triglycerides with fatty acids shorter than 14 carbon atoms.

Two well-established principles of general physiology are that membranes of cells behave as though composed of both protein and fat and that fat-soluble substances diffuse rapidly through a membrane that water-soluble substances penetrate slowly, if at all. These principles apply to the stomach.

A substance of low molecular weight such as ethanol, which is both water and fat soluble, is rapidly absorbed from gastric contents. Its rate of absorption is directly proportional to its concentration. If 350 ml of a 5.6% (w/v) solution of ethanol is placed in the stomach of a normal human subject, 5.4 gm of ethanol is absorbed through the gastric mucosa in 30 min. The rate of absorption of ethanol is entirely independent of the concentration of acid or of drugs (such as aspirin) simultaneously present in gastric contents.

Organic acids and bases may be water soluble at one pH and fat soluble at another. In general, organic acids are ionized and water soluble in a solution whose pH is above the acid dissociation constant (pK_a) of the compound and un-ionized and fat soluble in a solution whose pH is below the pK_a. If the water-soluble short-chain fatty acids—acetic, propionic, or butyric—are present in gastric contents at pH low enough (pH 4.0 or lower) for an appreciable fraction to be in the un-ionized, fat-soluble form, they rapidly diffuse through the lipoprotein membranes of the mucosal cells and are absorbed. Long-chain fatty acids are only slightly soluble in water. This limits their availability for absorption, and only minute amounts are absorbed through the gastric mucosa.

An important example of the influence of pH upon absorption is given by acetylsalicylic acid (aspirin), whose structure is shown in Figure 9–12. The pK_a of the acid group is 3.5. At neutral pH, all of the compound's molecules are in the ionized, water-soluble form; at pH below 2.0, more than 95% are in the un-ionized, fat-soluble form. When acetylsalicylic acid is present in acid gastric contents, its fat-soluble form diffuses across the mucosa into the cells. Because the cell interior is nearly neutral, the compound ionizes as soon as it crosses the membrane. The disappearance of the fat-soluble molecules from the intracellular side of the membrane prevents back-diffusion and maintains a steep gradient from lumen to cell. Eventually the compound escapes from the mucosal cells and is carried away by the blood. In man, acetylsalicylic acid is absorbed from the stomach at the rate of 50% of the amount in gastric contents in 30 min when the pH is 1.0. This is the same rate at which deuterium oxide (D_2O) is absorbed, and faster than that of ethyl alcohol. It is absorbed at about one tenth this rate from neutral solutions.

Organic bases are uncharged and fat soluble in a solution whose pH is on the alkaline side of their pK_a, but charged and water soluble in a solution whose pH is on the acid side of their pK_a. Therefore, they are absorbed in the stomach more readily from neutral than from acid solutions. Given the same pK_a, absorbability is a function of the solubility of the compound in oil. Barbital, whose pK_a is 7.8, is absorbed very slowly from an acid solution in the stomach, but thiopental, whose pK_a is 7.6 and which is 100 times more soluble than barbital in chloroform, is absorbed 10 times more rapidly. Caffeine is absorbed in the stomach from neutral but not acid solution, and the amount absorbed is directly proportional to the time caffeine solution remains in the stomach. Presence of caffeine has no effect upon the rate of emptying of liquids.

Exchange of Water and Electrolytes across the Resting Gastric Mucosa

Water molecules cross the gastric mucosa rapidly in both directions. The net rate of flow in either direction is the difference between the two rates:

$$\text{Flux rate}_{lumen\ to\ blood} - \text{Flux rate}_{blood\ to\ lumen}$$
$$= \text{Net flow} \#$$

The flux rate from lumen to blood can be estimated by placing water labeled with deuterium oxide or tritiated water (D_2O or THO) in the stomach and measuring the specific activity of the isotope in the lumen at frequent intervals. The results obtained when 50 ml of a solution containing D_2O was placed in contact with the oxyntic glandular mucosa of a dog are shown in Figure 9–13. Fifty percent of the isotopically labeled water was absorbed in less than 40 min; but because there was only a small net movement of wa-

---

#The direction of one-way fluxes and of net flow should always be specified in such a way as to make them instantly recognizable. Here the anatomical description, lumen to blood or mucosal surface to serosal one, will be used. Another system having wide, but not universal, currency is:

One-way flux, lumen to blood = Insorption
One-way flux, blood to lumen = Exsorption
Net flow, lumen to blood = Absorption
(insorption greater than exsorption)
Net flow, blood to lumen = Enterosorption
(exsorption greater than insorption)

See Code C.F.: The semantics of the process of absorption. *Perspect. Biol. Med.* 3:560, 1960.

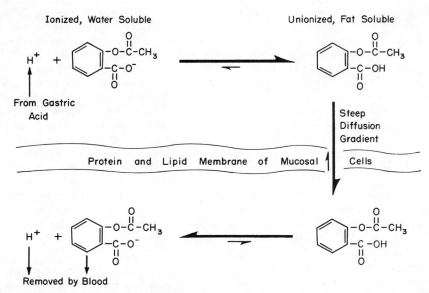

Fig 9–12.—Absorption of aspirin. Aspirin is acetylsalicyclic acid and is fat soluble. If it is ingested as acetylsalicylate, the ionized water-soluble form, and if it is inadequately buffered, acid secreted by the stomach converts it to acetylsalicylic acid.

ter, an almost equal number of water molecules must have entered the lumen as the labeled molecules left. These large fluxes of water in each direction across the gastric mucosa are slightly, or not at all, affected by the osmotic pressure of the contents of the lumen.

This accounts for the failure of hypo- or hypertonic gastric contents to come to isotonicity before a major part of gastric emptying occurs. Neither does water move readily across the mucosa along a hydrostatic pressure gradient. That there is rapid exchange of

Fig 9–13.—Fluxes across the oxyntic glandular mucosa of a dog's stomach. At the beginning of the experiment a solution containing D_2O, sodium labeled with ^{22}Na and potassium labeled with ^{42}K was placed in contact with the mucosa, and the rates of disappearance of the isotopes were measured. Then an infusion of histamine was started, and tritiated water was added to the solution. When acid secretion began, the rate of disappearance of ^{22}Na fell to zero, but fluxes of water and potassium into the mucosa were unaltered. (From Code C.F., et al.: J. Physiol. 166:110, 1963.)

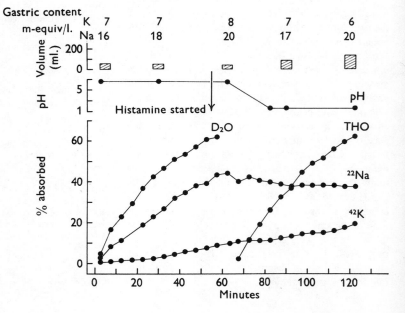

molecules of water across the mucosa but little bulk flow along osmotic or hydrostatic pressure gradients mean that there are many water-filled channels whose radius is very small. These are probably in the tight junctions between surface epithelial cells.

In the middle of the experiment described in Figure 9–13, an infusion of histamine was started, and copious secretion began. Tritiated water was placed in the lumen, and measurement of the specific activity of tritium showed that, despite a large net output of water from the mucosa in its secretions, flux of water from lumen to blood remained high.

Flux of sodium in either direction across the resting gastric mucosa is small. There is usually an electro-chemical gradient favoring net movement of sodium into the lumen, for interstitial sodium concentration is about 145 mN, and the surface of the mucosa is electrically negative with respect to the interstitial fluid. Nevertheless, net output of sodium is only a few microequivalents per cm^2 an hour. The unidirectional flux from lumen to blood is about one fifth the net output. Flux of sodium is reduced when the contents of the lumen are acidified, either by secretion by the mucosa itself (see Fig 9–13) or by exogenous HCl. The fluxes of potassium across the resting mucosa are still smaller than those of sodium, but they are unaffected by the acidity of the luminal contents.

When acid is present in the lumen, hydrogen ions diffuse slowly from the lumen into the mucosa; the rate is less than that at which sodium leaves the mucosa. If 250 ml of 100 mN HCl is placed in the stomach of a normal human subject, the concentration of acid recovered 15 minutes later may be 90–95 mN. Secretion of neutral fluid containing bicarbonate is responsible for part of the diminution in the concentration of acid, and back-diffusion of acid into the mucosa is responsible for the rest. In the dog, back-diffusion of acid into the pyloric glandular mucosa is faster than into the oxyntic glandular mucosa.

The Gastric Mucosal Barrier**

The property of the gastric mucosa that allows it to contain an acid solution and that prevents rapid penetration of itself by hydrogen ions is the *gastric mucosal barrier*. The ability of the mucosa to prevent rapid diffusion of sodium ions from its interstitial space into the lumen is likewise a property of the gastric mucosal barrier.

The gastric mucosal barrier is broken by many compounds; among them are aliphatic acids such as acetic, propionic and butyric acids, which, because they are un-ionized and fat soluble, diffuse through cell membranes into the gastric mucosa. Salicylic acid and acetylsalicylic acid in acid solution are likewise un-ionized and fat soluble, and they are rapidly absorbed from acid gastric contents into the mucosa. Once absorbed, they kill the surface epithelial cells, causing desquamation and barrier breaking. The salicylates in neutral solution are ionized and water soluble, and they are absorbed only slowly into the gastric mucosa from neutral gastric contents, causing no damage.‡ Other compounds breaking the barrier are those attacking the lipid components of the cell membranes: detergents, natural or synthetic. Among the natural detergents the most important are bile acids and lysolecithin, both present in normal duodenal contents. These destroy the barrier by dissolving lipids of the cell membrane. Patients with gastric ulcers have abnormally great regurgitation of duodenal contents, and many surgeons believe that gastric ulceration is caused by damage inflicted upon the barrier by regurgitated bile acids and lysolecithin. Ethanol is absorbed equally rapidly from neutral and acid solutions, and it also breaks the barrier.

Although it is reasonable to suppose that ischemia damages the gastric mucosa, there is no evidence to support the supposition. Prolonged ischemia resulting either from sympathetic discharge or the administration of large doses of vasoconstrictors does not break the barrier in the dog or in primates. Gastric blood vessels eventually escape from vasoconstrictor influences, and blood flow through the mucosa may return to normal in the face of continued sympathetic discharge or exogenous vasoconstrictors. When the mucosa is stimulated to secrete, oxygen demand resulting

---

**The *gastric mucosal barrier* is not to be confused with the *gastric barrier*. The latter is the ability of acid in the stomach to kill microorganisms, including ingested pathogens, so that chyme delivered to the small intestine is nearly sterile. The gastric barrier is particularly important in ruminants whose stomachs are a vast fermentation vat.

‡Gastric ulceration produced in cats and rats by IV administered acetylsalicylic acid is another matter. The gastric mucosal barrier is not broken. Because ulceration occurs only if the mucosa is secreting acid, it is prevented if acid secretion is inhibited by cimetidine. Acetylsalicylic acid inhibits prostaglandin synthesis, and ulceration is also prevented by treatment with several prostaglandins. Acetylsalicylic acid also interfers with blood clotting, and that may contribute to aspirin-induced gastric hemorrhage.

from increased metabolism overcomes vasoconstriction and results in active hyperemia. However, the combination of ischemia, acid in the lumen, and attack by bile acids has been found to break the barrier in experimental animals, and the combination may be equally deleterious in man. Other combinations are more damaging than their individual components: that of acid, aspirin, and alcohol is particularly effective in breaking the barrier and causing bleeding. If there is little or no acid in the lumen when the gastric mucosal barrier is broken, the consequences are trivial. Mucosal cells desquamate, and interstitial fluid leaks into the lumen. When the agent breaking the barrier is removed, the mucosa rapidly repairs itself, and the impermeability characteristic of an intact barrier is restored.

If there is acid in the lumen, acid diffuses back into the mucosa through the broken barrier, and there are many important pathophysiological consequences (Fig 9–14). Acid destroys mucosal cells, and by activating pepsinogen within the mucosa it contributes to further destruction of barrier function. Acid liberates histamine within the mucosa and accelerates histamine formation. Histamine stimulates acid secretion, and acid may be secreted into the lumen faster than it disappears through a highly permeable mucosa. Histamine causes vasodilatation, and increased blood flow removes the back-diffusing acid. Histamine also increases capillary permeability to plasma proteins, and plasma filters rapidly into the interstitium. The mucosa becomes edematous, and interstitial fluid containing plasma proteins leaks into the lumen. Acid stimulates intramural plexuses, and motility of the stomach and secretion of pepsinogen increase. Superficial capillaries and venous plexuses are destroyed, and petechial hemorrhages cover the damaged mucosa. If an arteriole is ruptured, copious bleeding occurs. Bleeding is more frequent and copious when there is concurrent cholinergic stimulation, probably because contraction of gastric muscle increases venous and capillary pressure.

The consequences of acid back-diffusion through a broken barrier can be prevented simply by removing acid from the lumen. This can be done by neutralizing the acid or by inhibiting its secretion. Really effective intragastric neutralization has been found to be better than treatment with cimetidine in prevention or control of the often exsanguinating gastric hemorrhage occurring in a severely ill or injured patient.

The rate at which plasma proteins normally leak through the gastric mucosa is of the order of a gram a day, but in some circumstances the gastric mucosa may shed such large amounts of plasma proteins, without any concurrent shedding of red blood cells, that hypoalbuminemia and anasarca result. Plasma proteins are shed in several forms of protein-losing gastropathy, one of which is caused by allergic response of the gastric mucosa to foreign proteins. Acetylcholine causes tight junctions between surface epithelial cells to become permeable to plasma proteins, and plasma shedding in protein-losing gastropathies can be reduced by treatment with atropine. Plasma shedding also occurs when the gastric mucosa is irrigated with ethanol solutions, either experimentally or convivially. Plasma shedding is also substantial when the mucosa is damaged by aspirin. Then the precipitated, acidified hemoglobin, the ''coffee grounds'' found in a bleeding stomach, may conceal leakage of plasma proteins. When frank bleeding occurs in hemorrhagic gastritis caused by alcohol or aspirin injury, the hematocrit of the shed blood may be as low as 10%.

Prior treatment of the stomach with a very small amount of any of a number of prostaglandins prevents damage to the mucosa by diverse agents such as absolute alcohol and boiling water. This is called *cytoprotection,* a word which will join *irritability* in the physiological limbo when we really understand the phenomenon.

Antacids

Antacids are bases which reduce gastric acidity for long periods if taken in substantial amounts after food. Antacid preparations differ greatly in their ability to neutralize acid in vitro and in vivo on account of their differing chemical composition, physical state, and reactivity. When 1 ml of a commercial preparation of aluminum phosphate was stirred with 100 ml of water, addition of only 4 ml of 0.1N HCl brought the pH to 3.0. In contrast, 35 ml of the same acid was required to bring 1 ml of a commercial mixture of aluminum hydroxide, calcium carbonate, and magnesium hydroxide to pH 3.0. For in vivo tests, patients with duodenal ulcers were given a standard meal of 150 gm of cooked and seasoned ground beef, two pieces of toast with butter, and 180 ml of water. One hour after the meal, the subjects were given either 60 ml of water or 60 ml of the same antacids cited above. The acidity of their gastric contents was measured an hour later. With water, the acidity aver-

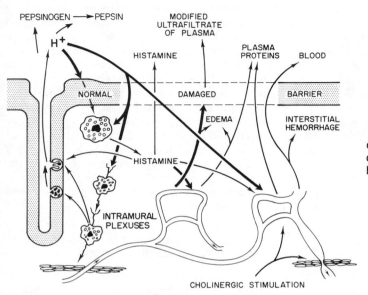

Fig 9–14.—Pathophysiological consequences of the back-diffusion of acid through the broken mucosal barrier.

aged 69 mN H$^+$. With aluminum phosphate, the acidity was 55 mN H$^+$, but with the aluminum-calcium-magnesium mixture it was 4 mN H$^+$. Results of in vitro and in vivo tests of many antacid preparations are available to the physician who wants to be sure that his prescribed therapy actually works.‡

Aluminum is absorbed from aluminum-containing antacids. A person taking such a preparation over many years may have 4–10 fold greater than normal urinary output of aluminum, and at autopsy his bone content of aluminum is found to be elevated. The brains of patients with premature senile dementia contain more than normal aluminum, and some physicians think that chronic ingestion of aluminum-containing antacids may contribute to that condition.

Temperature of the Stomach

The empty stomach is at the core temperature of the body. The temperature of swallowed food covers a range of about 80 C. Most persons like their coffee between 50 and 70 C, temperatures above the maximal tolerable temperature of hot water applied to the skin. At the other extreme is a mint julep whose temperature is −7.1 C.

‡Fordtran J.S., Morawski, S.G., Richardson C.T.: *In vivo* and *in vitro* evaluation of liquid antacids. *N. Engl. J. Med.* 288:923, 1973.

Temperature changes occurring in the stomach and the rate of return to core temperature depend upon the heat load, positive or negative, imposed. A 250-ml liquid meal taken at 57 C raises the temperature of the stomach to 44 C within 2 min, and the gastric temperature returns to core temperature in 18 min. Eating a dish of Philadelphia peach ice cream reduces gastric temperature to 22 C and often inverts the T wave of the ECG by reducing the rate of repolarization of the epicardial surface.

Gastric Blood Flow in Relation to Secretion

Arteries supplying the stomach pierce the muscular coat and form an arterial plexus, which gives small branches to the muscular layer and many branches to the mucosa. The transition from arterioles to capillaries occurs in the mucous membrane, and there are many anastomoses of arteries within the mucosa. The capillaries unite to form a superficial venous plexus around the orifices of the glands. The veins then form a more deeply placed venous plexus before going out through the muscularis. Because of the frequent arterial anastomoses, it is possible to ligate arteries to control hemorrhage without cutting off completely the blood supply to the mucosa. In fact, all major arteries to the stomach must be simultaneously ligated before the whole stomach is deprived of its blood supply.

Blood vessels of the stomach do not exhibit autoregulation, for blood flow increases linearly with increasing perfusion pressure over a wide range of pressures.

Effective blood flow through the gastric mucosa is measured by the clearance of aminopyrine. In man or the dog, ^{14}C-aminopyrine can be used in tracer amounts. Aminopyrine is a base that is both water and fat soluble. Its acid dissociation constant (pKa) is about 5, so that at the pH of blood it exists chiefly in the basic form. Because it is fat soluble, it diffuses rapidly through cell membranes from blood into tissues, and its concentration in cell water is, at equilibrium, equal to its concentration in water of the blood. In acid solution, aminopyrine picks up a proton; therefore, the concentration of its basic form is vanishingly small in acid solution, and a steep concentration gradient for the basic form is established between blood and the acid solution. Consequently, aminopyrine diffuses from blood into the acid solution, and this rate of diffusion is so fast that all or almost all of the aminopyrine contained in blood flowing through the mucosa is trapped in the acid solution. The acid solution may be either acid secreted by the mucosa or an exogenous solution placed in contact with mucosa.

The amount of aminopyrine brought to the mucosa per minute is equal to the blood flow through the mucosa per minute multiplied by the concentration of aminopyrine in the blood. If all the aminopyrine brought to the mucosa by the blood is cleared from the blood by being trapped in the acid solution, the amount trapped is equal to the amount brought to the mucosa by the blood. Since the amount trapped per minute and the concentration of aminopyrine in the blood can be measured, the blood flow can be calculated by dividing the amount trapped per minute by the blood concentration. In principle, this method of measuring effective mucosal blood flow is similar to the clearance method used to measure renal blood flow, and it is subject to the same theoretical and practical limitations.

Under resting conditions, effective mucosal blood flow measured by aminopyrine clearance is approximately equal to 50% of the total blood flow through mucosa and muscle measured by collection of venous effluent blood. Flow through mucosa and muscle are poorly correlated, for flow through the mucosa varies under many conditions without a corresponding change in flow through the muscle. In the nonsecreting stomach, effective mucosal blood flow is reduced

by infusion of vasopressin or norepinephrine, and it is increased by infusion of epinephrine. When acid secretion is stimulated by infusion of histamine, the mucosal blood flow increases parallel to the secretion rate. Infusion of gastrin also increases mucosal blood flow, but the increase per unit increment of secretion is less than that caused by histamine infusion. Unstimulated gastric mucosal blood flow is reduced by inhibitors of prostaglandin synthesis.

Increased blood flow in itself does not stimulate secretion, but blood flow is permissive in the sense that adequate flow is required if the stomach is to secrete. Profound vasoconstriction caused by sympathetic stimulation or by vasoconstrictor drugs is followed by depression of secretion.

Growth and Regeneration of the Gastric Mucosa

Epithelial cells of the gastric mucosa are in a dynamic state of growth, migration, and desquamation. Cells are shed by the normal human gastric mucosa at the rate of a half million cells a minute. The mucosa responds to mild injury by desquamation of its surface layer, followed by regeneration. The actual state of the mucosa, healthy or not, at any one instant or place is the result of a balance between protective and restorative forces and destructive forces.

The turnover time of cells in the epithelium of the lip, buccal mucosa, tongue, and esophagus, estimated by counting the fraction of nuclei of an excised sample that are in mitosis, ranges from 4 to 15 days. The human pyloric gland area renews itself much more rapidly, in 1.3–1.9 days, by division of cells in the basal parts of the glands followed by migration to the surface. In the oxyntic gland area, mitosis occurs in poorly differentiated surface epithelial cells at the base of the pits. The new cells, whose mucus content increases as they migrate, move up the sides of the pits to replace the old cells that have been lost from the surface. In normal circumstances, time required for renewal of surface epithelial cells is 1–3 days. The oxyntic and chief cells are derived from neck chief cells. In the mouse, the half-life of oxyntic cells is 23 days, and that of the chief cells still longer. In the embryo, all gastric glands are first lined with mucoid cells, some of which turn into oxyntic and chief cells; regeneration occurs in a similar order. If a piece of mucosa is excised down to the submucosa, an epithelial ledge composed of mucus-containing cells and derived from the surface epithelium appears at the rim of the lesion. It grows over the denuded submucosa, and from it cellular projections extend into the underlying granulation tissue. These ingrowths develop lumina, and mucus soon appears in the constituent cells, which become transformed through intermediate types, gradually losing mucus, into chief and oxyntic cells. Argentaffin cells also appear in the healing area, but transitional forms have not been seen. Mucous cells that remain in the neck are extruded within a week.

Cell division in the gastric mucosa has a circadian periodicity; in the rat, the highest rate of cell division occurs at 10 A.M. The cells are partially under the control of pituitary hormones, for the mucosa undergoes profound involution after hypophysectomy or adrenalectomy. The oxyntic cells are reduced in size, and their content of cytochrome c oxidase, DPN diaphorase, and other dehydrogenases, as measured by histochemical tests, declines. The mucosa's ability to secrete pepsinogen is reduced as much as 80%. Treatment of hypophysectomized animals with a mixture of somatotrophin, thyroid hormone, and cortisone enlarges the mucosal cells but does not return them to normal size or function.

Adrenal cortical hormones also affect the stomach by their influence on the response of the tissues to damage. Growth of the epithelium following chemically induced desquamation is not influenced by administration of ACTH or cortisone, but the healing of excision ulcers is delayed. These hormones also cause degranulation of the mast cells in the gastric mucosa.

Rapidity of repair depends on the extent of injury. When only restricted desquamation has occurred, healing is complete in 48 hours. Mucus first bridges the gap; and then new surface cells, migrating from the sides, close it. If injury has sliced off the surface layer, with almost complete destruction of foveolae, round or low cylindrical cells form a new surface within 1 hour; and by 36 hours the mucosa has partially recovered its barrier function. Columnar cells, already differentiated, migrate from the crypts; and if the crypts themselves have been destroyed, they are reformed. If injury has been severe, stripping off all but the deepest parts of the glandular tubules, complete regeneration and reorganization of the mucosa, with restoration of unimpaired function, may require 3–5 months.

Effects of Gastrectomy

Following partial gastrectomy, gastric digestion is never complete, and gastric absorption is minor; the major parts of digestion and absorption are left to the small intestine. Nevertheless, normal gastric function is necessary for normal nutrition.

In persons with partial gastrectomy, the usual consequence is weight loss, resulting not only from decreased food intake, although appetite may remain good, but from defective intestinal digestion and absorption. The absorption of fat measured in balance studies may be as low as 30% of the amount ingested, whereas normal absorption is over 90%. Both pancreatic secretion, measured by the concentration of pancreatic enzymes in intestinal contents, and biliary secretion, measured by bile concentrations, are reduced in partially gastrectomized subjects. The reason for this probably is that diminished gastric digestion and acid secretion provide subnormal stimulation of pancreatic secretion by way of pancreozymin and secretin. Reduced pancreatic secretion, poor mixing of intestinal contents with enzymes and increased bacterial decomposition of protein in the lower small intestine may account for poor assimilation of protein; digestion of carbohydrate is only slightly affected. Microcytic hypochromic anemia occurring after partial gastrectomy is attributed to faulty iron absorption, a consequence of reduced acid secretion, for acid promotes iron absorption.

REFERENCES

Allen A., Garner A.: Mucus and bicarbonate secretion in the stomach and their possible role in mucosal protection. *Gut* 21:249, 1980.

Andersson S.: Gastric and duodenal mechanisms inhibiting gastric secretion of acid, in Code C.F. (ed.): *Handbook of Physiology:* Sec. 6. *Alimentary Canal.,* vol 2. Washington, D.C., American Physiological Society, 1967, pp. 865–878.

Babkin B.P.: *Secretory Mechanisms of the Digestive Glands,* ed. 2. New York, Paul B. Hoeber, Inc., 1950.

Baron J.H.: *Clinical Tests of Gastric Secretion; History, Methodology and Interpretation.* New York, Oxford University Press, 1979.

Berglindh T., Dibona D.R., Ito S., et al.: Probes of parietal cell function. *Am. J. Physiol.* 238:G165, 1980.

Chowdhury A.R., Malmud L.S., Dinoso V.P. Jr.: Gastrointestinal plasma protein loss during ethanol ingestion. *Gastroenterology* 72:37, 1977.

Clemencon G.H.: Duodenogastric reflux. *Scand. J. Gastroenterol.* 16(suppl.):67, 1981.

Davenport H.W.: Physiological structure of the gastric mucosa, in Code C.F. (ed.): *Handbook of Physiology:* Sec. 6. *Alimentary Canal.,* vol. 2. Washington, D.C., American Physiological Society, 1967, pp. 759–780.

DiBona D.R., Ito S., Berglindh T., et al.: Cellular site of gastric acid secretion. *Proc. Natl. Acad. Sci. U.S.A.* 76:6689, 1979.

Durbin R.P.: Electrical potential difference of the gastric mucosa, in Code C.F. (ed.): *Handbook of Physiology:* Sec. 6. *Alimentary Canal,* vol. 2. Washington, D.C., American Physiological Society, 1967, pp. 879–888.

Engle G.L.: Memorial lecture: The psychosomatic approach to individual susceptibility to disease. *Gastroenterology* 67:1085, 1974.

Feldman M.: Comparison of acid secretion rates measured by gastric aspiration and by in vivo titration in healthy human subjects. *Gastroenterology* 76:954, 1979.

Gregory R.A.: *Secretory Mechanisms of the Gastrointestinal Tract.* London, Edward Arnold & Co., 1962.

Grossman M.I. (ed.): The biology of the oxyntic cell. *Gastroenterology* 73:873, 1977.

Guth P.H., Baumann H., Grossman M.I., et al.: Measurement of gastric mucosal blood flow in man. *Gastroenterology* 74:831, 1978.

Hirschowitz B.I.: Secretion of pepsinogen, in Code C.F. (ed.): *Handbook of Physiology:* Sec. 6. *Alimentary Canal.,* vol. 2. Washington, D.C., American Physiological Society, 1967, pp. 889–918.

Hunt J.N., Wan B.: Electrolytes of mammalian gastric juice, in Code C.F. (ed.): *Handbook of Physiology:* Sec. 6. *Alimentary Canal,* vol. 2. Washington, D.C., American Physiological Society, 1967, pp. 781–804.

Isenberg J.I., Maxwell V.: Intravenous infusion of amino acids stimulates gastric acid secretion in man. *N. Engl. J. Med.* 298:27, 1978.

Johnson L.R.: The trophic action of gastrointestinal hormones. *Gastroenterology* 70:278, 1976.

Johnson L.R. (ed.): Gastrointestinal hormones: Physiological implications. *Fed. Proc.* 36:1929, 1977.

Kelly D.G., Code C.F., Lechago J., et al.: Physiological and morphological characteristics of progressive disruption of the canine gastric mucosal barrier. *Dig. Dis. Sci.* 24:424, 1979.

Lanciault G., Jacobson E.D.: The gastrointestinal circulation. *Gastroenterology* 71:851, 1976.

Malagelada J.-R., Go V.L.W., Summerskill, W.H.J.: Different gastric, pancreatic, and biliary responses to solid-liquid or homogenized meals. *Dig. Dis. Sci.* 24:101, 1979.

Morris T., Rhodes J.: Antacids and peptic ulcer—a reappraisal. *Gut* 20:538, 1979.

Obrink K.J., Flenstrom G. (eds.): *Gastric Ion Transport. Acta Physiol. Scand. (special suppl)* 1978.

Priebe H.J., Skillman J.J., Bushnell L.S., et al.: Antacid versus cimetidine in preventing acute gastrointestinal bleeding. *N. Engl. J. Med.* 302:426, 1980.

Robert A. Nezamis J.E., Lancaster C., et al.: Cytoprotection by prostaglandins in rats; prevention of gastric necrosis produced by alcohol, HCl, NaOH, hypertonic NaCl, and thermal injury. *Gastroenterology* 77:433, 1979.

Rotter J.I., Sones J.Q., Samloff I.M., et al.: Duodenal-ulcer disease associated with elevated serum pepsinogen I. *N. Engl. J. Med.* 300:63, 1979.

Sachs G., Spenney J.G., Lewin M.: H^+ transport: Regulation and mechanisms in gastric mucosa and membrane vesicles. *Physiol. Rev.* 58:106, 1978.

Samloff I.M., Secrist D.M., Passaro E.P.: A study of the relationship between serum group I pepsinogen levels and gastric acid secretion. *Gastroenterology* 69:1196, 1975.

Smith B.M., Skillman J.J., Edwards B.G., et al.: Permeability of the human gastric mucosa; alteration by acetylsalicylic acid and ethanol. *N. Engl. J. Med.* 285:716, 1971.

Soll A.H.: The actions of secretagogues on oxygen uptake by isolated mammalian parietal cells. *J. Clin. Invest.* 61:370, 1978.

Soll A.H.: The interaction of histamine with gastrin and carbamylcholine on oxygen uptake by isolated mammalian parietal cells. *J. Clin. Invest.* 61:381, 1978.

Soll A.H., Grossman M.I.: Cellular mechanisms in acid secretion. *Ann. Rev. Med.* 29:495, 1978.

Soll A.H., Wollin A.: Histamine and cyclic AMP in isolated canine parietal cells. *Am. J. Physiol.* 237:E444, 1979.

Sugai N., Ito S.: Carbonic anhydrase, ultrastructural localization in the mouse gastric mucosa and improvements in technique. *J. Histochem. Cytochem.* 28:511, 1980.

Vagne M., Mutt V.: Extero-oxyntin: a stimulant of gastric acid secretion extracted from porcine intestine. *Scand. J. Gastroenterol.* 15:17, 1980.

Walsh J.H., Feldman M.: Distention-induced gastrin release. Effects of luminal acidification and intravenous atropine. *Gastroenterology* 78:912, 1980.

Walsh J.H., Grossman M.I.: Gastrin. *N. Engl. J. Med.* 292:1324–1377, 1975.

Williams S.E., Turnberg L.A.: Retardation of acid diffusion by pig gastric mucus: A potential role in mucosal protection. *Gastroenterology* 79:299, 1980.

Wolf S.: *The Stomach.* New York, Oxford University Press, 1965.

Zalewsky C.A., Moody F.G.: Mechanisms of mucus release in exposed canine gastric mucosa. *Gastroenterology* 77:719, 1979.

10

Pancreatic Secretion

THE PANCREAS SECRETES a juice having two major components, an alkaline fluid and enzymes, into the duodenum. The two components occur in variable proportions depending on the stimuli. The alkaline fluid component, ranging in volume from 200 to 800 ml/day, has a high concentration of bicarbonate, which neutralizes acid entering the duodenum and helps to regulate the pH of intestinal contents. The enzyme component contains the major enzymes necessary for the adequate digestion of carbohydrate, fat, and protein. Pancreatic secretion, by its influence on duodenal contents, also affects gastric secretion and emptying.

Structure and Innervation of Pancreas

Secretory acini of the pancreas are composed of cells containing zymogen granules; and, because of the roughly spherical shape of an acinus, the cells tend to resemble truncated pyramids. Those cells synthesize and secrete the enzymes of the juice. At rest each cell of an acinus is connected with every other cell by a low-resistance pathway; current injected into one cell by a microelectrode is conducted with decrement to all other cells. Small groups of neighboring acini are also electrically coupled. The acini are connected with excretory ducts by long intercalary and intralobular ducts whose walls are formed by cells that do not contain zymogen granules. Those cells contain carbonic anhydrase and secrete the aqueous component of the juice. Cells of the ducts frequently project into the acinus itself, where they are known as the centroacinar cells. As the intercalary ducts merge into excretory ducts the cells lining them become cuboidal epithelium. Cells lining the largest ducts stain for mucin, and small mucous glands open into the main pancreatic ducts. The walls of the largest ducts contain elastic fibers and smooth muscle

cells; when the walls contract, flow of juice into the duodenum is prevented.

Two or more ducts, their number and arrangement depending on the embryological history of the gland, enter the duodenum. In approximately half the human subjects examined at autopsy, the main pancreatic duct and the bile duct were joined to form an ampulla as they entered the duodenum. In the other half, the ducts entered separately or were divided by a septum running to within 2 mm of their common orifice.

The pancreas contains islets of α and β cells, which secrete glucagon and insulin internally. These islets and the acini, which form the external secretion, are served by separate arterioles. Those to the islets break up into large sinusoids and drain into smaller capillaries, which surround adjacent acini. Thus, blood flows from the islets to acini carrying blood containing high concentrations of hormones to cells synthesizing and secreting the enzymes of the external secretion of the pancreas. Insulin increases amylase synthesis and secretion by pancreatic acinar cells, and glucagon inhibits enzyme synthesis. Thus, the vascular arrangement may permit the endocrine secretions of the pancreas to have a local influence upon the exocrine secretions. Other acini receive blood from capillaries arising directly from arterioles.

The pancreas contains a network of postganglionic cholinergic neurones which stimulate secretion of enzymes. These postganglionic neurones are activated by preganglionic parasympathetic fibers of the vagal nerves. In addition, there are nervous connections between pancreatic postganglionic nerves and the intrinsic ganglia of the intestinal wall. These connections probably mediate an enteropancreatic reflex. Preganglionic sympathetic fibers synapse with postganglionic cells in the celiac and associated ganglia, and the postganglionic fibers are distributed to pancreatic blood vessels. In addition, sympathetic secretory fibers that do not synapse in the celiac ganglion occur in the splanchnic nerves. Myelinated afferent fibers from receptors analogous to those in the carotid pathetic nerves. Those in the vagus are the afferent limb of a vago-vagal reflex arc. Other fibers derive from receptors analogous to chose in the carotid sinus. Reduction of blood pressure within the pancreas reflexly evokes sympathetically mediated vasoconstriction and cardioacceleration. Afferent fibers from the pancreas mediate the sensation of pain; that from the head of the pancreas is localized in the midepigastrium and that from the tail in the left upper quadrant.

Methods of Collecting Pancreatic Juice

The best method of collecting pancreatic juice from experimental animals is to form a permanent duodenal fistula by inserting a metal and plastic tube through the body wall and into the duodenum, opposite the pancreatic duct. When not in use, the tube is closed, so that pancreatic juice and intestinal contents flow naturally through the intestine. Digestive functions and the acid-base balance remain normal. For collection of the juice, the cap is removed from the fistula, and an appropriately shaped glass cannula is passed through the fistula into the duct. At the same time substances can be placed in the duodenum to test their effect on secretion. For complete nervous isolation of the pancreas, a portion of it can be transplanted to the mammary gland of a lactating bitch, so that the duct drains through the nipple. In man, the current method of collecting pure pancreatic juice is by means of endoscopic retrograde cholangiopancreatography (ERCP) which has replaced collection of pancreatic juice from the lumen of the intestine.

Composition of the Bicarbonate-Containing Fluid

There are species differences in the secretion and electrolyte composition of pancreatic juice as the result of differences in secretion of two fluids which together make up the aqueous component.

In animals like man, the cat and the dog, who are intermittent feeders and whose pancreas secretes chiefly during the digestive phase, the major fraction of the aqueous fluid is the alkaline juice distinguished by its high concentration of bicarbonate. This juice is secreted by the centroacinar cells and by the cells of the ducts, and its secretion is stimulated by secretin and cholecystokinin. In the same animals, the zymogen cells secrete a fluid resembling an ultrafiltrate of plasma in response to stimulation by cholecystokinin and acetylcholine. This fluid accompanies the enzymes. In the cat and dog, and presumably in man, this second fluid is a minor constituent of the aqueous component of pancreatic juice, and it will be ignored in this section.

In animals like the rat, sheep, and rabbit, who are continuous feeders and whose pancreas secretes continuously, the fluid secreted by the zymogen cells is a more important part of pancreatic juice. Since it is secreted in company with the enzymes, it will be briefly described in the section dealing with enzyme secretion.

At the lowest rate of secretion, the concentration of bicarbonate is 20 mN or more in the alkaline juice. Its concentration rises as the rate of secretion increases and approaches a maximum characteristic of the particular man or animal. Curves illustrating the relationship in man are given in Figure 10–1. For man, the maximum bicarbonate concentration is from 120 to 140 mM. When plasma bicarbonate rises in metabolic alkalosis or falls in metabolic acidosis, pancreatic juice bicarbonate shows corresponding changes. Dissociation between rate of secretion and bicarbonate concentration occurs in chronic pancreatitis, in which the bicarbonate concentration of pancreatic juice is low at high rates of secretion, and during inhibition of secretion by anticholinergic drugs, in which bicarbonate concentration of the juice is high at low rates of secretion. In man, but not in the perfused cat pancreas, bicarbonate concentration tends to fall upon prolonged stimulation, although volume rate of secretion remains constant.

The P_{CO_2} of pancreatic juice is not far from 40 mm Hg, and consequently its pH ranges from 7.6 to 8.2.

The concentrations of sodium and potassium in pancreatic juice are independent of the rate of secretion and very nearly equal to their concentrations in plasma water. If plasma composition remains constant, both the sum of the cations and the sum of the anions in pancreatic juice are independent of the rate of secretion. Since chloride is, in addition to bicarbonate, the second major anion, chloride concentration falls as bicarbonate concentration rises. The osmotic pressure of pancreatic juice is identical with that of plasma. Experimental variations in the osmotic

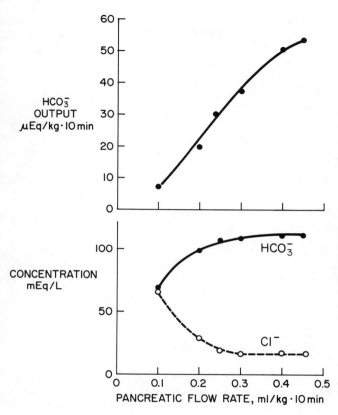

Fig 10–1.— *Top,* the relation between the bicarbonate output of the pancreas and the rate of secretion of pancreatic juice. Each point is the average of samples collected by endoscopic cannulation of the ampulla of Vater in five normal human subjects whose pancreas was stimulated by intravenous administration of secretin. *Bottom,* the relation between bicarbonate and chloride concentrations in the same samples of pancreatic juice. (Adapted from Domschke S., et al.: *Gastroenterology* 73:478, 1977. From Davenport H.W.: *A Digest of Digestion,* ed 2. Chicago, Year Book Medical Publishers, Inc., 1978.)

pressure of fluid perfusing the gland are promptly followed by corresponding changes in that of pancreatic juice.

Secretion of the Bicarbonate-Containing Fluid

The composition of samples of fluid collected by micropuncture from the cat's pancreas stimulated by secretin is shown in Figure 10–2. Fluid from the center of the lobule had a relatively high concentration of chloride, perhaps because the fluid secreted by the centroacinar cells was mixed with fluid secreted by the zymogen cells. Samples obtained from the extralobular ducts contained progressively higher concentrations of bicarbonate until the concentration in fluid from the main duct was 118 mM. Other experiments have shown that there is a passive 1:1 chloride-bicarbonate exchange in the more distal extralobular ducts. This exchange had little or no effect upon the com-

position of fluid described in Figure 10–2 for the reason that flow down the ducts was brisk, leaving little time for exchange. When the rate of flow is slower and there is more time for exchange, chloride concentration of fluid collected from the main duct rises, and bicarbonate concentration falls.

Not enough is known about the means by which the bicarbonate-containing fluid is secreted to justify a diagram purporting to describe the mechanism of secretion. What is known can be summarized as follows.

Apparent Bicarbonate Transport

During secretion bicarbonate ions disappear from the fluid perfusing the pancreas and appear in the secretion. This apparent bicarbonate transport is the result of an intracellular process separating H^+ for OH^-, transferring H^+ across the basal border of the cell and transferring OH^- in the direction of the luminal bor-

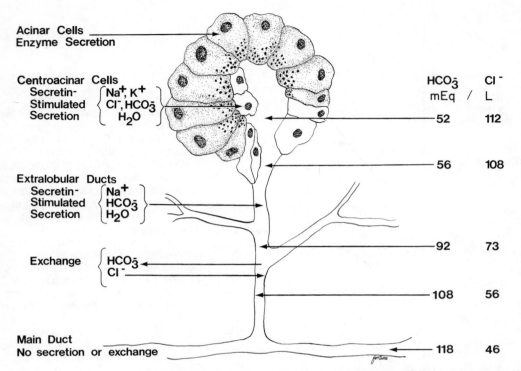

Fig 10–2.—Secretion by centroacinar cells and by cells of the extralobular ducts of the pancreas. Chloride concentrations (right) were determined on fluid collected by micropuncture, and the bicarbonate concentrations were inferred from the fact that the fluid is isotonic. These data are for the cat pancreas, but other species seem to be similar. (Adapted from Lightwood R., Reber H.A.: *Gastroenterology* 72:61, 1977.)

der. This process may be similar to that in the oxyntic cells responsible for the secretion of acid but oriented in the opposite direction. If so, it involves removal of electrons from a substrate and their transfer along a redox chain to oxygen.

H^+ is extruded into the perfusing fluid. This may be the result of a forced exchange with Na^+. If the concentration of Na^+ within the cell is low, energy derived from flow of Na^+ down its electrochemical gradient into the cell could be used to extrude H^+. In the perfusing fluid, H^+ reacts with HCO_3^-, thus causing HCO_3^- to disappear from the perfusing fluid. The subsequent dehydration of H_2CO_3, catalyzed by carbonic anhydrase, causes the partial pressure of CO_2 to rise in the perfusing fluid.

OH^- generated within the cell immediately reacts with CO_2 to produce HCO_3^-. Ninety-five percent of the CO_2 comes from the perfusing fluid and the rest from cellular metabolism. The reaction between OH^- and CO_2 is catalyzed by carbonic anhydrase, and inhibitors of carbonic anhydrase inhibit pancreatic secretion of bicarbonate-containing juice. HCO_3^- is extruded across the luminal border of the cell. This may or may not be an active process effected by a HCO_3^--activated ATPase. In any event, a small potential difference is generated, the luminal side of the cell being about 7 mV negative with respect to the basal side.

Thus, HCO_3^- secretion occurs for the reason that HCO_3^- is easily generated within the cell from ubiquitous CO_2.

NONSPECIFIC BUFFER ION TRANSPORT

If a cat's pancreas is perfused with fluid in which HCO_3^- is completely replaced by acetate ions, acetate substitutes almost completely for HCO_3^- in the secreted juice. (The small residual HCO_3^- in the juice is probably derived from CO_2 produced by cellular metabolism.) Thus, acetate as a permeant ion can carry the negative charge derived from OH^- into the juice, and acetate in the perfusing fluid can buffer H^+ extruded from the cell. Secretion of acetate-containing juice, because it does not involve reaction of OH^- with CO_2, is not inhibited by carbonic anhydrase inhibitors. Propionate, butyrate, and sulfamerazine, all permeant anions, can also substitute for HCO_3^- but impermeant anions cannot.

Nonspecific Cl^- transport.—Br^-, I^-, and NO_3^- can substitute for Cl^- in pancreatic secretion of alkaline juice as they can substitute for Cl^- in secretion of acid by the stomach. There is no evidence that the pancreas, like the stomach, contains a Cl^- pump.

Cation transport.—If an anion is secreted, electrical neutrality demands that a cation must be secreted as well. (The small potential difference across the cell represents only a miniscule separation of charge.) Na^+, as the most abundant cation, accompanies HCO_3^-, and Cl^- into pancreatic juice. Na^+ may travel by the paracellular pathway. The Na^+-K^+-activated ATPase which pumps Na^+ out of cells is located only on the basal and lateral borders of the secreting cells, not on the luminal border, and, therefore, Na^+ is probably not actively transported from the interior of the cell into the secreted fluid. How Na^+ pumping at the other borders of the cell contributes to secretion is not known. K^+ transport is probably passive.

Water transport.—Transport of osmotically active electrolytes from plasma to juice causes water to follow passively until isotonicity is achieved.

Enzyme Component

Enzymes of pancreatic juice are synthesized and secreted by the acinar cells which also secrete a fluid similar in electrolyte composition to an ultrafiltrate of plasma. In intermittent feeders such as man, the dog and cat, the volume of fluid secreted is very small, and it has little effect upon the volume and composition of pancreatic juice flowing in the main pancreatic duct. In continuous feeders such as the rat and rabbit, the volume secreted by the acinar cells is much greater, and by diluting the alkaline component it influences the composition of the ultimate pancreatic secretion. Enzymes are mixed in various proportions, depending upon the relative intensity of stimulation of the two components, with the alkaline component. Canine pancreatic juice contains protein ranging from

0.1 to 10%, and human pancreatic juice has the same range of protein content.

Most of the calcium in pancreatic juice accompanies the enzymes. Twenty percent is tightly bound to amylase in the amount of 25–30 micrograms per mg of protein. The protein-independent fraction is thought to diffuse into the juice from extracellular fluid.

Pancreatic juice contains three major enzyme groups: amylytic, lipolytic, and proteolytic. Most enzymes are secreted in multiple forms as isoenzymes. Details of enzyme structure and molecular weight, activation and inhibition, sites of action on substrates and specificities can be found in standard textbooks of biochemistry. Here the properties necessary for understanding their function will be summarized.

Crystalline amylase isolated from human pancreas is an α-amylase, which splits the α-1,4-glucosidic bond as does salivary amylase. The end-products of exhaustive digestion of unbranched starch are maltose and maltotriose. The products of digestion of a branched starch are glucose, maltose, maltotriose, and a mixture of dextrins containing α-1,6-branches. The enzyme, unlike salivary amylase, attacks raw, as well as cooked, starch. It is stable in the pH range 4–11, and its pH optimum is 6.9. Amylase is secreted in the juice in an active state.

Pancreatic juice contains at least three lipolytic enzymes. The first, usually called lipase, is a glycerol-ester hydrolase that hydrolyzes a wide variety of insoluble esters of glycerol at an oil-water interface; it requires cooperation of surface-active agents such as bile salts and of colipase which is also secreted by the pancreas. Its pH optimum depends on the substrate and ranges from 7.0 to 9.0. The second enzyme hydrolyzes esters of secondary and other alcohols, such as those of cholesterol at an optimum pH of 8.0, and it requires bile salts. The third enzyme hydrolyzes water-soluble esters. Pancreatic juice also contains prophospholipase A, which, when activated by trypsin, catalyzes the hydrolysis of lecithin to lysolecithin.

Fresh, uncontaminated pancreatic juice and extracts of pancreas have no proteolytic activity; but when suitably activated, they form several proteolytic enzymes. These are trypsin and a number of chymotrypsins. Their precursors, or zymogens, are trypsinogen and chymotrypsinogens. Trypsinogen is probably converted to trypsin in two ways: by action of trypsin itself and by action of enterokinase. At pH 7.0–9.0, activation of pure trypsinogen occurs autocatalytically, the activated enzyme converting more zymogen into the active enzyme. Eighty percent of the enzyme enterokinase in the human duodenal mucosa is firmly attached to the microvilli of the intestinal epithelial cells; the remaining 20% is either soluble or is easily detached. Enterokinase contains a large amount of polysaccharide and it resists digestion by secreted proteolytic enzymes, but it is destroyed by bacteria in the colon. In the pH range from 6.0 to 9.0, it catalyzes conversion of trypsinogen to trypsin. It is highly specific in that it activates only trypsin. Chymotrypsinogen is converted into the active enzyme only by trypsin. All these proteolytic enzymes attack the interior of peptide chains, and the end-products of action of pancreatic juice are a mixture of small polypeptides and amino acids. Previous action by pepsin is not necessary, and therefore protein digestion is unimpaired by absence of gastric juice.

Trypsin inhibitor, a polypeptide whose molecular weight is 5,000–6,000, is present in both pancreatic juice and extracts of pancreas. It combines with trypsin at pH 3.0–7.0 in the ratio of 1 molecule of trypsin to 1 of inhibitor; the product is enzymatically inactive. It also inhibits chymotrypsin, but less completely. Its concentration in human pancreatic juice is such that 1 ml of juice inhibits about 0.08 mg of trypsin. Its presence in the pancreas may protect the gland against autodigestion by small amounts of trypsin that may become active within it; but because its concentration in the juice or gland is much lower than that of trypsinogen, it does not prevent proteolytic activity by fully activated juice.

Pancreatic juice and extracts contain procarboxypeptidase, the zymogen precursor of carboxypeptidase. This enzyme attacks peptide chains at the end, liberating the amino acid with the free carboxyl group. Crude carboxypeptidase is activated by enterokinase; but this may be the result of contamination with trypsinogen, which, after activation by enterokinase, activates procarboxypeptidase. An elastase distinct from trypsin is also present, as are two nucleolytic enzymes, ribonuclease and deoxyribonuclease.

Question of Adaptation of Enzymes to Diet

The relative amounts of fat, protein, and carbohydrate in the diet vary greatly. In the short run (a matter of a day or so), the relative proportions of lipolytic, proteolytic, and amylytic enzymes in pancreatic juice do not vary as the diet is altered. Although there are minor deviations in the ratio of one kind of enzyme to another, the enzymes are secreted roughly in parallel. In long-term experiments, lasting weeks or months, manipulation of the proportions of foodstuffs in the diet does affect the enzyme content of the pancreas and pancreatic juice. For example, pancreatic glands of rats fed 18% casein contained twice as much proteolytic enzymes as glands of rats fed 6% casein. The glands of rats fed a high-starch diet synthesized amylase 3–4 times as rapidly as those fed a high-protein diet. The lipase activity per mg of protein in the dog's pancreatic juice rose 11% when the animal was fed a high-fat diet for 3 weeks, and the protease activity per milligram of protein rose 20% following a similar period on a high-protein diet. Adaptation may occur in man, but the evidence is not strong.

Synthesis and Secretion of Enzymes

Pancreatic enzymes are synthesized by the ribosomes of the rough endoplasmic reticulum in the acinar cells. There are at least two ways in which the enzymes reach pancreatic juice.

1. The enzymes are packaged in zymogen granules within the cell before being secreted. The granules are discharged by vagal stimulation and by agents releasing pancreozymin, and they have been seen to pass from the cells into the ducts. During fasting, when enzyme secretion is small, they increase in size and number. Zymogen granules have been isolated in pure state. They have been found to be stable at pH 5.5 and to dissolve at pH 7.2; this is evidence that intracellular pH in the neighborhood of the granules is near 5.5. Isolated bovine granules are composed of 95% protein: multiple forms of ribonuclease, three kinds of procarboxypeptidase, trypsinogen, chymotrypsinogen A and B, amylase, and some unidentified proteins. There is exact correspondence between the proteins of lysed granules and those of bovine pancreatic juice. Such an exact correspondence does not occur in all species; there are large differences between the enzyme content of zymogen granules and the secretions of rabbit and rat pancreases stimulated by pancreozymin or a cholinergic drug. Furthermore, in man as well as in other animals, the ratio of one enzyme to another changes in the course of prolonged secretion. This means that if the granules are the only route of protein secretion, different granules have dif-

ferent compositions, and the composition can change over time. Trypsin inhibitor is contained in the "cell sap," not in the granules.

As the enzymes are synthesized, they cross the membrane of the rough reticulum into the cysternae. Within the cysternae they are confined in small, smooth surfaced vesicles. These vesicles move to the Golgi complex where they are formed into zymogen granules. When exocrine stimulation is not occurring, zymogen granules accumulate in large numbers in the apical region of the cells. Following secretory stimulation, the membrane of the granules fuses with the plasma membrane on the apical border of the cells, and the contents of the granule are discharged into the lumen of the terminal pancreatic ducts. Each zymogen granule contains the whole array of digestive enzymes, and the enzymes are secreted when the contents of granules are discharged from the acinar cells.*

2. Strong and continuous stimulation depletes the cell's content of zymogen granules, and during subsequent rest they are reformed. However, when the granule content of the cell is reduced to zero during prolonged stimulation, synthesis and secretion still occur without segregation of the enzymes into granules. It is probable that even when enzymes are being packaged and transported in granules, enzymes are also traveling from their site of synthesis on ribosomes through the cytoplasm to the terminal pancreatic ducts without being sequestered in zymogen granules. In this case, the enzymes need not be secreted in parallel. Stimulation of the pancreas by a peptide extracted from the duodenal mucosa, the candidate hormone *chymodenin* (see chap. 2), selectively enhances the secretion of chymotrypsinogen while hardly, if at all, affecting the secretion of the other pancreatic enzymes.

When they are stimulated, acinar cells leak the digestive enzymes across their basal and lateral borders at a rate much lower than they secrete the enzymes across their luminal border. Enzyme traffic across the basal and lateral borders can be in the other direction as well. Acinar cells of the isolated perfused pancreas take up chymotrypsinogen from the perfusion fluid, and they secrete the absorbed enzyme into the juice. When trypsin labeled with ^{125}I was given IV to human subjects, between 13% and 38% of the dose was recovered from duodenal contents over the next 300 minutes. Separate collection of pancreatic juice and bile by means of ERCP showed that the enzyme was secreted by both pancreas and liver. In contrast, less than 1% of IV administered ^{131}I-labeled albumin was recovered from the duodenum. Normal intestinal epithelium absorbs many intact protein molecules (see chap. 15), and pancreatic enzymes are among them. When labeled trypsin was placed in the duodenum, 11% of the total dose was found in the circulation over 75 minutes. There is recirculation of some pancreatic enzymes.

Control of Pancreatic Secretion at the Cellular Level

Secretin is the major stimulant for secretion of bicarbonate-containing fluid by the centroacinar and duct cells. Secretin is present in S cells throughout the duodenum, but it is most concentrated in the duodenal bulb. Over the range of pH from 0 to 3, release of secretin into the blood is independent of the pH of the fluid entering the duodenum; the amount of secretin released is directly proportional to the amount of acid entering the duodenum, not to the concentration of the acid. Above pH 3.0 there is a sharp decline in the amount of secretin released as the pH rises, and above pH 3.5 little is released. There is an absolute threshold at pH 4.5–5.0. Stimulation of the pancreas by secretin depends entirely upon the presence of acid in the duodenum, and consequently secretion of bicarbonate-containing pancreatic fluid is secondary to any stimulus which results in delivery of acid to the duodenum. Among such stimuli are insulin hypoglycemia and the interdigestive myoelectric complex.

Cyclic AMP in the pancreas rises within 30 sec after the gland has been stimulated by secretin, and it remains high as long as secretin is present. The nonsecreting cat's pancreas is stimulated to secrete by addition of cyclic AMP, and secretion is enhanced by theophylline, the methyl xanthine that inhibits the enzyme-destroying cyclic AMP.

Secretion of bicarbonate-containing fluid is weakly stimulated by cholecystokinin, but cholecystokinin augments the action of secretin. Peptides sharing the active C-terminal sequence of amino acids, gastrin and caerulein among them, also stimulate bicarbonate secretion and augment the action of secretin.

---

*This story is not universally accepted. See Rothman S.S.: *Science* 190:747, 1975, for a dissenting opinion, and Scheele G.A., Palade G.E., Tartakoff A.M.: *J. Cell Biol.* 78:110, 1978, for a reply.

Although acetylcholine is a poor stimulant of bicarbonate secretion, cholinergic nerves to the pancreas appear to have an important permissive influence. Secretion of bicarbonate-containing fluid in response to a meal is depressed by atropine and reduced as much as 90% by truncal vagotomy.

Cholecystokinin is the major stimulant for secretion of enzymes by the acinar cells. This property is shared by gastrin and caerulein. For clinical and experimental work, a commonly used stimulant is the synthetic C-terminal octopeptide of cholecystokinin designated CCK-OP. Cholecystokinin is present in the mucosa of the duodenum and upper jejunum, and it is released into the blood by amino acids, the most potent being L-phenylalanine and L-tryptophan. It is also released by oligopeptides containing one of those amino acids.

Acetylcholine is also a major stimulant for enzyme secretion. It is released near acinar cells from postganglionic fibers of the pancreatic plexus. These in turn are activated by preganglionic vagal fibers and by fibers connecting the enteric plexuses with the pancreatic plexus. Acetylcholine augments the action of cholecystokinin, and cholecystokinin augments the action of acetylcholine. Consequently, enzyme secretion in response to a meal is reduced as much as 50% by truncal vagotomy.

Secretin is only a weak stimulant of enzyme secretion, but it augments the response to other stimuli.

Cholecystokinin and acetylcholine depolarize an acinar cell's membrane and uncouple it from neighboring cells. They release bound calcium from the cell membrane and permit entry of calcium from extracellular fluid. Their second messenger is cyclic GMP.

Perfusion of the ileum or colon with a solution of oleic acid inhibits pancreatic secretion, perhaps by liberating an unidentified hormone. Enkephalins and pancreatic polypeptide also inhibit pancreatic secretion, but their role in the normal control of the pancreas is unknown.

Control of Pancreatic Secretion During Digestion

During the cephalic phase of digestion of a meal, pancreatic secretion is directly stimulated through vagal impulses to the pancreas. It is also stimulated indirectly by gastrin released and acid secreted during the cephalic phase of gastric secretion. During the gastric phase of digestion, pancreatic secretion is further stimulated by gastrin. Distention of the stomach also initiates a vago-vagal reflex stimulating pancreatic secretion.

The most important stimulation of pancreatic secretion occurs during the intestinal phase of digestion. Enzyme secretion begins at once and continues steadily throughout the whole course of gastric emptying (Fig 10–3). Protein and fat digestion products arrive in the duodenum as soon as the stomach begins to empty, and they release cholecystokinin. In addition, a vago-vagal reflex originating in the duodenum and an enteropancreatic reflex in which impulses are carried from the enteric plexuses directly to the pancrease bring cholinergic stimulation to the acinar cells.

Although gastric secretion of acid begins early in digestion and reaches its highest rate within the first hour, food dilutes and buffers the acid. Consequently, gastric contents delivered to the duodenum early in digestion have a pH higher than 4.5 (see Fig 9–11). No secretin is released, and there is only a trivial secretion of bicarbonate-containing fluid. Later, gastric contents become acid, and chyme in the duodenum does release some secretin. In the experiments whose results are shown in Figure 10–3, gastric contents did not become so acid as pH 3.5 until 60 min after feeding. Even when gastric contents are acid late in digestion of a meal, delivery of acid to the duodenum occurs in spurts. As Figure 4–11 shows, contents of the duodenal bulb and the first part of the duodenum are only intermittently acid. As a result, only a small amount of secretin is released, and it is released in bursts. The amount of secretin released is, in itself, not enough to drive the pancreas to secrete the bicarbonate-containing fluid the pancreas is observed to secrete. Nevertheless, acid in the duodenum does drive pancreatic secretion of bicarbonate through secretin. The explanation is that cholecystokinin, which is simultaneously released, augments the action of secretin. The two hormones together in relatively small amounts cause the pancreas to secrete at a rate 70% of that occurring during maximal stimulation with exogenous secretin.

Control of pancreatic secretion by secretin is a negative feedback loop: acid releases secretin which stimulates secretion of bicarbonate which neutralizes acid. Cholecystokinin is also part of a negative feedback loop. Trypsin in intestinal contents suppresses release of cholecystokinin and thereby removes the

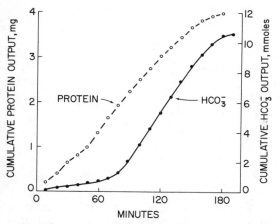

Fig 10–3.—Cumulative protein and bicarbonate outputs by the pancreas of two conscious dogs fed meals containing 15% liver extract. Pancreatic juice was collected by cannulation of the pancreatic duct. The juice was not returned to the duodenum, and consequently the negative feedback loops controlling secretion were broken. (Adapted from Moore E.W., et al.: *Acta Hepat.-Gastroenterol.* 26:30, 1979.)

stimulus for trypsinogen secretion. In man, perfusion of the duodenum with a solution of trypsin or with a mixture of pancreatic juice and bile reduces pancreatic secretion of amylase and lipase, but if trypsin inhibitor is mixed with the perfusate, secretion of the enzymes is not inhibited. If a rat or a chicken is fed the trypsin inhibitors present in soybean meal and egg white, it develops a hypertrophic, hypersecretory pancreas. The inhibitors bind trypsin, and this prevents trypsin from suppressing the release of cholecystokinin. Continuously high circulating levels of cholecystokinin, through the trophic action of the hormone, cause the pancreas to grow and to secrete.

Pancreatic secretion rises as the interdigestive myoelectrical complex passes over the gastric antrum and the duodenum, probably as the result of delivery of acid from stomach to duodenum.

Pancreatic Secretion in Man

Pancreatic secretion of bicarbonate in man is measured by collecting pancreatic secretions after secretin is given. When juice is collected from the duodenum, it risks being modified by duodenal or biliary secre-

tions. The results of studies on 47 normal human subjects obtained by duodenal collection are given in Table 10–1. Many studies have produced similar results.

Uncontaminated juice can be collected by the more difficult endoscopic retrograde cholangiopanreatography (ERCP). Data obtained in one such collection from a normal subject are given in Figure 10–4. Patients with pancreatitis studied by the same means were found to secrete juice containing bicarbonate at a concentration over 100 mM.

The ability of the pancreas to secrete enzymes can be tested by infusing cholecystokinin IV or by perfusing the duodenum with test solutions. When the duodenum is perfused with a glucose solution, there is washout of trypsin, but sustained enzyme output occurs only with perfusion with a mixture of amino acids or with a micellar solution of fat (Fig 10–5). The output attained during perfusion with amino acids is equal to that following the maximal tolerated IV dose of cholecystokinin. Only the essential amino acids are effective, and a mixture of phenylalanine, valine, and methionine is more effective than the individual amino acids. Lipase, trypsin, and amylase are secreted in parallel by the normal human pancreas, but in patients with acute or chronic pancreatitis, there is greater impairment of secretion of proteolytic than of nonproteolytic enzymes.

Rapid hydrolysis of proteins in the duodenum during digestion of a meal liberates amino acids that stimulate pancreatic secretion of enzymes. When three meals are eaten during the day, the rate of pancreatic secretion equals the maximum attained after infusion of cholecystokinin, and the high rate of secretion is sustained for 12 hours.

TABLE 10–1.—PANCREATIC SECRETION IN 47 NORMAL HUMAN SUBJECTS GIVEN 2 UNITS OF SECRETIN PER KG OF BODY WEIGHT*

Volume, ml/kg · hr	2.68	± 0.24†
Volume, ml/hr	176	± 20
Total bicarbonate output, µEq/kg · hr	199	± 22
Total bicarbonate output, mEq/hr	13.5	± 1.8
Peak volume, ml/10 min	50	± 15‡
Peak bicarbonate output, µEq/kg · 10 min	60	± 6
Peak bicarbonate concentration, mEq/L	99	± 2

*From Hartley R.C., Gambill E.E., Summerskill W.H.J.: *Gastroenterology* 48:312, 1965.
†Mean ± 2 SE.
‡Ten subjects only.

Pancreatic Enzymes in Plasma and Urine

Small amounts of pancreatic enzymes normally escape from the gland into plasma, and those absorbed across the intestinal epithelium likewise enter the plasma. Because pancreatic enzymes are proteins of low molecular weight, they appear in urine. Injury to the pancreas results in a rise in plasma concentration and urinary excretion.

Normal human plasma slowly digests starch, but less than one fourth of plasma amylase comes from the pancreas. Elevation of intraductal pressure promotes escape of amylase from the pancreas into the blood; when the pancreas secretes against obstruction, ductules rupture, fluid containing amylase escapes through the capsule, and the amylase is carried by lymph to the plasma. Stimulation of the pancreas by parasympathomimetic drugs, palpation of the gland, or eating of a heavy meal accompanied by liberal amounts of alcohol raises the plasma amylase level; and increasingly severe insults raise it much higher. The enzyme is filtered through the glomerular membrane and, along with other proteins, is partially reabsorbed by the renal tubules. Consequently, a rise in urinary amylase can result from increased liberation of the enzyme from the pancreas or from reduced reabsorption by the renal tubules. To obviate the latter in making a diagnosis of pancreatitis, some clinicians regard an increase in the ratio of the renal clearance of amylase to the renal clearance of creatinine (a crude measure of glomerular filtration) as being diagnostically significant. During severe injury to the pancreas, plasma amylase rises 15–40 times, and urinary

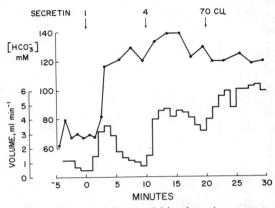

Fig 10–4.—Volume and bicarbonate concentration of pancreatic juice collected from a normal human subject by ERCP. Purified secretin was given by bolus injection in the clinical units (CU) indicated. One CU caused a rise in plasma immunoreactive secretin concentration similar to that occurring when the duodenum is perfused with acid. There was a washout of amylase when 1 CU was given, and the concentration of amylase in the juice remained low until a small peak appeared after 70 CU. (Adapted from Cotton P.B.: *Gut* 18:316, 1977.)

amylase may rise 100 times. The plasma level falls again as recovery ensues or as fibrotic tissue replaces acinar cells.

Lipase is also liberated into plasma, and its concentration roughly parallels that of amylase. Trypsinogen is present in normal plasma, and in acute pancreatic inflamation its concentration rises as much as 15

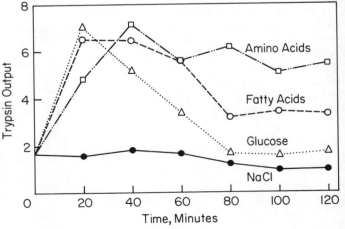

Fig 10–5.—Rate of trypsin output in normal human subjects during duodenal perfusion with isotonic NaCl, a glucose solution, a micellar solution of fatty acids, or a solution of a mixture of essential amino acids. Pancreatic secretion of bicarbonate-containing juice was maintained by continuous intravenous infusion of secretin. (Adapted from Go V.L.W., Hofmann A.F., Summerskill W.H.J.: *J. Clin. Invest.* 49:1558, 1970.)

times. Doubtless, all other pancreatic enzymes could be found in plasma and urine if tests for them were sufficiently sensitive.

Pancreatic and Biliary Duct Pressures

Forceful retrograde injection of bile into the pancreas causes immediate coagulation of lobules. Fatty necrosis and dissolution of necrotic cells with hemorrhage follow in 24 hours. Reflux of bile into the pancreas, particularly when their common duct has been obstructed, has frequently been blamed for naturally occurring pancreatitis, but that plausible explanation of the etiology of pancreatitis is probably not correct for the reason that pressure in the bile ducts is probably always lower than pressure in the pancreatic duct. Pressures in the two ducts have been measured in fasting human subjects by ERCP. Common duct pressure was found to range from 7 to 17 mm Hg, whereas main pancreatic duct pressure ranged from 20 to 40 mm Hg. In all subjects, pancreatic duct pressure was two to three times higher than the corresponding common bile duct pressure. Similar observations have not been made in subjects digesting a meal, but in dogs pancreatic duct pressure is not increased when the pancreas is stimulated to secrete by food or secretin. In man, administration of secretin causes pancreatic duct pressure to fall.

Consequences of Loss of Pancreatic Juice

When pancreatic juice flows to the outside through a fistula, the negative feedback loop controlling its secretion is broken, and the pancreas secretes a larger than normal volume of bicarbonate-containing fluid. The consequences are dehydration and metabolic acidosis. In the absence of pancreatic juice, the pH of duodenal contents averages about one unit lower than normal. This means that the hydrogen ion concentration has increased from 0.001 to 0.01 mN. Obstruction of the pancreatic duct, with subsequent fibrotic destruction of acinar cells, results in deficiency of digestive enzymes whose consequences are considered in chapter 16.

REFERENCES

Babkin B.P.: *Secretory Mechanism of the Digestive Glands,* ed 2. New York, Paul B. Hoeber, Inc., 1950.

Case R.M.: Pancreatic secretion: cellular aspects, in Duthie H.L., Wormsley K.G. (eds.): *Scientific Basis of Gastroenterology.* Edinburgh, Churchill Livingstone, 1979, pp. 163-198.

Case R.M., Hotz J., Hutson D., et al.: Electrolyte secretion by the isolated cat pancreas during replacement of extracellular bicarbonate by organic anions and chloride by inorganic anions. *J. Physiol.* 286:563, 1979.

Denyer M.E., Cotton P.B.: Pure pancreatic juice studies in normal subjects and patients with chronic pancreatitis. *Gut* 20:89, 1979.

Gregory R.A.: *Secretory Mechanisms of the Gastrointestinal Tract.* London, Edward Arnold & Co., 1962.

Hallenbeck G.A.: Biliary and pancreatic intraductal pressures, in Code C.F. (ed.): *Handbook of Physiology:* Sec. 6. *Alimentary Canal,* vol. 2, Washington, D.C., American Physiological Society, 1967, pp. 1007-1026.

Harper A.A.: Hormonal control of pancreatic secretion, in Code C.F. (ed.): *Handbook of Physiology:* Sec. 6. *Alimentary Canal,* vol. 2. Washington, D.C., American Physiological Society, 1967, pp. 969-996.

Henderson J.R., Daniel P.M., Fraser P.A.: The pancreas as a single organ: The influence of the endocrine upon the exocrine part of the gland. *Gut* 22:158, 1981.

Janowitz H.D.: Pancreatic secretion of fluid and electrolytes, in Code C.F. (ed.): *Handbook of Physiology:* Sec. 6. *Alimentary Canal,* vol. 2, Washington, D.C., American Physiological Society, 1967, pp. 925-934.

Lake-Bakaar G., Rubio C.E., McKavanagh S., et al.: Metabolism of ^{125}I-labelled trypsin in man: Evidence for recirculation. *Gut* 21:580, 1980.

Malagelada J-R., DiMagno E.P., Summerskill W.H.J., et al.: Regulation of pancreatic and gallbladder function by intraluminal fatty acids and bile acids in man. *J. Clin. Invest.* 58:493, 1976.

Moore E.W., Verine H.J., Grossman M.I.: Pancreatic bicarbonate response to a meal. *Acta Hepato-Gastroenterol.* 26:30, 1979.

Palade G.E.: Intracellular aspects of the process of protein synthesis. *Science* 189:347, 1975.

de Reuck A.V.S., Cameron M.P. (eds.): *The Exocrine Pancreas.* London, J. & A. Churchill, Ltd., 1962.

Rinderknecht H., Renner I.G., Douglas A.P., et al.: Profiles of pure pancreatic secretions obtained by direct pancreatic duct cannulation in normal healthy human subjects. *Gastroenterology* 75:1083, 1978.

Rothman S.S.: Protein transport by the pancreas. *Science* 190:747, 1975.

Rothman S.S.: Passage of proteins through membranes—old assumptions and new perspectives. *Am. J. Physiol.* 238:G391, 1980.

Scheele G.A.: Biosynthesis, segregation, and secretion of exportable proteins by the exocrine pancreas. *Am. J. Physiol.* 238:G467, 1980.

Singh M., Webster P.D. III: Neurohormonal control of pancreatic secretion. *Gastroenterology* 74:294, 1978.

Swanson C.H., Solomon A.K.: Micropuncture analysis of the cellular mechanism of electrolyte secretion by the in vitro rabbit pancreas. *J. Gen. Physiol.* 65:22, 1975.

Webster P.D. III: Hormonal control of pancreatic and acinar cell metabolism, in Botelho S.Y., Brooks F.P., Shelley

W.B. (eds.): *Exocrine Glands*. Philadelphia, University of Pennsylvania Press, 1969.

Webster P.D. III, Black O. Jr., Mainz D.L., et al.: Pancreatic acinar cell metabolism and function. *Gastroenterology* 73:1434, 1977.

11

Secretion of the Bile

BILE IS CONTINUOUSLY secreted by the liver into bile capillaries from which it flows into the hepatic ducts. There are two major components of bile. A *bile-acid independent fraction* has a composition similar to that of pancreatic juice, and its rate of secretion is largely controlled by the hormones secretin and cholecystokinin. The *bile-acid dependent fraction* contains newly synthesized bile acids and bile acids returned to the liver in portal blood. These bile acids are secreted in micelles, which also contain sodium, lecithin, and cholesterol. The rate of secretion of the bile-acid dependent fraction is governed by the rate of return of bile acids to the liver. In man, the two components together are secreted at the rate of 250–1,100 ml/day. During the interdigestive period in man and other animals having a gallbladder, the resistance to bile flow offered by the sphincter at the duodenal termination of the common bile duct shunts much of the bile into the relaxed gallbladder. Bile acids, bile pigments, and other organic constituents of the bile are concentrated as much as 20 times by the gallbladder's rapid reabsorption of water and electrolytes. Within 30 min of a meal, chyme in the duodenum releases the hormone cholecystokinin, which causes the gallbladder to discharge its contents into the duodenum. Bile acids assist in emulsification, hydrolysis, and absorption of fats. Some bile acids are deconjugated and dehydroxylated in the small intestine, and some are absorbed by passive diffusion the length of the small intestine. Conjugated bile salts, and some deconjugated bile acids, are actively absorbed in the terminal ileum, and most reabsorbed bile acids are extracted from blood by the liver and are again secreted into the bile. Deconjugated bile acids are reconjugated before being secreted. Approximately one quarter of the bile acid pool escapes absorption in the small intestine and enters the colon. In the colon, bile acids are deconjugated and dehydroxylated, and some, together with their degradation products, are passively absorbed through the colonic mucosa.

Pressure of Bile Secretion

The secretion of bile is an active process requiring expenditure of energy by liver cells; it is not simply ultrafiltration. If the bile flows into a vertical tube, its rate of flow is constant until a pressure *above* the pressure of blood in the liver is reached; then flow stops abruptly. In man, this limiting pressure is about 23 mm Hg.

After bile has ceased to flow against a high pressure, a further increase in pressure distends the biliary tree and results in irreversible loss of bile from the tree until the pressure of bile returns to the limiting pressure. The loss occurs because bile leaks out of the bile capillaries; and the rate of leakage is higher, the greater the bile pressure. Therefore, the limiting pressure is the one at which continuous secretion of bile into the capillaries equals the rate at which bile leaks from them.

Extrahepatic cholestasis occurs when flow of bile is obstructed for any reason. Plasma concentration of bile acids increases as bile leaks into the capillaries, and the concentration of bile acids in hepatocytes rises.

Pressure in the portal vein of a man with a normal liver is about 10 mm Hg; pressure in the unobstructed hepatic vein, measured by means of a catheter threaded into it by way of the antecubital vein and inferior vena cava, is 3–4 mm Hg. Pressure in liver sinusoids must lie between these two values. An estimate of sinusoidal pressure is obtained by wedging a catheter with an end-tip into the hepatic vein until it obstructs a small branch. Because the hepatic vein is a valveless end-vein distal to freely communicating hepatic sinusoids, a catheter with an end-hole, wedged into and blocking a division of the hepatic vein, measures pressure in the sinusoids. There is, then, direct connection between the tip of the catheter and the sinusoid bed by way of a static column of blood, but blood can continue to flow through the sinusoids because it can escape into other branches of the hepatic vein. The "wedge pressure" so obtained is 4–5 mm Hg. Direct measurement of pressures within the liver vessels of the rat, made by inserting a microneedle attached to a manometer, has shown the mean pressure in the intralobular portal venules to be 4–5 mm Hg and in the central veins to be only 1–2 mm Hg.

Bile flow depends on liver blood flow chiefly because the oxygen supply to liver cells, and therefore their energy supply, depends on blood flow. If abundant oxygen is insured by raising the quantity of oxygen in blood going to the liver, bile flow becomes nearly independent of blood flow.

Electrolytes of Bile; The Bile Acid-Independent Fraction

Bile flow increases directly in proportion to the rate of secretion of bile acids. The increment in flow resulting from an increment in bile acid secretion is the *bile acid-dependent* fraction of bile. It has been customary to calculate the linear regression of flow (the *y* variable) upon bile acid secretion (the *x* variable) and to extrapolate the line derived to zero bile acid secretion. The intercept on the *y* axis is always positive, and the rate of flow it represents has been called the *bile acid-independent fraction*. Some hepatic synthesis and secretion of bile acids always occurs, and zero bile acid secretion is experimentally unattainable. At low bile acid secretion rates, the calculated intercept varies widely. As a result, the reality of a

truly independent fraction has been questioned. Nevertheless, when the rate of bile acid secretion is constant, administration of the hormone secretin causes a large increment in volume flow with no accompanying increase in bile acid secretion (Table 11–1; Fig 11–1). The incremental flow is a bile acid-independent fraction. A similar increment in secretion of the bile acid-independent fraction is caused by hydrocortisone or phenobarbital.

The electrolyte composition of human bile secreted at basal rate, except for the presence of bile salts, resembles an ultrafiltrate of plasma (see Table 11–1). Stimulation of bile flow by secretin raises the concentration of bicarbonate and lowers that of chloride. In this respect the response of the liver to secretin is similar to that of the pancreas. The carbonic anhydrase inhibitor acetazolamide reduces secretin-stimulated bicarbonate output and volume flow of bile as it reduces those of pancreatic juice. The intrahepatic bile ducts and canaliculi, rather than the parenchymal cells of the liver, are probably the sites at which the major fraction of water and electrolytes of the bile are secreted, and their cells are the ones that respond to secretin and that are inhibited by acetazolamide.

The bile acid-independent fraction of bile has a high bicarbonate concentration and is under the same control as is the aqueous fraction of pancreatic juice, and therefore what is known about the function and control of pancreatic bicarbonate secretion can be applied to the function and control of bile acid-independent secretion by the liver. Secretin stimulates bile flow without increasing bile acid secretion, and all stimuli for secretin release elicit bile secretion. Cholecystokinin augments the effect of secretin, and so do peptides sharing cholecystokinin's C-terminal sequence of amino acids, the gastrins, caerulein, and CCK-OP. All stimuli of acid secretion indirectly

TABLE 11–1.—Mean and Ranges of Basal Flow and Electrolyte Composition of Human Bile Obtained from 7 Subjects Compared with Flow and Composition at the Peak of Response to Intravenous Secretin*

	FLOW (ML/10')	HCO$_3$ (MEQ/L)	CL$^-$ (MEQ/L)	NA$^+$ (MEQ/L)	K$^+$ (MEQ/L)	BILE SALTS
Basal:						
Mean	3.5	26	102	151	4.6	27
Range	2.7– 4.1	19–29	83–112	144–159	3.6–5.2	19–55
Secretin:						
Mean	11.9	46	95	146	4.0	9
Range	5.8–28.5	31–81	62–111	140–153	3.7–4.7	3–12

*From Waitman A.M., et al.: *Gastroenterology* 56:286, 1969.

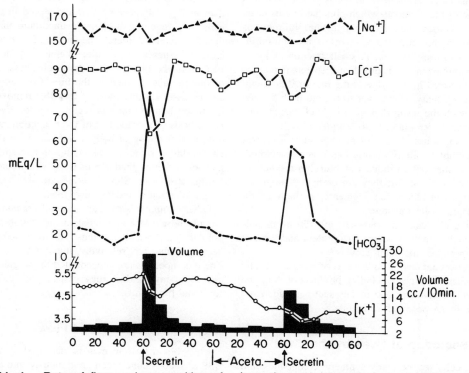

Fig 11–1.—Rate of flow and composition of hepatic bile in a human subject. At the vertical *arrows* secretin (1 unit/kg of body weight) was rapidly injected intravenously. During the third hour the carbonic anhydrase inhibitor acetazolamide was infused in a dose of 50 mg/kg of body weight. (From Waitman A.M., et al.: *Gastroenterology* 56:286, 1969.)

stimulate bile secretion. Delivery of acid to the duodenum during the interdigestive myoelectrical complex probably accounts for the fact that bile flow increases more than four times as the activity front passes through the duodenum. Stimulation of the vagus, as by insulin hypoglycemia, increases bile secretion, and the response is blocked by anticholinergic drugs. It is not known whether the effect of vagal stimulation is directly upon the liver or indirectly through humoral mediators. Histamine stimulates bile secretion independently of its effect upon acid secretion, and the effect is blocked by H_2 antagonists.

Afferent nerves from the gallbladder and biliary tree are stimulated by tension in the walls of those vessels; these nerves mediate pain, which is referred to the epigastrium with radiation around the costal margins. Stimulation of the splanchnic nerves to the liver causes vasoconstriction and decreases bile flow.

Systemically released epinephrine has two effects: (1) direct vasoconstriction of hepatic vasculature, causing increased impedance; and (2) a redistribution of cardiac output, producing greater blood flow in the splanchnic area with a corresponding increase in pressure gradients through the liver. The latter effect is the more important.

Bile secretion is not stimulated by calomel or by the proprietary preparations of vegetable drugs sold as liver pills.

Other Constituents of the Bile

Bilirubin, the metabolic product of hemoglobin porphyrin degraded in the reticuloendothelial system, is normally present in the plasma at about 0.2–0.8 mg/100 ml. It is conjugated in the liver with glucuronic acid and secreted into the bile at up to 1,000 times this concentration. Bilirubin is yellow, but it is only one of the pigments which give the bile its characteristic golden hue. In the intestine, bilirubin is converted by bacteria first to colorless mesobilirubinogen and to urobilinogen. A small fraction of intestinal urobilinogen is absorbed and secreted into the bile or excreted in the urine at the rate of 0.5–2 mg/day. The capacity of the liver to secrete urobilinogen is low, with the result that in mild liver damage urinary urobilinogen may rise to 5 mg/day.

The stool is usually one of many shades of brown, and in complete biliary obstruction, it is clay-colored. However, metabolic products of bilirubin are not responsible for the color; they are orange, blue-green, or colorless. Unidentified pigments, perhaps of dietary origin, color the stool.

Hepatic bile contains 280–410 mg/100 ml of protein, a small fraction being mucoprotein and the rest plasma proteins, including amylase.

Bile also contains up to 3 gm/L of neutral fat, and many other substances including products of drug and hormone metabolism, and metals such as lead, zinc, and copper are secreted in the bile.

Secretion of Dyes by the Liver

Liver cells secrete many synthetic compounds into the bile. Two dyes, sulfobromophthalein (Bromsulphalein, BSP) and indocyanine green, are used for measurement of hepatic blood flow, rate of bile secretion, and liver function. Indocyanine green is more convenient than BSP for the reason that its absorption curve does not overlap that of hemoglobin. It is secreted unchanged into the bile, and it has no extrahepatic metabolism or enterohepatic circulation.

The handling of BSP by the liver can be divided into two separate processes: (1) removal of BSP from the plasma at a rate that depends on its plasma concentration, and (2) secretion of BSP into the bile. The latter process is also a function of plasma concentration; it is a rate-limited process that becomes constant when the plasma concentration of BSP is greater than 3 mg/100 ml. The maximal rate of BSP secretion into the bile therefore has a T_m (transport maximum) analogous to that of the renal tubules. In 12 men without liver disease, the T_m for BSP secretion was found to be 9.5 mg/min, with a standard deviation of 1.9. In 7 women who were also without liver disease, the value was 7.1 ± 0.8. The T_m is reduced, sometimes nearly to zero, in patients with liver disease.

If BSP is infused into the blood at a rate greater than its T_m, it is accumulated and stored in the liver cells; this function, too, has a maximum that is reduced by liver disorders. The uptake of BSP by extrahepatic tissues is negligible; likewise, absorption of BSP from the intestine is small, being less than 5% of the amount secreted in the bile. Consequently, it can be assumed that most of the BSP removed from the blood is taken up and secreted by the liver. This is the basis of a crude test of liver function, in which a dose of 5 mg/kg of body weight is given IV and a single plasma sample is taken 45 min later. In normal persons, no more than 5% of the dose is found in the plasma at 45 min. Similar results are obtained with a single injection of indocyanine green.

Other compounds secreted by the liver into the bile include tetraiodophenolphthalein and tetrabromophenolphthalein, which, because of their content of heavy atoms, are relatively impervious to x-rays. These compounds are concentrated in the gallbladder, and the x-ray shadow they cast reveals gallbladder form and function.

Liver Blood Flow

Liver blood flow in man is measured by injecting BSP or indocyanine green continuously into a vein at the rate of Q mg/min, and the rate is adjusted so that the arterial plasma concentration of the dye remains constant, as determined by frequent analysis. This rate Q is, of course, below the T_m of the liver. A catheter is inserted, so that a sample of blood draining the liver through the hepatic veins can be secured. It is assumed that this sample is representative of the total venous drainage of the liver. It is also assumed that the concentration of the dye in all plasma reaching the liver by way of the hepatic artery and portal vein is represented by the concentration in the systemic artery from which the blood is drawn.

If the concentration of the dye in arterial plasma is A mg/L and if the estimated hepatic plasma flow is $EHPF$ in liters per minute, then the rate at which the dye is brought to the liver is $A(EHPF)$ mg/min. The rate at which the dye leaves the liver in the same volume of blood is $V(EHPF)$ mg/min, where V is the concentration of the dye in hepatic venous blood. The rate at which the dye is removed by the liver is equal to the difference between that entering and leaving, or $(EHPF)(A - V)$ mg/min. If the only dye that leaves the plasma is that removed by the liver, the rate at which the liver removes it is equal to the rate Q, at which it must be infused to keep the plasma level constant. Then $Q = (EHPF)(A−V)$, or $EHPF = Q/(A−V)$. The estimated hepatic blood flow *(EHBF)* is determined by dividing this value by the fraction of plasma in arterial blood. With this method, normal values of $1,000−1,800$ ml/1.73 m² body surface area are obtained. The mean is about 1,500, about 25% of the cardiac output at rest.

Estimated hepatic blood flows measured simultaneously with BSP and indocyanine green correlate reasonably well (r = 0.80).

In unanesthetized dogs, the hepatic artery supplies about one third of the total hepatic blood flow. There are no reliable estimates of the proportion in man. Blood reaching the liver by the hepatic artery and the portal vein is apparently thoroughly mixed within the liver sinusoids. Of the total oxygen uptake of $40−50$ cc/min by a 1,500-gm liver, about half is supplied through the hepatic artery.

Lymph flow from a 1,500-gm liver ranges from 0.4 to 0.8 ml/min, and its protein content is almost equal to that of plasma.

Bile Acids and Bile Salts*

Bile acids are synthesized in the liver from cholesterol (Fig 11-2). The limiting reaction is hydroxylation catalyzed by 7-α-hydrolase, and this step in synthesis is the one controlled in the negative feedback regulation of bile acid synthesis.

The major bile acids formed in the human liver are the trihydroxy cholic acid (Greek *chole* = bile) and the dihydroxy chenodeoxycholic, or chenic, acid (Greek *chenos* = goose). Because these are the ones synthesized, they are the *primary bile acids*.

Primary bile acids are dehydroxylated by bacteria in the digestive tract. The major products are deoxycholic acid, a dihydroxy acid derived from cholic acid, and lithocholic acid (Greek *lithos* = stone), a

---

*Bile acids is the generic term for the steroid compounds first isolated in protonated form by organic chemists. The papers by O. Rosenheim and H. King (The ring system of sterols and bile acids, *J. Soc. Chem. Ind.* 51:464 and 554, 1932), in which their structure was first correctly described, are masterpieces of armchair chemistry. *Bile salts* is a generic term referring to the conjugates of bile acids with glycine or taurine in peptide linkage or bile alcohols linked in ester bond with sulfate. The term bile salts is used here to name the conjugated compounds, whether or not they are ionized. The term bile acids is used as an all-inclusive name for all compounds containing the cholane nucleus, whether or not they are conjugated or ionized. The nomenclature of these compounds is not settled, and the reader can expect to encounter changes. For example, taurine is linked to cholic acid in a peptide bond, and the name for the compound that is likely to be adopted is *cholyltaurine,* replacing *taurocholic acid.*

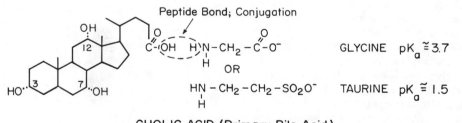

CHOLIC ACID (Primary Bile Acid)
All -OH's in α configuration

−OH on 3 and 7, not on 12 = Chenodeoxycholic Acid − Primary

−OH on 3 and 12, not on 7 = Deoxycholic Acid − Secondary

−OH on 3, not on 7 or 12 = Lithocholic Acid − Secondary

(−OH on 3 sulfated) = Sulfolithocholic Acid

−OH on 3α, −OH on 7β = Ursodeoxycholic Acid − Tertiary

Fig 11−2.—The structure of the common bile acids. The bile acid is conjugated with either glycine or taurine by elimination of water to form a peptide bond. The approximate ionization constants of glycocholic and taurocholic acids are given on the right. (Adapted from Hofmann A.F.: *Gastroenterology* 48:484, 1965.)

monohydroxy acid derived from chenodeoxycholic acid. These are *secondary bile acids*.

The bond connecting carbon atoms at positions 3, 7, and 12 with hydroxyl groups can have one of two configurations. The hydroxyl group may project above the plane of the molecule as it is represented in Figure 11–2. In that case the hydroxyl is said to be in the α-configuration, and the bond is conventionally represented by an interrupted line. If the hydroxyl projects below the plane of the molecule, it is said to be in the β-configuration, and the bond is represented by an uninterrupted line. All hydroxyls of the primary and secondary bile acids are in the α-configuration.

During the enterohepatic circulation of chenodeoxycholic acid, the hydroxyl group at the 7-position is oxidized to a ketone. The ketone-containing acid is then a secondary bile acid. The ketone group is subsequently reduced in the liver, and the resulting hydroxyl group is in either the α- or the β-configuration. If it is β, the molecule is ursodeoxycholic acid (Latin *ursus* = bear). Ursodeoxycholic acid, being derived from a secondary one, is a *tertiary bile acid*. It occurs in human bile. Some bacteria are apparently capable of converting chenodeoxycholic acid directly to ursodeoxycholic acid. In that case the product would be classified as a secondary bile acid.

The *bile acid pool* is the total amount of primary and secondary bile acids present in the body at any one time.

Primary and secondary bile acids are secreted as conjugated bile salts by the liver. Either taurine (Greek *tauros* = bull) or glycine is added in peptide bond to the carbonyl group of the side chain of the cholane nucleus, making the corresponding taurocholic or glycocholic bile salts. These occur in human bile in the ratio of about 1:3, because taurine is in short supply. If taurine is fed, the fraction secreted as taurocholic acids rises, and when taurine is lost as the result of imperfect reabsorption, the ratio falls to 1:20.

Bile acids are planar molecules. Hydrophobic groups project on one side of the molecule, and hydrophilic groups project on the other side. The hydrophilic groups are the hydroxyl groups of the cholane nucleus, the peptide bond on the side chain, and the carbonyl or sulfate group of glycine or taurine. Bile acids accumulate at oil-water interfaces and decrease interfacial tension. The effectiveness of individual bile acids in reducing interfacial tension depends upon

the number and character of their hydrophilic groups. Thus, taurocholic acid emulsifies fat and forms micelles far more readily than does unconjugated lithocholic acid.

Taurine conjugates of bile acids are anions at the pH (6.0–7.7) of bile and intestinal contents, and consequently they always exist as salts. Glycine conjugates of bile acids are less ionized at intestinal pH, and consequently glycocholates in the intestine are a mixture of ionized and un-ionized molecules. The charges of bile anions are balanced by cations, chiefly sodium. The osmotic pressure, measured directly by the freezing point depression method of 67 samples of bile drawn from the common ducts of cholecystectomized dogs, was found to be 299 ± 11 mOsm, plasma osmolality of the same dogs was 303 ± 4 mOsm. However, the sum of the cations and anions of bile is always greater than 299 mOsm. The reason for this discrepancy is that the bile salts form osmotically inactive micelles (Fig 11–3) within which some cations are sequestered. In samples of dog bile whose

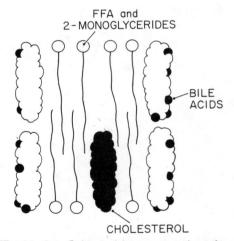

Fig 11–3.—Schematic representation of a micelle in longitudinal section. Bile acids form a cylindrical shell with their hydrophilic groups outward. The core consists of interdigitating lipid molecules. During digestion these are free fatty acids and 2-monoglycerides, as depicted here. In the bile, the lipids are lecithin and neutral fats. Cholesterol is dissolved in the lipid phase of the micelle. The micelle is surrounded by cations, not shown here, which balance the anionic charges on the bile acids.

sodium concentration determined chemically was 245–270 mN, the sodium activity, measured with a sodium-sensitive electrode, was only 148–186 mN. The potassium concentration of the same samples was 8.8–10.2 mN, but the potassium activity was 2.7–3.9 mN. Osmotic coefficients of sodium and potassium in bile are even lower than their activity coefficients.

Ionized bile salts, being large, negatively charged ions at the pH prevailing in the intestinal lumen, are not absorbed by passive diffusion in the duodenum or jejunum. Consequently, their concentration remains high until fat digestion and absorption are complete. Some small fraction of the un-ionized glycocholates are absorbed by diffusion in the duodenum and jejunum; but the remainder, plus the ionized bile salts,

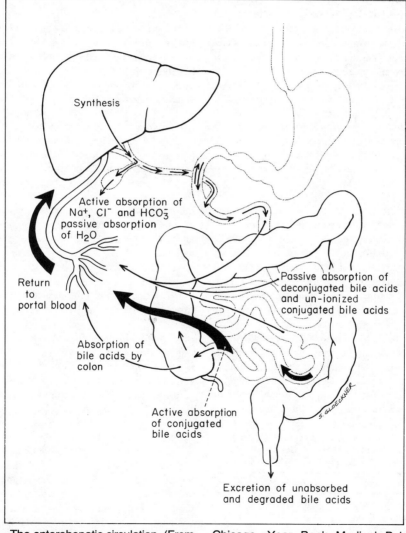

Fig 11–4.—The enterohepatic circulation. (From Davenport H.W.: *A Digest of Digestion,* ed. 2. Chicago, Year Book Medical Publishers, Inc., 1978.)

are absorbed by active transport in the lower ileum. Upon reaching the liver, bile salts in the portal blood are immediately taken up by liver cells and again secreted into bile. This cycle is called the *enterohepatic circulation* (Fig 11–4).

Bile-Acid Dependent Bile

A substance increasing bile flow is called a *choleretic*. Bile salts themselves are powerful choleretics, the most effective being dehydrocholate. Bile salts act directly on the parenchymal cells of the liver, and their IV injection is promptly followed by their secretion into the bile. The active secretion of bile salts is accompanied by an apparently passive flow of water and electrolytes into the bile, with the result that the volume, as well as bile salt concentration, of the bile increases. When bile salts are continuously infused, the rate of bile secretion is equal to, or only slightly greater than, the rate of infusion. When the rate of infusion is increased, the rate of flow of bile likewise goes up. The secretion of bile salts is interrupted by their storage in the gallbladder during the interdigestive period.

The rate of secretion of the bile acid-dependent fraction is governed by the amount of bile acids available to the hepatocytes, and this in turn is the sum of newly synthesized bile acids and of bile acids returned to the liver in portal blood. Under normal circumstances, the rate of synthesis is nearly constant, but the rate of return of bile acids varies with the digestive cycle. In the interdigestive period, particularly overnight, bile acids are sequestered in the gallbladder, but when digestion begins, the gallbladder empties its contents into the small intestine. Then bile acids participating in digestion are absorbed throughout the small intestine, returned to the liver, and secreted once more. An individual bile acid molecule may circulate several times during the digestion of a

meal. Consequently, the rate of bile secretion is a function of the frequency and nature of meals. Table 11–2 shows that doubling the caloric intake increases bile secretion and the frequency of the enterohepatic circulation. The rate of bile secretion also depends on small intestinal transit time, for the more rapidly bile acids move down the intestine, the more rapidly are they absorbed and returned to the liver.

Bile Acid Pool Size

The method of measuring bile acid pool size is illustrated in Figure 11–5. A nasogastric tube is passed with its opening in the subject's duodenum so that freshly excreted bile can be sampled. A small amount of isotopically labeled bile is given. In this example, glycocholic acid doubly labeled with ^3H on the cholic acid and ^{14}C in the glycine was used. The administered radioactivity in terms of disintegrations per min (dpm) of each isotope is accurately known. Bile samples are collected over 168 hours. The radioactivity and the cholic acid concentrations are determined, and the specific activity of the cholic acid in the sample is calculated. This is dpm per milligram or per millimole of cholic acid. Likewise, the specific activity of the glycine conjugated to cholic acid is calculated. This is dpm per milligram or per millimole of glycine.

The rate of disappearance at any instant of each of the labeled components of glycocholic acid is proportional to the amount present at that instant. This is a first-order reaction, and the logarithm of the specific activity of the serial samples plotted against time gives a straight line with negative slope. The line extrapolated to zero time gives the specific activity of the bile acid pool if mixing of the administered bile acid with the rest of the pool were instantaneous. The zero-time specific activity, dpm per milligram or millimole, divided into the number of dpm's originally administered gives the size of the pool of glycocholic acid in milligrams or millimoles.

Other components of the pool can be measured in the same way. Alternatively, if all the components of the pool are measured in an early sample, their individual pool sizes can be calculated once the size of the pool of any component is known.

Representative data on the size and composition of the bile acid pool in normal subjects are given in Table 11–3. The lower limit of the normal pool in adult men is about 1.8 gm.

TABLE 11–2.—BILE ACID AND CHOLESTEROL SECRETION IN BILE OF NORMAL SUBJECTS FED A LOW-CALORIE AND HIGH-CALORIE DIET*

	20 CAL/KG	40 CAL/KG
Bile acid secretion, gm/day	16.0–37.0	19.0–72.0
Cholesterol secretion, gm/day	1.6– 3.0	2.4– 3.7
Bile acid cycles/day	3.4– 5.7	5.0–13.5

*Adapted from Brunner H., Hofmann A.F., Summerskill W.H.J.: Gastroenterology 62:188, 1972.

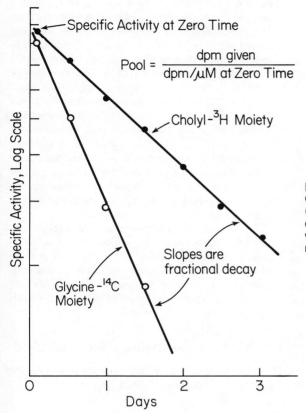

Specific Activity at Zero Time

$$\text{Pool} = \frac{\text{dpm given}}{\text{dpm}/\mu M \text{ at Zero Time}}$$

Cholyl-^{3}H Moiety

Specific Activity, Log Scale

Slopes are
fractional decay

Glycine -^{14}C
Moiety

Days

Fig 11–5.—Measurement of bile acid pool size, fractional decay, and deconjugation. In this example, cholylglycine (glycocholic acid) labeled with ^{3}H in the cholane nucleus and with ^{14}C in the glycine was given. Dpm = disintegrations per minute.

The slope of the lines plotted in Figure 11–5 is the fractional decay, or rate constant, of the labeled compound. The data given for the cholyl moiety show that half the activity disappeared in 1.56 days. The rate constant is 0.321 days^{-1}. If the pool size remains constant, the amount of cholic acid disappearing each

TABLE 11–3—Representative Figures for Composition of Bile Acid Pool in Normal Adult Human Subjects

	GM	RANGE, GM
Primary bile acids		
Total cholic conjugates	1.45	0.55–1.90
Cholylglycine	1.05	
Cholyltaurine	0.40	
Total chenic acid conjugates	0.68	0.13–0.77
Secondary bile acids		
Total deoxycholic conjugates	0.35	0.01–0.72
Total lithocholic conjugates	0.05	0.01–0.07
Total pool size	2.53	1.86–3.24

day is replaced by newly synthesized cholic acid. With this rate constant, synthesis of 0.40 gm/day of cholic acid would be required to maintain a cholic acid pool of 1.45 gm.

During its enterohepatic circulation, some cholic acid is converted to deoxycholic acid by 7-α-dehydroxylation. The newly produced secondary bile acid enters a pool of deoxycholic acid and participates in the enterohepatic circulation.

Both pool size and rate constant for chenodeoxycholic acid are smaller than those of cholic acid. Chenodeoxycholic acid circulates 1.34 times more rapidly than cholic acid, because more chenodeoxycholic acid is absorbed in the proximal small intestine and thus reaches the liver more rapidly than does cholic acid. Synthesis of about 0.18 gm/day maintains a pool size of about 0.68 gm. Bacterial dehydroxylation of chenodeoxycholic acid produces lithocholic acid, which is discussed below.

The bile acid pool of patients with cholelithiasis is

usually below normal, and the pool remains small after cholecystectomy.

The damaged liver of cirrhotic patients produces cholic acid at 25% of the normal rate as the result of impaired conversion of cholesterol to cholic acid. The rate of production of chenodeoxycholic acid is 70% of normal. As a result, the bile acid pool of patients with cirrhosis is about half that of normal persons, and its fraction of cholic and deoxycholic acids is reduced.

The enterohepatic circulation of cholic, deoxycholic and chenodeoxycholic (or chenic) acids is summarized in Figure 11−6.

Deconjugation During Enterohepatic Circulation

Bacteria in the small intestine deconjugate bile acids during their enterohepatic circulation, and bacteria in the colon deconjugate bile acids that have escaped reabsorption in the terminal ileum. The concentration of deconjugated bile acids rises from zero in the duodenum to about 1 mM in the lower ileum during digestion of a meal.

The data in Figure 11−5 were obtained when doubly labeled glycocholic acid was given; the ^{3}H label on the cholic acid and the ^{14}C in the glycine were chemically linked by a peptide bond. Nevertheless,

the fractional decay of the glycine label is about 3 times as fast as the cholic acid label. Eighteen percent of glycocholic acid is deconjugated during each enterohepatic cycle. The deconjugated bile acid is reabsorbed to the extent of 95% and, being reconjugated in the liver, is again secreted in the bile. Glycine liberated from the bile acid enters the large glycine pool of the body, and its carbons are almost entirely oxidized to carbon dioxide. Carbon dioxide is eventually breathed off in the lungs, and this is the basis of a simple test of the rate of deconjugation. A bile acid conjugated with ^{14}C-glycine is given, and the rate at which ^{14}C-CO_2 appears in the breath is a measure of the rate of deconjugation. Replacement of glycine oxidized after deconjugation requires 450 mg of glycine a day.

The rate of deconjugation of taurine conjugates is less than half that of glycine conjugates. A small fraction of the taurine removed from bile acids is reabsorbed and enters the taurine pool; little is reincorporated into conjugated bile acids. The taurine not absorbed is degraded by bacterial action, and its sulfur appears in the urine as sulfate.

Sulfation of Bile Acids; Lithocholic Acid

Lithocholic acid is derived from chenodeoxycholic acid by 7-α-dehydroxylation in the cecum. In normal

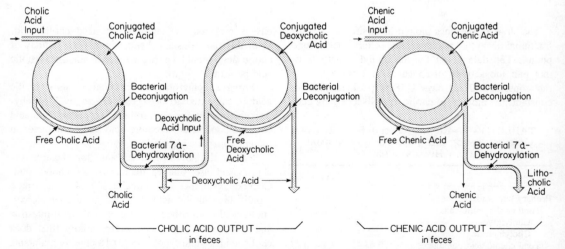

Fig 11−6.—Enterohepatic circulation of the major primary and secondary bile acids in man. The circulation of lithocholic acid, which is normally less than 5% of the total bile acid pool, is shown in Figure 11−7. (Adapted from Hoffman N.E., Hofmann A.F.: *Gastroenterology* 67:887, 1974.)

man, 60–150 mg of chenodeoxycholic acid, or one third to one half of the chenodeoxycholic acid synthesized each day, is converted to lithocholic acid. The same proportion of chenodeoxycholic acid is converted to lithocholic acid in a patient whose chenodeoxycholic acid pool has been greatly expanded by ingestion of chenodeoxycholic acid. Despite its rapid formation, lithocholic acid is only a trace constituent of normal human bile or of bile of a patient fed chenodeoxycholic acid. Lithocholic acid is toxic to the liver, but it does not rise to toxic levels in man for the reason that it is cleared as rapidly as it is formed.

Lithocholic acid is cleared by two means. It is poorly absorbed in the lower small intestine so that only about one fifth of the lithocholic acid produced in the small intestine returns to the liver. The rest is excreted in the stool. Lithocholic acid reaching the liver is, in part, sulfated by addition of sulfate to the hydroxyl group in the 3-position. The glycyl-conjugate, the predominating form, is more readily sulfated than the less abundant tauryl-conjugate. The result of sulfation is that about 75% of the lithocholic acid reaching the liver is secreted into the bile as the sulfate. Most of the sulfated lithocholic acid reaching the intestine is excreted in the stool. The process of lithocholic acid clearance is illustrated in Figure 11–7.

The liver of the rhesus monkey or of the baboon has a lesser ability to sulfate lithocholic acid, and when an animal of either species is fed chenodeoxycholic acid, a hepatotoxic amount of lithocholic acid accumulates in its bile acid pool. The chimpanzee is like man in readily sulfating lithocholic acid, and it

avoids liver damage when fed chenodeoxycholic acid.

Sulfation becomes an important metabolic fate of bile acids in cholestasis when the enterohepatic circulation is interrupted and bile acids are retained in the liver. Because there is decreased passage of bile acids through the distal intestine, primary bile acids are not converted to secondary ones, and secondary bile acids disappear from the bile acid pool. The concentration of bile acids in the plasma rises with the concentration of cholic acid being greater than that of chenodeoxycholic acid. About 7% of the chenodeoxycholic acid in plasma is sulfated. Although there are virtually no bile acids in the urine of a healthy person, urine becomes the major route of excretion of bile acids in a patient with cholestasis. Renal excretion occurs, because the primary bile acids, in particular chenodeoxycholic acid, are sulfated, and renal clearance of sulfated bile acids is 10–100 times greater than renal clearance of unsulfated bile acids. Sulfated chenodeoxycholic acid is the predominant bile acid in the urine. A third is sulfated in the 3α-position, and nearly half has an additional sulfate at the 7α-position.

Intestinal Absorption of Bile Acids

Bile acids are absorbed the whole length of the small intestine by passive diffusion and in the terminal ileum by active transport. Absorption is highly efficient, and more than 95% of the bile acids entering the small intestine is absorbed there. Nevertheless, enterohepatic cycling is so frequent that about

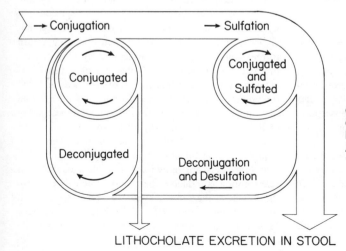

Fig 11–7.—The enterohepatic circulation of lithocholic acid and its clearance from the bile acid pool by sulfation. (Adapted from Cowen A.E. et al.: *Gastroenterology* 69:67, 1975.)

LITHOCHOLATE EXCRETION IN STOOL

one quarter of the bile acid pool escapes into the colon each day. Some of those bile acids are absorbed in the colon by passive diffusion, and the rest, after bacterial deconjugation and modification, is excreted in the stool. Excretion averages 500 mg/day.

Because bile acids have molecular weights greater than 500, their passive absorption depends chiefly upon their solubility in the lipid portion of the membrane of intestinal epithelial cells. Therefore, factors decreasing water solubility and increasing fat solubility promote passive absorption. Chenodeoxycholic acid, having only two hydroxyls, is absorbed in the upper small intestine, and its enterohepatic cycle is shorter and more rapid than that of cholic acid. Glycine conjugates, because their pK_a is high, are only slightly ionized in intestinal contents. They are more readily absorbed by passive diffusion than are taurine conjugates whose pK_a is low and which are ionized in intestinal contents. Deconjugation increases passive absorption by 9 times, and dehydroxylation increases passive permeation by more than 4 times.

Conjugated and ionized bile acids are rapidly absorbed by active transport by a mechanism confined to the terminal ileum. The transport process exhibits saturation kinetics, and there is competition among bile acids for absorption. A small amount of deconjugated bile acids may also be absorbed by active transport in the terminal ileum.

The fraction absorbed by each process is unknown for man, but abolition of active absorption in the monkey by removal of the last third of the small intestine reduces total bile acid absorption by 45%. Perhaps in man, half of the absorption is active, half passive.

In the normal course of digestion, most fat is absorbed by the time intestinal contents have passed through the jejunum, and micelles reaching the terminal ileum are composed chiefly of bile salts. If, however, digestion and absorption of fat are defective, the fat digestion products in mixed micelles reaching the terminal ileum interefere with absorption of bile salts, and an abnormally large amount of bile salts escape into the colon.

In man, resection or disease of the terminal ileum reduces absorption of bile acids, and diarrhea may ensue for one of two reasons. If bile acid malabsorption is mild, synthesis of bile acids by the liver is able to maintain pool size, and no defect in fat digestion and absorption occurs. However, unabsorbed bile acids

escape into the colon, where they or their degradation products inhibit sodium and water absorption. This form of diarrhea can be prevented by feeding cholestyramine, a resinous polymer of aminated styrene, which absorbs and thereby sequesters bile acids. If bile acid malabsorption is severe, digestion and absorption of long-chain triglycerides is defective, and unabsorbed fatty acids or their derivatives cause diarrhea. This form can be prevented by substituting medium-chain for long-chain triglycerides.

Bile Acids in Blood

Absorbed bile acids are carried from the intestine to the liver in the plasma of portal blood. Dihydroxy bile acids are bound to plasma proteins to the extent of 96%–98%, and trihydroxy acids are bound 83%–91%. Bile acids are rapidly removed from portal blood by the liver, and the rate of removal is slightly less for those more tightly bound to plasma proteins. Consequently, the concentration of chenodeoxycholic acid in peripheral blood is higher than that of cholic acid, although chenodeoxycholic acid is a smaller fraction of the bile acid pool.

The plasma concentrations of conjugates of chenodeoxycholic acid and cholic acid over a 24-hour period in normal persons, in cholecystectomized persons, and in patients following ileal resection are shown in Figure 11–8.

Control of Bile Acid Synthesis

Under normal circumstances, 500–900 mg of bile acids is lost each day, and pool size is kept constant by equal synthesis of new bile acids. Rate of synthesis is governed by the rate at which bile acids return to the liver, and control is exercised by feedback inhibition of 7-α-hydrolase. If the rate of return is increased by feeding bile acids, pool size rises and rate of synthesis falls. If the rate of return is reduced, rate of synthesis is increased up to the maximum capacity of the liver. In the monkey, the liver can synthesize bile acids up to 10 times the normal rate. The same is probably true for the human liver.

If bile acids are drawn off through a fistula or if reabsorption is reduced, the rate of synthesis rises. In the monkey, the rate of synthesis can keep up with loss if no more than 20% of the bile acids secreted fails to be reabsorbed. However, if 33% of the bile

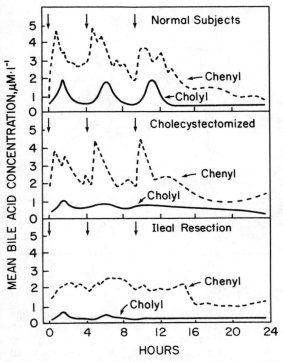

Fig 11–8.—The concentration of conjugates of chenodeoxycholic (chenic) acid and cholic acid in peripheral plasma over 24 hours. The subjects were given liquid meals at the time indicated by the arrows. *Top:* Five representative healthy subjects. *Middle:* Five subjects six or more months after cholecystectomy. *Bottom:* Five patients with ileal resection and documented bile acid malabsorption. (Adapted from Schalm S.W., et al.: *Gut* 19:1006, 1978.)

acids secreted is lost, synthesis cannot increase further, and pool size falls. With greater losses, pool size falls further. If the enterohepatic circulation is abolished in man by total failure of reabsorption, the rate of secretion of bile acids falls to 3–5 gm/day; this rate of secretion is equal to the maximal capacity of the liver to synthesize bile acids. Thus, in patients incapable of reabsorbing bile acids, only 3–5 gm of bile acids reaches the duodenum each day, in contrast with the normal 16–72 gm/day. In such patients, most of the bile is delivered in the morning when the gallbladder evacuates bile it has stored overnight.

Function of the Gallbladder

In animals possessing a gallbladder* that concentrates and stores bile, the entry of bile into the duodenum does not parallel its secretion by the liver. In the interdigestive period, the sphincter of the ampulla offers resistance to flow. The gallbladder is relaxed, and hepatic bile enters it. The major storage of bile occurs overnight during the interdigestive period (Fig 11–9), but even then approximately half of the bile secreted by the liver reaches the duodenum. In man the gallbladder's capacity ranges from 14 to 60 ml.

The wall of the gallbladder rapidly absorbs water and electrolytes, leaving bile salts, bile pigments, and cholesterol behind. The solutes become 5–20 times more concentrated in bladder bile than in hepatic bile (Table 11–4). In the rabbit and in fish, the primary process responsible for concentrating bile is the active transport of sodium and chloride from lumen to blood. No potential difference is generated, and water follows the ions in isotonic proportion. The human gallbladder develops a potential difference of about 8 mV, serosal surface positive, when it is absorbing, and it can maintain a short circuit current. The current is not present if choline or potassium is substituted for sodium; it is abolished by ouabain; and it is almost independent of anions. In man, absorption is apparently effected by an electrogenic sodium pump. Chloride ions follow passively. Secretion of acid by the gallbladder epithelium removes bicarbonate and slightly acidifies the bile. Of water-soluble substances, only the small ones whose molecular diameter is less than 6–8 A can penetrate the gallbladder mucosa. The concentration of K^+ tends to rise as bile is concentrated, but because it can penetrate the mucosa, K^+ diffuses down its concentration gradient until it reaches electrochemical equilibrium. Probably

---

*In animals without a gallbladder (e.g., pigeon, rat, pocket gopher, and horse), the sphincter of the hepatopancreatic ampulla has little or no resistance, and the liver secretes large amounts of bile. Some animals (e.g., guinea pig, rabbit, and bush rat) whose gallbladders have a low ability to concentrate bile also have a large continuous secretion. The pig, goat, sheep, and cow have gallbladders with low concentrating capacity, but their livers secrete relatively small amounts of bile. The duck, man, mouse, chicken, dog, cat, and striped gopher secrete bile at low rates; but their gallbladders have high concentrating capacity, with the result that the volume of bile reaching the intestine is small.

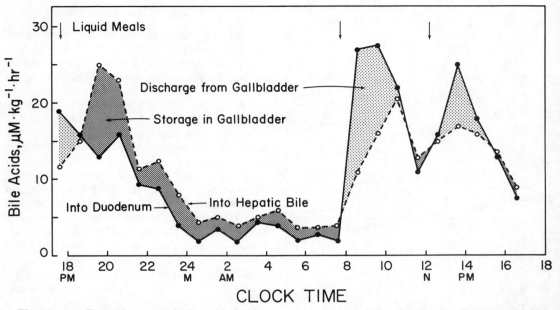

Fig 11–9.—Rate of output of bile acids by the liver into hepatic bile and rate of delivery of bile acids into the duodenum from both liver and gallbladder. When the rate of output into hepatic bile exceeds the rate of delivery into the duodenum, bile acids are being stored in the gallbladder. When the rate of delivery into the duodenum ex- ceeds the rate of output by the liver, stored bile acids are being discharged from the gallbladder. The data are mean values for 5 healthy young men. (Adapted from van Berge Henegouwen G.P., Hofmann A.F.: *Gastroenterology* 75:879, 1978.)

Ca^{2+} is also at its electrochemical equilibrium in gallbladder bile. The high total concentration of Na^+ in gallbladder bile is accounted for by the fact that a large fraction of the ion is held in association with bile salts in osmotically inactive micelles.

The mucosa of the gallbladder consists of a single layer of tall, columnar epithelial cells, each bound to its neighbors at the apical end by tight junctions. The tight junctions are channels through which water and ions can diffuse. When the mucosa is reabsorbing fluid, long lateral channels appear between the cells; they are the site of a standing osmotic gradient responsible for the reabsorption of water (Fig 11–10).

The gallbladder mucosa secretes mucin, the concentration of which in gallbladder bile is about 1.6%.

The organic constituents of bile are concentrated by being left behind during absorption of water and electrolytes. About 10% of the lecithin in the gallbladder is absorbed, and a small amount of deconjugated bile acids, when present in gallbladder bile, is absorbed by passive diffusion. Iodine-containing, synthetic organic compounds are secreted by the liver into bile, and they are not absorbed by the gallbladder epithelium. Consequently, they become concentrated in gallbladder bile. Because they are radiopaque, they serve as contrast media.

Contraction of the Gallbladder

Within the first half hour after a meal, the human gallbladder begins to contract. Emptying occurs irregularly over a period of 20–105 min, and it is seldom complete. In 15 normal human subjects whose gallbladder size was measured by x-ray photographs taken in two directions, the greatest volume expelled was found to be 27 ml and the least 8 ml. In another 15 subjects, the percentage emptying ranged from 51 to 99 and averaged 84. In the interdigestive period, the sphincter of the ampulla maintains a basal pressure about 4 mm Hg higher than that within the com-

TABLE 11–4.—COMPARISON OF LIVER AND
GALLBLADDER BILE*

	UNITS	LIVER BILE	GALLBLADDER BILE
Na$^+$	mEq/L	140–159	220– 340
K$^+$	mEq/L	4– 5	6– 14
Ca^{2+}	mEq/L	2– 5	5– 32
Cl$^-$	mEq/L	62–112	1– 10
Bile salts	mEq/L	3– 55	290– 340
pH		7.2–7.7	5.6– 7.4
Cholesterol	mg/100 ml	60–170	350– 930
Pigment	mg/100 ml	50–170	200–1,500

*Compiled in part from data on human bile contained in Waitman A.M., et al.: *Gastroenterology* 56:286, 1969, and Dittmer D.S. (ed.): *Blood and Other Body Fluids.* Washington, D.C., Federation of American Societies for Experimental Biology, 1961, and from data on canine bile contained in Ravdin I.S., et al.: *Am. J. Physiol.* 99:317, 1932, and Wheeler H.O., Ramos O.L.: *J. Clin. Invest.* 39:161, 1960.

mon bile duct or the pancreatic duct. Phasic increases of pressure are superimposed upon the basal pressure. The contractions responsible for the phasic pressure increases are peristaltic, and they cause bile to spurt into the duodenum. During the digestive period, when the gallbladder contracts, the basal pressure within the ampulla falls, but the pumping action of the sphincter increases.

A substance promoting discharge of gallbladder bile is a *cholecystagogue,* and the effect is mediated by two pathways—nervous and humoral. The efferent nerves to both gallbladder and sphincter are in the vagus; their mediator is acetylcholine. Gallbladder contraction is part of the cephalic phase of digestion. Besides being activated by the process of eating, this reflex is subject to influences having emotional concomitants. The drinking of olive oil by an Italian may be promptly followed by gallbladder contraction, whereas the gallbladder of an Irishman, to whom olive oil may be repulsive, may be unaffected by the same stimulus. Afferent nerves from the duodenum and other organs may carry impulses arousing or inhibiting vagally mediated gallbladder and sphincter movements, but such reflexes have not been fully studied.

The gallbladder is under humoral control, for an extrinsically denervated gallbladder contracts in response to chyme in the duodenum. The most effective stimuli are fat in any form, egg yolk, and meat. The hormone released from the duodenal mucosa is cholecystokinin. This hormone has the same C-terminal

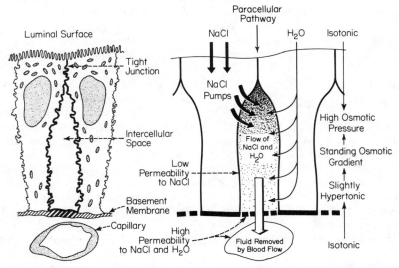

Fig 11–10.—Scheme showing fluid transport by the gallbladder. In the rabbit gallbladder, Na$^+$ and Cl$^-$ pumps may be coupled. In the human gallbladder there may be only a Na$^+$ pump with Cl$^-$ following passively. Water and ions can pass in both directions through the paracellular pathway provided by the tight junctions. (Adapted from Dietschy J.M.: *Gastroenterology* 50:692, 1960. From Davenport H.W.: *A Digest of Digestion,* ed 2. Chicago, Year Book Medical Publishers, Inc., 1978.)

sequence of amino acids (Trp-Met-Asp-Phe-NH$_2$) as gastrin and caerulein. When compared with cholecystokinin on a molar basis, the tetrapeptide itself has 1/143rd the potency of cholecystokinin in stimulating gallbladder contraction. Gastrin I and II are equal and are 1/22nd as effective. Because caerulein is 16 times more effective than cholecystokinin, it has been used clinically as a cholecystagogue.

Cholecystokinin has no effect upon electrolyte and water transport by the gallbladder.

The wall of the gallbladder is richly supplied with nerves containing vasoactive intestinal peptide (VIP). When infused experimentally, VIP relaxes the gallbladder and changes sodium and bicarbonate absorption into secretion into the lumen. The role of VIP in controlling gallbladder function is unknown.

In patients with gluten enteropathy, the release of cholecystokinin is reduced, and bile remains sequestered in the gallbladder. The rate of synthesis of bile acids is normal, but their half-life is prolonged. As a result, the bile acid pool size increases and may become as large as 14 gm.

The gush of gallbladder bile into the duodenum early in gastric emptying raises the concentration of bile salts in duodenal contents to 10–46 mM. Thereafter, as gastric contents empty into the duodenum, bile salts remain nearly constant at 2–7 mM throughout the period of intestinal digestion and absorption. Therefore, bile salt concentration in duodenal contents is always above the critical micellar concentration.

Cholesterol in Bile

Bile is a five-component system of water, phospholipids, bile acids, cholesterol, and bilirubin. Phospholipids, bile acids, and cholesterol are considered in terms of their mole fraction or mole percent in bile. The amount of each component in a sample of bile is determined. The sum of the three components, in moles, is set equal to 1, and the fraction that each particular component contributes to the sum is the mole fraction. The mole percent is 100 times the mole fraction.

Phospholipids and bile acids form micelles in which cholesterol dissolves. The molar ratio of bile acids to lecithin in the micelles is about 3:1. The amount of cholesterol dissolved depends on the amount of lecithin in a micelle, for lecithin provides solvent capacity. The function of the bile acids is to keep lecithin dispersed in micellar form. There are four possible relations between cholesterol concentration in bile and the phospholipid-bile acid micelles.

1. The amount of cholesterol present is not enough to saturate the micelles. Unsaturated bile does not form cholesterol gallstones; it is *nonlithogenic*. Nonlithogenic bile, in fact, dissolves cholesterol gallstones, and to make the liver secrete nonlithogenic bile is the aim of dietary modification of bile composition.

2. Micelles in bile are saturated with cholesterol, but there is no additional cholesterol in the bile sample.

3. Supersaturated bile contains micelles saturated with cholesterol, and, in addition, there is cholesterol in various aggregates in the liquid phase. There are no cholesterol crystals in the bile.

4. If nuclei are present, cholesterol in the liquid phase of supersaturated bile first forms microcrystals of cholesterol monohydrate, and eventually the crystals grow to macroscopic size. Nuclei may be aggregates of cholesterol itself or any of a large number of substances whose nature is not understood.

If the total concentration of phospholipids, bile acids, and cholesterol in a solution is constant, the solubility of cholesterol is a function of the ratio of phospholipids to the sum of phospholipids plus bile acids (Fig 11–11). In addition, the solubility of cholesterol at any one ratio of phospholipids to the sum of phospholipids plus bile acids increases with the total lipid concentration. Figure 11–11 shows the solubility of cholesterol in solutions having four different total lipid concentrations.

Total lipid concentration is important in considering the cholesterol saturation, and therefore the lithogenicity, of bile. Samples of hepatic bile have been found to have total lipid concentrations ranging from 0.2 to 4.2 gm/dl and gallbladder bile concentrations ranging from 8.7 to 24.9 gm/dl. In Figure 11–11 the point representing a cholesterol mole fraction of 0.076 at a PL/(PL + BA) ratio of 0.3 lies above two lines and below two others. If the total lipid concentration had been 10 gm/dl, cholesterol saturation would be at 0.082. The *lithogenic index* of that bile would be 0.076/0.082 or 0.95. The bile would be nonlithogenic. On the other hand, had the total lipid concentration been 5 gm/dl, saturation would occur at a cholesterol mole fraction of 0.072. The lithogenic index would then be 0.076/0.072 or 1.05, and the sample would be lithogenic.

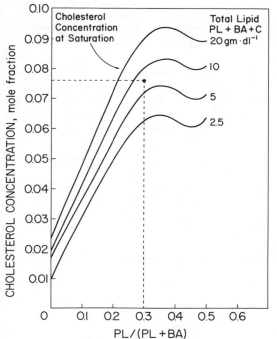

Fig 11–11.—The solubility of cholesterol at 37 C in 0.15N NaCl as a function of the ratio of phospholipid to the sum of phospholipid plus bile acids. Curves are shown for four different total lipid concentrations. (Adapted from Carey M.C.: *J. Lipid Res.* 19:945, 1979.)

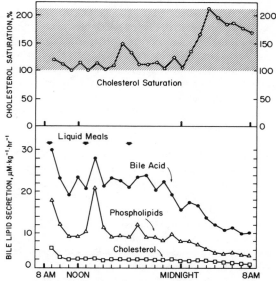

Fig 11–12.—Average rates of secretion of bile acids, phospholipids, and cholesterol in the bile of six patients with cholesterol gallstones. (Adapted from the data of LaRusso N.F., et al.: *Gastroenterology* 69:1301, 1975. Cholesterol saturation was calculated by the method of Thomas P.J., Hofmann A.F.: *Gastroenterology* 65:698, 1973.)

The composition of bile varies diurnally. The rate of secretion of cholesterol is almost constant over 24 hours (Fig 11–12). However, at night when digestion is abated, much of the bile acid pool is sequestered within the gallbladder. The rate of secretion of both phospholipids and bile acids falls off with the result that, even in persons without cholesterol gallstones, bile secreted at night is frequently saturated or supersaturated with cholesterol.

Gallstones

All gallstones have a mixed composition, but they fall into two general classes: pigment stones and cholesterol stones.

The chief component of pigment stones is the calcium salt of unconjugated bilirubin, and on account of their calcium content they are radiopaque. Small amounts of other calcium compounds, apatite, pal-

mitate, carbonate, and phosphate may be present. Pigment stones contain less than 3% bile acids and less than 4% cholesterol.

Calcium in hepatic bile is bound 80% to lipids, and in gallbladder bile it is 50% bound. Bilirubin conjugated with glucuronic acid is soluble, but unconjugated bilirubin precipitates with calcium as a pigment stone. Gallbladder bile of patients with pigment stones has been found to be supersaturated with unconjugated bilirubin, whereas bile of control subjects contained less than 1.2% unconjugated bilirubin. In pigment stone patients, some unconjugated bilirubin is secreted by the liver along with conjugated bilirubin. Conjugated bilirubin is deconjugated in the gallbladder by action of the enzyme β-glucuronidase. The enzyme is released from the wall of a damaged gallbladder, and it is present in high concentration in *Escherichia coli,* which sometimes infects the gallbladder.

Cholesterol stones contain 50%–75% cholesterol in three crystal forms and about 5% of other sterols derived from the diet or cholesterol metabolism. Although precipitated pigments make up less than 1% of the stones, 87% of cholesterol stones have a bilirubin center. On account of their low calcium content, cholesterol stones are radiolucent. Cholesterol stones precipitate from supersaturated bile, and therefore supersaturation of gallbladder bile is necessary but not sufficient for stone formation.

About 50% of the bile secreted by persons without cholesterol stones is supersaturated with cholesterol. Without exception, patients with cholesterol stones secrete supersaturated bile. In nonobese stone patients, saturation occurs, because lipid secretion is reduced, but in obese patients, cholesterol secretion is increased. Both classes of patients have a bile acid pool about half of the normal size both before and after cholecystectomy. Consequently, gallbladder bile is supersaturated with cholesterol in those patients whose genetic or metabolic defect leads to hepatic secretion of supersaturated bile.

Supersaturated bile is an unstable system in which microscopic crystals grow and agglomerate as macroscopic crystals. Crystal growth is slow at low degrees of saturation which occur in normal subjects, but it accelerates exponentially at the high saturation of cholesterol stone patients. Cholesterol crystals gather around a crystal nucleus. In a homogeneous system, cholesterol molecules themselves make up the nuclei. This occurs only in a very supersaturated solution, one whose lithogenic index is greater than 3, and such nucleation is probably unimportant in stone formation in man. Heterogeneous nucleation results in easier crystalization at lower degrees of supersaturation. Nuclei 10–100 nm in diameter are provided by pigments, fatty acids, or carbonate precipitated with calcium, by coagulated mucus and by bacteria. Difference in nucleation may account for difference in stone formation. Supersaturated but crystal-free bile from control subjects has been found, on standing, to wait 15 days before forming crystals, whereas crystal-free bile from patients with stones formed crystals in 3 days. The nature of the nuclei in the latter samples is unknown.

Behavior of the gallbladder contributes to stone formation. More frequent emptying results in a smaller bile acid pool size, which may promote secretion of more saturated bile. When bile is concentrated in a normal gallbladder, the ratio of phospholipids to the sum of phospholipids plus bile acids remains constant, and the increase in concentration of total lipids

actually increases the solubility of cholesterol. In a gallbladder whose mucosa has been damaged, bile acids, but not cholesterol, are absorbed. Cholesterol saturation may then increase.

Obstruction of the bile ducts by stones or tumors results in bile deficiency and its effect on digestion. Absence of bile results in incomplete activation of pancreatic lipase and in reduced absorption of fat. Because the absorption of vitamin K requires bile salts, the prothrombin content of plasma is low and hemostasis becomes inadequate. In the absence of bile pigments, the stool is clay-colored. Although secretion of bile by the liver continues, the high biliary pressure caused by obstruction results in dilatation of the ducts and leakage of the bile back into the circulation. In effect, the liver fails to clear the blood of bile acids and bilirubin, and jaundice results.

Dietary Modification of Bile Composition

If bile composition can be favorably modified by diet, the liver will secrete bile that is less than saturated with cholesterol. Unsaturated bile does not form cholesterol gallstones, and it slowly dissolves cholesterol stones already present in the gallbladder.

Feeding lecithin does not increase the lecithin content of bile. The components of lecithin—glycerol, fatty acids, phosphate, and organic bases—enter their respective metabolic pools and have other fates. Only a very small fraction of the choline fed in lecithin reappears in biliary lecithin. The only effect upon bile of feeding lecithin is to increase the cycling frequency of bile acids. Feeding choline itself does not decrease cholesterol saturation of bile.

When safflower oil or triolein is taken in a large amount, biliary lipid secretion and bile acid pool size increase, but the composition of the bile is unchanged.

Conversion of lithogenic bile to nonlithogenic bile can be accomplished by feeding chenodeoxycholic acid or ursodeoxycholic acid, and when nonlithogenic bile is secreted, cholesterol gallstones dissolve at the rate of about 1mm in diameter per month. The dose of chenodeoxycholic acid is 15–20 mg/kg day, and the dose of ursodeoxycholic acid is 8–10 mg in the same units. When either of those bile acids is fed, the bile acid pool expands, its composition changes (Table 11–5), and bile becomes unsaturated (Table 11–6). The effect of chenodeoxycholic acid in causing secretion of unsaturated bile takes some time, but the

TABLE 11–5.—Average Bile Acid
Composition in Percent of Bile of Ten
Cholesterol Gallstone Patients Fed Either
Chenodeoxycholic Acid or Ursodeoxycholic
Acid in Daily Doses of 12.7–17.9 mg/kg for 3
Months*

BILE ACID FED	NONE	CHENO	URSO
Primary bile acids			
Cholic acid	37.3	5.4	15.7
Chenodeoxycholic acid	40.7	84.3	17.9
Secondary bile acids			
Deoxycholic acid	20.0	2.1	8.0
Lithocholic acid	1.1	2.0	0.6
Lithocholic acid sulfate	0.7	1.6	0.5
Tertiary bile acid			
Ursodeoxycholic acid	1.4	4.6	57.1

*Adapted from Stiehl A., et al.: *Gastroenterology*
75:1016, 1978.

TABLE 11–6.—Average Bile Composition
in Moles Percent of Ten Patients
with Cholesterol Gallstones Fed Either
Chenodeoxycholic Acid or Ursodeoxycholic
Acid in Daily Doses of 12.7–17.9 mg/kg
for 3 Months*

BILE ACID FED	NONE	CHENO	URSO
Cholesterol	10.5	6.2	4.3
Total bile acids	62.7	65.6	67.1
Phospholipids	26.8	28.9	28.7
Cholesterol saturation index	1.4	0.8	0.6

*Adapted from Stiehl A., et al.: *Gastroenterology*
75:1016, 1978.

effect of ursodeoxycholic acid is immediate. Bile becomes unsaturated, because the rate of cholesterol secretion decreases. The means by which chenodeoxycholic acid or ursodeoxycholic acid decreases cholesterol secretion is unknown.

Feeding cholic acid or deoxycholic acid does not convert lithogenic bile to nonlithogenic bile.

REFERENCES

van Berge Henegouwen G.P., Brandt K-H., Eyssen H., et al.: Sulfated and unsulfated bile acids in serum, bile, and urine of patients with cholestasis. *Gut* 17:861, 1976.
Boyer J.L.: New concepts of mechanisms of hepatocyte bile formation. *Physiol. Rev.* 60:303, 1980.
Carey M.C.: Critical tables for calculating the cholesterol saturation of native bile. *J. Lipid Res.* 19:945, 1979.
Carey M.C., Small D.M.: The physical chemistry of cholesterol solubility in bile. *J. Clin. Invest.* 61:998, 1978.
Cowen A.E., Korman M.G., Hofmann A.F., et al.: Metabolism of lithocholate in healthy man. *Gastroenterology* 69:77, 1975.
Diamond J.M.: Transport mechanisms in the gallbladder, in Code C.F. (ed.): *Handbook of Physiology:* Sec. 6. *Alimentary Canal*, vol. 5. Washington, D.C., American Physiological Society, 1968, pp. 2451-2482.
Forker E.L.: Mechanisms of hepatic bile formation. *Ann. Rev. Physiol.* 39:323, 1977.
Gilmore I.T., Hofmann A.F.: Altered drug metabolism and elevated serum bile acids in liver disease: A unified pharmokinetic explanation. *Gastroenterology* 78:177, 1980.
Hallenbeck G.A.: Biliary and pancreatic intraductal pressures, in Code C.F. (ed.): *Handbook of Physiology:* Sec. 6. *Alimentary Canal*, vol. 2. Washington, D.C., American Physiological Society, 1967, pp. 1007-1026.
Hanson R.F., Pries J.M.: Synthesis and enterohepatic circulation of bile salts. *Gastroenterology* 73:611, 1977.
Hofmann A.F.: Functions of bile in the alimentary canal, in Code C.F. (ed.): *Handbook of Physiology:* Sec. 6. *Alimentary Canal*, vol. 5. Washington, D.C., American Physiological Society, 1968, pp. 2507-2533.
Hofmann A.F., Poley J.R.: Role of bile acid malabsorption in pathogenesis of diarrhea and steatorrhea in patients with ileal resection. *Gastroenterology* 62:918, 1972.
Hoffman N.E., Hofmann A.F.: Metabolism of steroid and amino acid moieties of conjugated bile acids in man. V. Equations for the perturbed enterohepatic circulation and their application. *Gastroenterology* 72:141, 1977.
LaMorte W.W., Schoetz D.J. Jr., Birkett D.H., et al.: The role of the gallbladder in the pathogenesis of cholesterol gallstones. *Gastroenterology* 77:580, 1979.
Northfield T.C., LaRusso N.F., Hofmann A.F., et al.: Biliary lipid output during three meals and an overnight fast. *Gut* 16:12, 1975.
Richardson P.D.I., Withrington P.G.: Liver blood flow. I. Intrinsic and nervous control of liver blood flow. *Gastroenterology* 81:159, 1981.
Reuss L.: Mechanisms of sodium and chloride transport by gallbladder epithelium. *Fed. Proc.* 38:2733, 1979.
Shaffer E.A., Small D.M.: Biliary lipid secretion in cholesterol gallstone disease: The effect of cholecystectomy and obesity. *J. Clin. Invest.* 59:828, 1977.
Small D.M.: Cholesterol nucleation and growth in gallstone formation. *N. Engl. J. Med.* 302:1305, 1980.
Spring K.R., Hope A.: Dimensions of cells and lateral intercellular spaces in living *Necturus* gallbladder. *Fed. Proc.* 38:128, 1979.
Weiner I.M., Lack L.: Bile salt absorption; enterohepatic circulation, in Code C.F. (ed.): *Handbook of Physiology:* Sec. 6. *Alimentary Canal*, vol. 3. Washington, D.C., American Physiological Society, 1968, pp. 1439-1456.
Wheeler H.O.: Water and electrolytes in the bile, in Code C.F. (ed.): *Handbook of Physiology:* Sec. 6. *Alimentary Canal*, vol. 5. Washington D.C., American Physiological Society, 1968, pp. 2409-2432.

12

Intestinal Secretion

THE INDISPENSABLE FUNCTION of the small intestine is absorption of water, salts, and foodstuffs. Net amounts of approximately 9 L of water with accompanying salts plus the digestion products of several hundred grams of food move each day across the intestine from the mucosal side to the blood. In the case of water and salts, this net transfer is the resultant of very large flow in the direction of serosal to mucosal side, offset by still larger flow in the opposite direction. Transfer in both directions during absorption is described in chapter 13. In this chapter, only the volume and composition of secretions delivered into the empty bowel will be considered.

Secretory Structure of the Small Intestine

The small intestine is divided functionally into three parts—duodenum, jejunum, and ileum—but the microscopic structures of the mucosa of all three parts are similar. The surface, which is thrown into folds, is covered with finger-like or ridge-shaped villi 0.5–1.5 mm long, at a density of 10–40/mm^2. The surface of the villi is a layer of columnar epithelial cells whose free border is composed of hundreds of minute processes called microvilli. Mucus-containing goblet cells are scattered among the columnar epithelial cells. At the base of the villi, and lined with epithelium continuous with them, are the crypts, simple tubes 0.3–0.5 mm deep. There are three crypts to every villus.

Epithelial cells are formed by mitosis which is restricted to the crypts. After two or more mitoses, daughter cells migrate in columns up a villus as old cells are extruded from the tip (Fig 12–1). Migration is complete earlier in the ileum than in the jejunum, because villi are shorter in the ileum. Extrusion occurs as the result of intercellular vaculation, which detaches a cell from those surrounding it. Although most extrusion occurs at the tip of a villus, it can occur along the whole length and even in the crypts. When a cell is extruded, the gap is not immediately filled by apposition of five or six neighboring cells, and the gap is a site at which large molecules can pass from lumen to blood and at which plasma proteins are shed into the lumen. Shedding increases during absorption of fluid or during protozoal infection or secretion induced by cholera toxin. Under normal circumstances, cells are shed at the rate of 100 million per minute, and in this process 30–50 gm of endogenous protein, together with other cell constituents, is delivered to the lumen every day. The entire epithelium is replaced in 3–6 days. In the steady state, cells are formed in the proliferative zone at the same rate they are shed. Cell production is decreased by x-radiation, methotrexate, folic acid and vitamin B$_{12}$ deficiencies, starvation, and old age. Cell production is increased by presence of nutrients, bile, and pancreatic juice in the lumen; by cortisol administration; by bacterial overgrowth, and by surgical shortening of the bowel.

Cells change as they leave the crypts. They are sensitive to radiation while in the crypts but not after migration to the villi. In the crypts, the cells have low concentrations of phosphatases, esterase, and succinic dehydrogenase, but they acquire the enzymes on leaving; there is a sharp line at the base of the villus where precursor cells lacking phosphatases abut on cells rich in those enzymes.

Other enzymes develop more gradually. Differentiation continues during migration to the apex, and apical cells have the greatest capacity for absorption. Morphological as well as chemical differentiation occurs as the cells migrate. Free ribosomes disappear as endoplasmic reticulum is formed. The brush border becomes a tightly packed array of microvilli, each of which contains a core of 20–30 actin filaments whose basal ends are embedded in the newly formed termi-

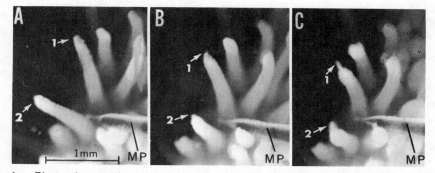

Fig 12–1.—Photomicrographs showing cell extrusion from the tip of a villus of the dog's intestine. **A,** a piece of mucosa has been bathed in physiological salt solution for 10 min. Villi *1* and *2* were separated by a micropipette *(MP).* No cell extrusion occurred. **B,** after 20 min the cell extrusion in villi *1* and *2* became visible *(arrows).* **C,** after 40 min there was a large accumulation of extruded cells at the tips of villi *1* and *2,* and extruded cells are visible at the tips of other villi. Villi *1* and *2* also show contraction. (From Lee J.S.: *Am. J. Physiol.* 217:1528, 1969.)

nal web containing heavy-chain subunits of myosin. As a result, the microvilli can contract. The apical surface of the cell increases 7–8 times, and this change is accompanied by an increase in the concentration of brush-border enzymes. The total protein of the brush border turns over in 14 hours. A fuzzy coat, or glycocalyx, is added to the microvilli. The fuzzy coat is composed of discrete filaments of sulfated glycoprotein aligned in parallel with each other and with the long axis of the microvilli. It is intensely hydrophylic and the site of at least part of the unstirred layer of water.

If the mucosa is destroyed in one spot, epithelialization occurs first by migration of cells from adjacent areas followed by increased rate of cell division in the crypts at the margin of the injured area; later, new crypts form on its floor and villi are regenerated. If the injury penetrates the muscularis mucosae, that layer is not re-formed during repair.

Duodenal Glands

In the duodenum, there is a further anatomical division, for special glands extend from an irregular border at the pyloric ring down the duodenum. In more than 85% of men, the distal border of the area of the glands is marked by islands of glands below the infrapapillary section of the duodenum; and in nearly one third of the men, glandular islands extend into the jejunum. Among other animals, the carnivores have the shortest distribution of these glands,

the omnivores a medium one, and the herbivores the longest.

The duodenal glands consist of branched and coiled tubules arranged in lobules lying in the submucosa. Their ducts pierce the muscularis mucosae to empty into the bottoms or sides of the crypts. In most species, the glands are composed of one type of cuboidal cell, and they stain deeply for mucus.

Duodenal Secretions

The secretion collected from the upper duodenum is a mixture of that from the duodenal glands and the crypts, but the major components appear to come from the duodenal glands. The juice is highly viscid, being similar to egg white, although its mucin content is only 0.5%. The viscosity of the juice declines rapidly on standing, and there is an increase in nonprotein nitrogen, both effects probably being caused by a proteolytic enzyme contained in the juice. Most samples of juice from the upper duodenum contain amylase, peptidases, enterokinase, and other enzymes. Most or all of the enzymes, with the exception of enterokinase, are derived from desquamated cells.

The aqueous portion of the secretion is probably isotonic with the blood. Its mean electrolyte composition is Na^+ 145 (range 136–150), K^+ 6.3 (4.5–8.0), Cl^- 136 (130–140), and HCO_3^- 17 (14–22) mEq/L. Pepsin is present in low concentration.

Rate of secretion by duodenal glands during fasting

is very low, but secretion increases after feeding. It is not known whether the bicarbonate secretion in man is sufficient to make a substantial contribution to neutralization of gastric contents as they pass through the duodenum. The response to food persists after extrinsic denervation. Gastrin, secretin, cholecystokinin, and glucagon, all of which are liberated during digestion of a meal, stimulate duodenal secretion when given in large doses, but whether these or other hormones are responsible for the observed increase in secretion following feeding is unknown. Mucin lubricates the area and is a mechanical barrier holding neutral or alkaline fluid against the mucosal surface. The upper duodenum is more resistant to attack by acid chyme than is the jejunum; if in the dog the duodenum is bypassed surgically so that gastric contents empty directly into the jejunum, ulceration of the jejunal mucosa almost invariably occurs. However, duodenal ulcers in man occur exclusively in the duodenal gland area.

Intestinal Secretion

Most digestive enzymes of the small intestine are integral parts of the membrane of the brush border; they are not secreted. When epithelial cells desquamate, intracellular enzymes as well as the enzymes of the brush border are shed into the lumen, and they are solubilized by bile.

Water and electrolytes are absorbed from the intestinal lumen by processes described in chapter 13. Water and electrolytes also flow from intestinal interstitial fluid into the lumen as the result of hydrostatic pressure differences and active secretion of electrolytes.

If intestinal interstitial fluid pressure exceeds luminal fluid pressure by about 3–5 mm Hg, interstitial fluid containing plasma proteins is filtered across the mucosa into the lumen. Such a rise in interstitial pressure is caused by a rise in portal venous pressure above 25 mm Hg. Increased interstitial pressure appears to open intercellular channels and to cause filtration of fluid through leaky tight junctions. Fluid also escapes through gaps left by desquamated epithelial cells. An excess of luminal pressure over interstitial pressure does not force fluid from the lumen across the intestinal mucosa.

In addition to its ability to absorb by active transport mechanisms, the intestinal mucosa is capable of active secretion. It is probable that most absorption

occurs at the tips of the villi and that active secretion occurs in the crypts. Since cells on the villi are derived from cells in the crypts, intestinal epithelial cells appear to change their transport properties as they migrate up the villi.

Intestinal secretion is chiefly the result of active secretion of anions: chloride and bicarbonate. Sodium follows to maintain electrical neutrality, perhaps through the paracellular pathway, and water follows to maintain isotonicity. Secretion of anions is stimulated by an increase in intracellular cyclic AMP or by entry of ionized calcium. Cholera toxin promotes the former, and 5-HT and acetylcholine promote the latter. Acetylcholine, which stimulates secretion, is delivered to intestinal epithelial cells by ending of nerves of the intrinsic plexuses, and therefore the neural apparatus of the intestine controls secretion as well as motility.

Jejunal secretion occurs during digestion of a meal, and it is probably hormonally mediated. In man, jejunal absorption of water in the interdigestive state has been found to average 0.43 ml/min in a length of 25 cm. Intravenous infusion of a mixture of gastrin, secretin, cholecystokinin, glucagon, and gastric inhibitory polypeptide in concentrations which cause a rise in blood levels similar to those seen during digestion of a meal converted jejunal absorption to secretion at the rate of 0.32 ml/min in the same length of intestine. A burst of secretion accompanies contraction of the small intestine and is suppressed by atropine.

Intestinal Secretion in Cholera

The toxin produced by *Vibrio cholera* during its growth phase stimulates the jejunum and to a lesser extent the ileum of man and animals to secrete massive amounts of fluid. The toxin has no direct effect upon the colon, but because the colon's reabsorptive capacity is overwhelmed, fulminant diarrhea results. Enterotoxins elaborated by noninvasive strains of *E. coli* also stimulate intestinal secretion, and they cause many of the cases of severe, acute diarrhea suffered by travelers.

The fluid secreted by the intestine is protein-free, and its major cations are sodium and potassium. Its concentration of bicarbonate approaches 80 mM, and its concentration of chloride is correspondingly low. Loss of many liters of this fluid results in dehydration and metabolic acidosis. Secretion of aldosterone is stimulated, and reabsorption of sodium and secretion of potassium by the colon are increased. If dehydration and metabolic acidosis can be corrected by adequate IV infusions, the disease is painless and self-limiting. Administration by mouth of electrolyte fluids not containing glucose or amino acids is ineffective, for the administered fluid merely adds to that secreted by the intestine and gushes from the anus.

Cholera toxin is a protein of approximately 84,000 MW, which can be separated into subunits. One fragment, which is biologically inert, binds to a specific ganglioside on the cell surface and perhaps partitions itself into the membrane matrix. The smaller fragment, which is highly hydrophobic, penetrates the membrane and activates adenylcyclase, thereby greatly increasing the cell's content of cyclic AMP. Fluid shed by the mucosa is actively secreted, not filtered. There are no changes in capillary or intestinal permeability or in intestinal hemodynamics which can account for the effects of the toxin. There is increased blood flow through the intestine and a slight increase in capillary filtration coefficient equal to that expected on account of vasodilatation. The toxin produces no cytotoxic effects, and it does not cause desquamation. Consequently, plasma proteins are not lost in greater than normal amounts in the diarrhea fluid. However, the cholera-stimulated mucosa does secrete glycoprotein.

The toxin probably stimulates secretion of bicarbonate and chloride ions. Bicarbonate secretion is actually the result of the secretion of hydroxyl ions, which, upon entering the lumen, combine with ubiquitous carbon dioxide. This is demonstrated by the fact that the partial pressure of carbon dioxide in the luminal fluid falls when bicarbonate secretion is stimulated by the toxin. Sodium ions to maintain electrical neutrality and water to maintain isotonicity follow passively, perhaps through the paracellular pathway. Potassium appears to be actively secreted, for its concentration in the fluid is about 3 times that which can be accounted for by its electrochemical gradient. Reabsorption of sodium may be simultaneously reduced. However, reabsorption of sodium can be strongly stimulated by glucose or amino acids present in the luminal fluid, and consequently water is absorbed as well. Glucose and amino acids may exert this effect by solvent drag rather than by direct coupling of their absorption with that of sodium. A combination of glucose and glycine is more effective than either alone. Administration by mouth of an electrolyte solution containing glucose or amino acids mitigates the diarrhea, and this is enormously important in the treatment of a patient with cholera. Such a solution can easily be prepared and need not be made sterile and pyrogen-free as one given IV should be.

Other Stimuli of Intestinal Secretion

Some non-β-cell tumors of the pancreas contain and secrete hormones that stimulate secretion of bicarbonate-containing fluids by the pancreas, liver, and small intestine, and the resulting diarrheal state is called *pancreatic cholera*. The hormone mainly responsible is vasoactive intestinal peptide (VIP, see chapter 2), and the concentration of VIP in the plasma may be very high. In some instances, secretin or a secretin-like hormone may also be secreted by the tumor, and other hormones including 5-HT have been found in tumor tissue. The tumors do not secrete gastrin, and secretion of acid is not increased. VIP and secretin stimulate secretion of bicarbonate-containing fluid by the pancreas and liver, and the gallbladder of a patient with pancreatic cholera may be grossly distended with dilute bile. However, the volume of fluid passing the duodenal-jejunal junction is normal, and therefore the diarrhea must result from excessive secretion of fluid by the jejunum and perhaps by the ileum. In man, IV infusion of VIP at a rate which elevates the plasma concentration of VIP to that found in pancreatic cholera causes a dose-dependent decrease in water and sodium absorption and secretion of chloride against its electrochemical gradient. Bicarbonate is unaffected. Reabsorption of fluid by the colon is increased as the result of secondary aldosteronism.

Methyl xanthines inhibit phosphodiesterase and thereby cause increased intracellular accumulation of cyclic AMP, which in turn stimulates intestinal secretion. Caffeine in the amount ordinarily taken (75–300 mg) stimulates net intestinal secretion lasting about 15 min. Secretion is correlated with the appearance of caffeine in the jejunum or ileum, not with the blood level. Such stimulation may account for the urge to defecate which frequently follows coffee drinking.

The process of absorption reduces the volume of intestinal contents essentially to zero. A major constituent of the diet of herbivorous animals is cellulose, which is digested by microorganisms in the cecum. As it passes through the small intestine, it contributes nothing to the osmotic pressure of luminal contents. To provide adequate volume, the small intestine of herbivorous animals regularly secretes fluid into the lumen. In man, intestinal secretion is stimulated by oleic and ricinoleic acid, and reabsorption is inhibited by bile acids; in this way, adequate luminal volume is maintained during digestion and absorption of fat.

The Appendix

Among animals having a cecal appendix, only in man, the rabbit, and the chimpanzee does the appendix secrete a significant volume of fluid spontaneously or in response to pilocarpine; the appendix of the dog, red fox, howling gibbon, and tiger does not. If the secretion drains freely, secretion can continue indefinitely; but if the lumen is obstructed, continued secretion results in a rise in intraluminal pressure, which in 4 hours may come to equal systolic blood pressure. Focal ischemic necrosis occurs, followed by rupture, with infection in the scattered contents. Lymphoia tissue of the entire small intestine, the colon, and perhaps the stomach is capable of forming antibodies, but the lymphoid tissue of the appendix appears to be an especially lively site of antibody production.

REFERENCES

Alpers D.H., Kinzie J.L.: Regulation of small intestinal protein metabolism. *Gastroenterology* 64:471, 1973.

Banwell J.G., Sheer H.: Effect of bacterial enterotoxins on the gastrointestinal tract. *Gastroenterology* 65:467, 1973.

Binder H.J. (ed.): *Mechanisms of Intestinal Secretion*. New York, Alan R. Liss, Inc., 1979.

Cooke A.R.: The glands of Brunner, in Code C.F. (ed.): *Handbook of Physiology:* Sec. 6. *Alimentary Canal,* vol. 2. Washington, D.C., American Physiological Society, 1967, pp. 1087-1095.

Eastwood G.L.: Gastrointestinal epithelial renewal. *Gastroenterology* 72:962, 1977.

Field M.: Intestinal secretion. *Gastroenterology* 66:1063, 1974.

Field M., Fordtran J.S., Schultz S.G. (eds): *Secretory Diarrhea*. Washington, D.C., American Physiological Society, 1980.

Gregory R.A.: *Secretory Mechanisms of the Gastrointestinal Tract*. London, Edward Arnold & Co., 1962.

Lipkin M.: Proliferation and differentiation of gastrointestinal cells. *Physiol. Rev.* 53:891, 1973.

Loehry C.A., Creamer B.: Three-dimensional structure of the human small intestinal mucosa in health and disease. *Gut* 10:6, 1969.

Moss J., Vaughan M.:Activation of adenyl cyclase by choleragen. *Ann. Rev. Biochem.* 48:581, 1979.

Potten C.S., Allen T.D.: Ultrastructure of cell loss in intestinal mucosa. *J. Ultrastruct. Res.* 60:272, 1977.

Williamson R.C.N.: Intestinal adaptation. *N. Engl. J. Med.* 298:1393, 1444, 1978.

Sandow M.J., Whitehead R.: The Paneth cell. *Gut* 20:420, 1979.

PART IV

Digestion and Absorption

13

Intestinal Absorption of Water and Electrolytes

THE SMALL INTESTINE is the indispensable site of absorption, and its surface area is very large. The macroscopic mucosal area of six human small intestines obtained at autopsy and handled so as to minimize postmortem distortion was found to be 1.9–2.7 m². Mucosal folds form transverse ridges projecting into the lumen; and because these are less prominent in the distal part of the intestine, the mucosal area per unit length falls off sharply from proximal to distal intestine. Almost half the total mucosal surface occurs in the first quarter of the small intestine. Presence of villi multiply the total absorptive area about 8 times, and microvilli on the absorptive cells increase the area a further 14–24 times, so that the total absorptive area of the small intestine is 200 to 500 m².

In the small intestine, the columnar epithelial cells cover the villi. One or more central arterioles carry arterial blood to the tip of a villus, and near the tip the arterioles break into a rich capillary network that descends the villus just beneath the bases of the epithelial cells. Thus, blood shoots up the villus as in a fountain and then falls back around the central stream. This arrangement permits countercurrent exchange of absorbed substances between the descending network and the ascending vessels. The collecting veins drain into the portal vein; all blood and substances absorbed into the portal vein normally pass through the liver before reaching the systemic circulation. Blood flow through the capillaries of the small intestine is about 1 L/min in the normal man at rest. Flow decreases following sympathetic stimulation, as in exercise; and it increases in the hyperemia that accompanies digestion. Intestinal blood flow is sufficiently fast so that it does not limit the rate of absorption. Every villus also contains a central lacteal emptying into the lymphatic channels, which drain eventually into the left thoracic duct. Total left thoracic duct flow in man is between 100 and 200 ml/hr. The volume of lymph coming from the intestine is not known, but it cannot be more than a few milliliters per minute. Lymph flow increases during absorption, as the result of the pumping action of intestinal contraction and the increase in volume of intestinal interstitial fluid.

Only a few of the absorptive processes are regulated. The intestine indiscriminately absorbs water, the major electrolytes, and the products of digestion of foodstuffs, but absorption of calcium and iron is adjusted to the body's needs. The absorbing capacity of the small intestine is normally far in excess of need, for 50% or more of the intestine can be removed without deleterious effect. Yet in conditions preventing adequate digestion and in some diseases of intestinal mucosa or muscle, the absorption of water, fat, or protein may be so deficient as to endanger life.

Fundamental Properties of the Small Intestinal Mucosa

Three properties of the intestinal mucosa determine its handling of water and electrolytes.

1. *Gradient of permeability:* Permeability of the intestinal mucosa to water and electrolytes is high in the duodenum and jejunum. It decreases almost linearly along the length of the intestine until permeability in the ileum is low. The tight junctions between intestinal epithelial cells are the site of most permeation by water and electrolytes. In the duodenum and jejunum they are "leaky" tight junctions. Water and electrolytes can pass through them in either direction, from mucosa to serosa or from serosa to mucosa, depending upon the effective osmotic gradient for wa-

ter or the electrochemical gradient for an electrolyte. These tight junctions form the "paracellular pathway." In the ileum the tight junctions are "tight," relatively, but not absolutely impermeable to water and electrolytes. Along the length of the intestine, the tight junctions are cation-selective; that is, cations such as Na^+ and K^+ pass through the junctions more readily than do anions such as Cl^- and HCO_3^-. The junctions are not, however, absolutely cation-selective. Divalent ions such as Mg^{2+}, Ca^{2+}, and SO_4^{2-} penetrate the junctions only slightly, if at all, and uncharged molecules above 200 mW are excluded.

2. *Na^+-H^+ exchange:* Throughout the length of the small intestine there is exchange of Na^+ from the lumen for H^+ from the cell interior across the apical membrane of the intestinal epithelial cells. Na^+ is pumped out of a cell across its basolateral border, and the concentration of Na^+ within the cell is relatively low. In addition, the potential difference across the apical membrane is positive outward and negative inward. Consequently, there is an electrochemical gradient driving Na^+ from luminal fluid into the cell. Exit of Na^+ from cell to interstitial fluid is accomplished by an energy-consuming pump in the basolateral membrane. Although a number of H^+ ions is extruded equivalent to the number of Na^+ ions entering the cell, the gradient of H^+ concentration across the apical membrane of the cell is very small; extrusion requires very little energy, and that energy may be derived from the Na^+ gradient. As Na^+ is absorbed, Cl^- follows passively.

When Na^+ is extruded into the lateral space between cells, it may diffuse back into the lumen through the tight junction at the luminal end of the space; whether and to what extent it does diffuse depends upon the leakiness of the junction, the electrochemical gradient for Na^+, and concurrent flow of water which may carry Na^+ by solvent drag.

3. *Cl^--HCO_3^- exchange:* In the ileum, but not in the jejunum, there is equal exchange of Cl^- from luminal fluid for HCO_3^- from the cell interior across the apical membrane of the epithelial cells. Cl^- entering the cell is extruded across the basolateral border into interstitial fluid. Little is known about the energetics of this exchange, and it apparently does not require the intervention of carbonic anhydrase.

Cl^--HCO_3^- exchange may not be confined exclusively to the ileum. HCO_3^- is secreted by the first 2 cm of the bullfrog intestine, a process which disposes of acid arriving from the stomach. In the mammalian duodenum, acid is likewise, in part, neutralized by HCO_3^- from the mucosa, but whether HCO_3^- is actively secreted is not known.

Water Fluxes Across the Intestine

The intestinal mucosa is permeable to water molecules, but the permeability declines from duodenum to ileum. Consequently, water molecules move in both directions across the intestinal epithelium, more readily proximally than distally. Bulk flow of water in either direction occurs along an effective osmotic pressure gradient.*

The net flow of water across the intestine is equal to the difference between the unidirectional flux from intestinal lumen to interstitial fluid (and eventually into blood and lymph) and the opposite flux from interstitial fluid to intestinal lumen. When 40 ml of isotonic NaCl solution containing a trace of D_2O was placed in a 16–18 cm length of dog duodenum, the isotope was found to move from lumen to blood at a rate directly proportional to its concentration in the lumen contents. From the measured rate of D_2O movement, the flux of water from lumen to blood was calculated to be 2 ml/min; and because there was no net volume change, the flux from blood to lumen must also have been 2 ml/min. Similar measurements made using 21- to 25-cm lengths of ileum filled with 40 ml of isotonic NaCl showed that the flux from lumen to blood was 2.2 ml/min, and the flux from blood to lumen was 1.8 ml/min. The net rate of absorption, 0.4 ml/min, was one fifth the rates of the one-way movements of water.

---

*The *effective osmotic pressure difference* depends upon the reflection coefficient of the solute at the membrane. If the membrane is relatively impermeable to the solute responsible for the osmotic pressure of a solution on one side of the membrane—that is, if the solute's reflection coefficient approaches 1—an effective osmotic pressure difference can be established. If the membrane is highly permeable to the solute—that is, if the solute's reflection coefficient at the membrane approaches zero—an effective osmotic pressure difference cannot be created. The osmotic pressure of a mint julep, measured by freezing point depression, is enormously high, but because the reflection coefficient of ethanol at the gastric or intestinal membrane is very low, a mint julep does not establish an effective osmotic pressure gradient and cause water to flow from the gastric or intestinal mucosa to dilute it.

Fluxes across the human intestine are correspondingly rapid. The results obtained when 50 gm of D$_2$O was placed in the upper intestine of normal adult men are shown in Figure 13–1, where they can be compared with similar results showing water flux in the stomach. The D$_2$O was given alone, or it was made isotonic at 154 mN NaCl or hypertonic at 250–1,000 mN NaCl. Fluxes from hypotonic or isotonic solutions were the same; those from hypertonic solutions were slightly slower. Fifty percent of the D$_2$O reached the blood in only 2–3 min, and 95% in 3–11 min. These fluxes are much faster than from the stomach, one reason being the larger mucosal area to which water in the intestine is exposed. When barium sulfate was mixed with the water, the x-ray shadow of barium was distributed throughout most of the small intestine within 2–3 min after the fluid was given. An IV dose of the parasympatholytic drug methantheline bromide (0.3 mg/kg of body weight) caused the water to remain in a shorter segment of the bowel, and flux from lumen to blood was reduced by about half.

When 1,000 ml of water is drunk, about 50% is absorbed in 30 min. Since 50% of isotopic water enters the blood in 2–3 min, flux from lumen to blood is about 10 times as fast as net absorption. This means that during absorption there must be a flux of water from blood to lumen only slightly slower than that in the opposite direction. Net absorption is there-fore the difference between two very large, oppositely oriented fluxes; a small change in either individual flux can cause a large change in net flow.

Net Water Absorption

Between 5 and 10 L of water, derived from food and drink and from salivary, gastric, pancreatic, biliary, and intestinal secretions, enters the small intestine of the normal adult man in a day. Only about 1 L leaves the intestine for the colon; therefore, the small intestine absorbs water at an average rate of 200–400 ml/hr. This is the minimum rate. The maximum rate, if any upper limit exists, is imposed not by any property of the intestinal mucosa but by the body's ability to bear a water load. The steady state of maximal excretion of water in the urine when water is drunk at quarter-hour intervals is about 1% of the body weight per hour. For a 70-kg man, this equals absorption at the rate of 700 ml/hr. Absorption of a single drink of 1 L of water occurs within an hour; and because the maximal dilution of plasma during absorption occurs 30–60 min after the drink is taken, the peak rate of absorption must be greater than 1 L per hour. In the cat, rat, and rabbit, water can be absorbed so fast that intravascular hemolysis occurs; as much as 3 gm/100 ml of hemoglobin has been found in portal venous plasma during water ab-

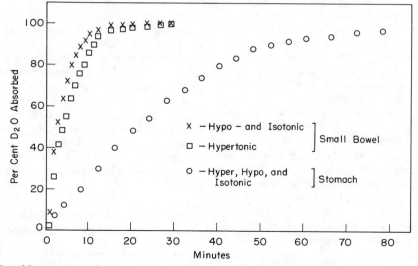

Fig 13–1.—Mean rates of absorption of D$_2$O from small intestine of normal fasting human sub-jects. (Adapted from Lee P.R., et al.: *Gastroenterology* 29:1008, 1955.)

sorption in these animals. A dehydrated camel can drink water equal to 30%–33% of its body weight in 10 min.

In the experimental process of whole intestinal perfusion, a physiological salt solution is pumped through a tube into the stomach of a human subject at the rate of 75 ml/min. After an hour a steady state is reached in which 200–300 ml is passed through the anus at 3- to 4-min intervals. All food particles, fecal masses, and 99% of colonic bacteria are flushed out. Water is absorbed from the perfusate at the rate of 600 ml/hr.

Water is absorbed throughout the small intestine, but the chief locus of absorption following a meal is the upper part of the tract. When the unabsorbable substance polyethylene glycol is added to a meal, its concentration at various parts of the digestive tract shows how much the meal has been diluted or concentrated in its passage through the gut. After a meal of steak, bread and butter, salad, and tea whose initial volume is 645 ml and which contains a known amount of polyethylene glycol has been eaten, the concentration of polyethylene glycol in samples recovered from the upper duodenum is 20%–40% of its initial value. The meal, as it enters the duodenum, has been diluted by several times its volume of gastric secretions. The volume passing the midduodenum is 1,500 ml; that in the proximal jejunum is 750 ml; and that in the lower small bowel is 250 ml. Since biliary and pancreatic secretions of about 1,000 ml have been added in the duodenum, their volume together with water contributed by the stomach has been absorbed. The net movement of water in the duodenum is in the direction that tends to keep its contents isotonic, and this large absorption of water is the result of absorption of osmotically active components of the meal, chiefly glucose and amino acids, and the neutralization of acid, with consequent reduction of osmotic pressure. Absorption of 55 gm of glucose, or 306 mOsm, would allow 1,000 ml of water to be absorbed. Addition of 1 part of isotonic pancreatic juice to 1 part of isotonic HCl gives, roughly, 1 part of isotonic NaCl and 1 part of free water, which must be absorbed if the duodenal contents are to remain isotonic. By the time the intestinal contents have reached the ileum, the concentration of polyethylene glycol is greater than that in the original meal, showing that all the water added by secretion, as well as some ingested, has been absorbed.

When water containing solutes flows through aqueous channels in a membrane, frictional forces act on the solute; and the solute, especially if it is uncharged, tends to move with the water. This phenomenon is called solvent drag, and it accounts for a fraction of the absorption of urea and other small nonelectrolytes.

Most or all water, and solutes as well, are absorbed by the cells at the tips of the villi.

Intestinal Lymph Flow

Lymph flow from the intestine increases as water is absorbed. The greatest flow occurs during absorption of isotonic NaCl. The percentage of absorbed water transported through lymph channels increases with increasing venous pressure; when venous pressure is below 5 mm Hg, lymph accounts for 10%–25% of water transport; but when venous pressure is greater than 20 mm Hg, the rate of lymph flow equals the rate of water absorption. Absorption through lymph channels is assisted by rhythmic contraction of mesenteric lymph ducts and by contraction of the villi.

The lymph flowing in the large ducts is not the actual water and electrolytes absorbed from the intestinal lumen. The reason for this is that water and solute molecules rapidly exchange among capillary blood, interstitial fluid, and lymph. If water being absorbed is labeled with tritium, 97% of the tritium is recovered in portal blood and only 3% in lymph. Only fat droplets, once in lymph capillaries, cannot escape.

The concentration of plasma proteins in gastrointestinal interstitial fluid is high, and, consequently, their concentration in intestinal lymph is high at low lymph flow rates. Their concentration in lymph falls when lymph flow increases as the result of water absorption. Lymph is the major route of transport of IgA synthesized in the intestinal mucosa. Physiologically insignificant amounts of digestive hormones and modest amounts of digestive enzymes occur in intestinal lymph, but in acute pancreatitis, intestinal lymph carries large quantities of digestive enzymes and toxic products.

Osmotic Pressure and Water Movement; The Dumping Syndrome

Water flows in either direction across the intestinal mucosa along an effective osmotic pressure gradient. Under normal circumstances, the duodenum and upper jejunum are the major sites of osmotic flow of water. Gastric contents emptied into the duodenum may have any osmotic pressure. They may be hypotonic if water is drunk, or they may be hypertonic. Hydrolytic action of pancreatic enzymes upon protein and starch rapidly produces a large number of small molecules from a few large ones. If the rate of absorption is slower than the rate at which osmotic equilibrium is reached, the volume of fluid in the intestinal lumen increases.

Solutions in the duodenum and upper jejunum rapidly come into osmotic equilibrium with blood, for the mucosa of the upper small intestine is highly permeable to water and electrolytes. Chyme remains isotonic as it moves down the rest of the small intestine. As osmotically active particles, electrolytes, or the products of digestion, are absorbed, an equivalent amount of water is likewise absorbed.

If the solute responsible for the osmotic pressure of intestinal contents cannot be absorbed, sufficient water to dilute it to isotonicity must remain in the lumen. The ions Mg^{2+} and $SO_4^=$ are slowly absorbed, and their presence in the lumen prevents water absorption. The usual dose of 15 gm of Epsom salts ($MgSO_4 \cdot 7H_2O$) retains between 300 and 400 ml of water; this is the basis of the action of this saline cathartic.

It is important that hypertonic solutions become isotonic before they reach the jejunum, for the presence of hypertonic solutions in the jejunum causes nausea, a sense of epigastric fullness with pain, pallor, sweating, vertigo, and eventually fainting. This concatenation of signs and symptoms following meals in persons who have been subjected to partial gastrectomy or gastroenterostomy is known as the dumping syndrome, and it results from the rapid arrival of hypertonic fluid in the jejunum. Its physiological expression is very similar to the vasovagal syndrome that may follow venisection or sudden emotional shock: great vasodilatation in the muscles and moderate vasodilatation elsewhere, leading to a profound fall in arterial blood pressure and fainting. Great variability in the dumping syndrome among patients or in one patient from time to time is probably related as much to differences in cardiovascular responses as it is to variations in the cause.

The dumping syndrome may be produced in normal persons by intrajejunal instillation of 250 ml of hypertonic (about 2,000 mOsm) glucose, fructose, or amino acids. All of these solutions, when placed in the jejunum, increase rapidly in volume; and the fluid entering the intestine is, of course, drawn from extracellular fluid and plasma. If a patient subject to dumping is fed 130–200 ml of a hypertonic meal, his plasma volume begins to fall within 10 min and has fallen 20%–30% in 30–40 min. Recovery occurs in 100–120 min. An initially hypotonic diet containing starch has the same effect, for intestinal starch hydrolysis is so rapid that osmotically active oligosaccharides are released within a few minutes. Although distention of the jejunum does not in itself cause a fall in plasma volume, it induces the sense of fullness, nausea, and hypermotility leading to diarrhea.

The Unstirred Layer

Bulk contents of the lumen of the intestine are mixed by segmentation and peristalsis, and water and solutes are brought to the surface of the mucosa by convection. However, the fluid trapped between the microvilli and within the glycocalyx is stationary, and a series of thin layers, each progressively more stirred, extends from the surface of the epithelial cells to the bulk phase in the lumen. Together, these are the *unstirred layer,* whose effective thickness ranges from 0.01 to 1 mm.

Because there is no convection in the unstirred layer, molecules move within it only by diffusion. The rate at which a substance diffuses is a function of the concentration gradient down which it diffuses and of its diffusion coefficient. The coefficient is, among other things, roughly inversely proportional to the square root of the molecular weight below 450 and to the cube root above that weight. Therefore, during absorption two processes are in series: diffusion from the bulk phase through the unstirred layer to the cell membrane and translation through the cell membrane to the cell's interior. The latter process may be passive or active. If the substance being absorbed is fat soluble, or if its diffusion through the membrane is facilitated, its translation into the cell may be so rapid that the substance's concentration at the bottom of the unstirred layer is nearly zero. Then diffusion through the unstirred layer is the rate-limiting step in absorption.

Absorption of Sodium

Net absorption of sodium results from the difference between two opposing unidirectional fluxes.

Throughout the length of the small intestine, sodium is actively absorbed by the process of Na^+-H^+ exchange, which has been described above.

Sodium also passes through tight junctions, and it can pass in either direction. Its ability to pass decreases from proximal to distal intestine.

Passage of sodium through permeable tight junctions is governed by fluid movement and by sodium's electrochemical gradient. If there is substantial bulk flow of water from lumen to intercellular spaces through tight junctions, sodium (and chloride) is pulled along by solvent drag. If, on the other hand, there is a large potential difference across the mucosa with lumen negative with respect to serosa, sodium moves from intercellular spaces through tight junctions into the lumen. This occurs when active sodium absorption is stimulated by glucose, and a potential difference of about 30 mV is established. Then flow of sodium into the lumen through the tight junctions along sodium's electrochemical gradient equals active absorption of sodium, and net sodium flux is, in effect, independent of glucose stimulation.

The pioneer work demonstrating bidirectional fluxes of sodium and differences between segments of the gut is shown in Figure 13–2. Isotonic solutions of NaCl containing a trace of radioactive sodium ($^{22}Na^+$ or $^{24}Na^+$) were placed in loops of the intestine of chronically prepared dogs. The data for flux from lumen to blood are shown on the left of the figure. Sodium marked with an asterisk (Na*) is the labeled

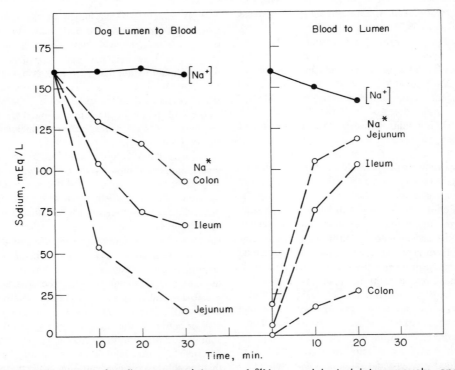

Fig 13–2.—Movement of sodium across intestinal mucosa of unanesthetized dogs with surgically prepared, chronic loops of jejunum, ileum, and colon. **Left,** intestinal loops were filled with isotonic NaCl containing a trace amount of ^{24}Na. The total sodium concentration [Na$^+$] remained constant, but the labeled sodium Na* (see text) originally present in the loop rapidly disappeared. **Right,** intestinal loops were filled with approximately isotonic Na$_2$SO$_4$ solutions. A trace amount of ^{24}Na was injected intravenously, and its appearance in the contents of the luman observed. Total sodium concentration fell, but the sodium in the lumen of the loop was rapidly replaced by labeled sodium from the blood. Movement of labeled sodium in both directions is in the following order: jejunum > ileum > colon. (Adapted from Visscher M.B., et al.: *Am. J. Physiol.* 141:488, 1944.)

sodium concentration, which at zero time is the total sodium concentration. In 30 min, there was no change in the total NaC1 concentration in the lumen, but the labeled sodium disappeared rapidly; at the end of the period, only 10% of the sodium originally present in the loop was still there. Since the total concentration of sodium was constant, a quantity of sodium equal to that leaving the lumen, minus the net amount of sodium absorbed, must have entered it from the blood. The exchange of sodium was slower in the ileum and still slower in the colon. Movement in the opposite direction was measured by placing unlabeled sodium in the lumen, injecting a tract of $^{24}Na^+$ into the blood and following the appearance of the isotope in the contents of the lumen. The results given in Figure 13–2 (right) show that, although there was a slight fall in total sodium concentration, which indicates net absorption of sodium, there was rapid movement of labeled sodium from blood to lumen.·

As the result of active transport, solvent drag, and leak through tight junctions, the concentration of sodium in human jejunal fluid remains at about 135 mM during digestion. In contrast, active absorption of sodium in the ileum can reduce sodium's concentration in ileal fluid below 100 mN. This is because the relative impermeability of ileal tight junctions restricts flux of sodium from interstitial fluid to lumen. For the same reason, solvent drag plays little part in ileal absorption of sodium. In the ileum of men and dogs, solute absorption can be faster than water absorption, for when the ileum is perfused at a constant rate with isotonic NaC1, the fluid recovered from the lumen has a lower osmolality than that instilled. Under normal circumstances, ileal contents remain in the lumen long enough for osmotic equilibration to occur.

The minimal concentration of sodium reached in the ileal contents of severely salt-depleted human subjects is between 75 and 95 mEq/L, and the quantity of sodium delivered per day by the ileum to the colon far exceeds sodium intake. During salt depletion, mineralocorticoid secretion by the adrenal cortex increases, and urinary excretion of sodium falls. However, the small intestine exhibits little or no adaption to sodium load. The variations in sodium absorption brought about by changes in mineralocorticoid levels, if they exist, are very small. In contrast, sodium absorption by the colon definitely increases when aldosterone rises as the result of salt depletion or administration of the hormone.

Countercurrent Flow and Sodium Concentration

Arterial blood is carried up a villus in 1–3 arterioles, which run unbranched toward the tip along the lamina propria of the villus. On their way they lose their smooth muscle coat. At the villous tip the arterioles branch into a dense capillary network, which descends surrounding the central vessels (see Fig 5–9). The capillaries of the network eventually unite into one or a few venules. This arrangement permits countercurrent mulitplication of diffusible substances absorbed by cells at the tip of the villus.

In experiments on anesthetized cats, the intestine has been quickly frozen in liquid nitrogen. While the tissue was held at very low temperature, the mucosa was sectioned and placed upon a microscope slide. The temperature of the slide was slowly raised, and a villus was seen to thaw first at its tip. From the freezing point depression, the osmolality of the tip, and then of the rest of the villus, could be calculated.

When the lumen of the intestine was perfused with physiological salt solution containing sodium and glucose, and the blood flow was at its resting value of 28ml/min·100 gm, osmolality at the tip of jejunal villi was found to be greater than 1,100 mOsm/kg H_2O (Fig 13–3). Osmolality fell off sharply toward the base of the villus.

Countercurrent multiplication depends upon appropriate balance between blood flow and solute absorption. When blood flow was increased to 146 ml/min·100 gm, the gradient of osmolality was partially washed out, and it almost disappeared when blood flow was stopped. These relations between blood flow and hyperosmolality are the same as those found in renal papillae.

The gradient is reduced if glucose is omitted from the solution perfusing the lumen, and it is almost completely abolished if sodium is omitted. The osmotic gradient established by the countercurrent multiplier facilitates water absorption.

Absorption of Potassium

As with sodium, the net movement of potassium across the intestinal mucosa is the difference between two opposed unidirectional fluxes. Potassium fluxes are very much smaller than sodium fluxes, and most or all of the fluxes are through the tight junctions. Net movement throughout the small intestine and colon occurs only down the ion's electrochemical gradient;

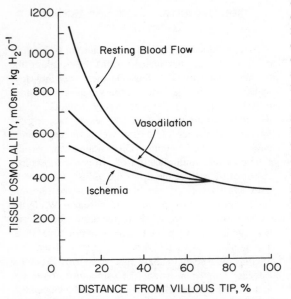

Fig 13–3.—Tissue osmolality in the villi of jejunal segments of the cat intestine during three different circulatory conditions as a function of distance from the villous tip. Vasodilatation was induced by close intra-arterial infusion of isopropyl norepinephrine, and ischemia was induced by occlusion of the mesenteric artery. (Adapted from Jodal M., et al.: *Acta Physiol. Scand.* 102:94, 1978.)

there is no evidence of active transport. With a plasma concentration of 4.8 mEq/L and a potential difference of 5 mV, mucosa negative, the equilibrium concentration in the lumen is 6 mEq/L, and absorption occurs when the luminal concentration is 2 mEq/L higher than the plasma concentration. When the potential difference is increased, as it may be to 30 mV during glucose absorption, the potassium equilibrium concentration is much higher. Then there is net movement of potassium from blood to lumen; potassium appears to be secreted, although it only moves toward its equilibrium concentration.

Chloride and Bicarbonate

The fact that a potential difference is generated during Na^+ absorption means that charge separation occurs. Because the value of the faraday (the charge carried by one gram equivalent) is so large and the distance that the charges are separated is so small, only a very small amount of Na^+ need be separated from anions to produce the potential difference found during Na^+ absorption. In the jejunum, most of the charge on absorbed Na^+ is balanced by (1) passive absorption of Cl^-, (2) extrusion of H^+ in exchange for Na^+, and (3) absorption of HCO_3^-.

During active Na^+ absorption, the electrochemical gradient for Cl^- is in the direction of absorption, and the passive flow of Cl^- down its electrochemical gradient accounts for the bulk of Cl^- absorption in the ileum. How much travels through the cells and how much by the paracellular pathway is unknown. In its passive absorption of Cl^-, the intestine does not distinguish among Cl^-, Br^-, I^-, SCN^-, and NO_3^-; all are absorbed at approximately equal rates.

H^+ extruded into the lumen in exchange for Na^+ combines with HCO_3^- to give H_2CO_3. Carbonic acid is dehydrated to CO_2, and the partial pressure of CO_2 in the ileal fluid is high. These reactions result in disappearance of HCO_3^-.

When H^+ is extruded across the apical membrane, an equivalent amount of HCO_3^- is exchanged across the basolateral membrane. Thus, H^+ secretion does effect the apparent reabsorption of HCO_3^-. In the intestine, as in the pancreas and kidney, which also appear to transport HCO_3^-, it is difficult to determine whether HCO_3^- ions are being transported as such or whether apparent transport is the result of H^+ secretion. Nevertheless, there is some evidence that HCO_3^- is actively reabsorbed as such in the human jejunum.

Jejunal contents recovered from the human intestine usually have HCO_3^- concentration of 6–20 mM and a pH of 6.6–6.8. Cl^- concentration is about 120 mM. The sum of HCO_3^- and Cl^- does not equal the sum of Na^+ (135 mM) and K^+ (8mM); the difference, or anion gap, is filled largely by bile salts.

In the ileum, in addition to Na^+-H^+ exchange at the apical border of the epithelial cells, there is exchange of Cl^- and HCO_3^-. Cl^- can be actively absorbed against its electrochemical gradient. If an isotonic solution of $NaCl$ and Na_2SO_4 is placed in a dog's ileum, the presence of the slowly absorbed $SO_4^=$ keeps the Na^+ concentration in the luminal fluid approximately constant, but in 90 min the concentration of Cl^- falls to less than 1 mEq/L. Secretion of HCO_3^- in exchange for Cl^- raises the HCO_3^- concentration in ileal fluid. A typical value is 40 mEq/L, and the pH is correspondingly high, being about 7.8.

Water and electrolytes not absorbed are passed on to the colon. Data on their fate in human subjects are given in chapter 17.

Effects of Obstruction

Obstruction of the small bowel is rapidly followed by distention above the point of obstruction. Distention reduces blood flow through the segment, and this may result in necrosis and rupture. Only a small amount of the distention is caused by the accumulation of fluid delivered from above, for distention inhibits gastric and intestinal motility. Most of the fluid comes from the distended part itself.

In the unobstructed intestine, insorption, the flux of water from lumen to blood, is usually greater than exsorption, the flux of water from blood to lumen. As pressure rises in an obstructed segment, insorption decreases, and exsorption increases. At an intraluminal pressure of 20 mm Hg, insorption is only about one third of exsorption, and at a pressure of 30 mm Hg insorption is almost abolished, whereas exsorption is greatly increased. Fluxes of electrolytes parallel those of water, with the result that the obstructed segment of bowel is progressively distended with isotonic fluid secreted by its own mucosa. Sequestration of this fluid in the lumen reduces plasma and interstitial fluid volume, sometimes as much as 35%, with consequent hemoconcentration, renal insufficiency, vascular collapse, and death.

Other sequelae are anorexia and vomiting, which exacerbate the fluid loss. These are mediated by afferent nerves, and they occur in otherwise normal dogs following distention of blind pouches made from the jejunum but not in dogs whose jejunal pouches have been denervated. Bilateral splanchnicotomy and excision of the lumbar chain prevents the response of vomiting; but, in addition, bilateral vagotomy is required to abolish anorexia.

Other Factors Influencing Fluxes Across the Intestine

Pentagastrin infused IV into dogs in a dose submaximal for acid secretion causes a 15%–20% reduction in the rate of absorption of water by both jejunum and ileum. A similar reduction follows release of endogenous gastrin. Intravenous infusion of secretin or cholecystokinin reduces the absorption of sodium, potassium, and chloride in the most proximal part of the human jejunum. VIP and prostaglandin E_1 induce net secretion, probably by first raising intracellular concentration of cyclic AMP. Serotonin and calcitonin induce secretion, not by raising intracellular cyclic AMP but by increasing influx of calcium. Acetylcholine, released from nerve endings running from the intrinsic plexuses to the epithelium, likewise stimulates secretion.

Free or conjugated dihydroxy bile acids (deoxycholate, chenodeoxycholate, and their conjugates) and a free trihydroxy bile acid (cholic) at concentrations of 3–5 mM inhibit net water and electrolyte absorption, and at higher concentration, up to 10 mM, stimulate net secretion. Other compounds normally present in the intestinal lumen during digestion also inhibit water and electrolyte absorption. These include lysolecithin and oleic acid. If the effects of hormones, bile acids, and fatty acids have any role in the normal digestive process, it may be to maintain fluidity in the lower small intestine where, otherwise, absorption would dry the lumen.

Ricinoleic acid, the dihydroxy fatty acid which is the active ingredient of castor oil, inhibits absorption of water, electrolytes, glucose, and amino acids and thereby produces net secretion. This, and not a supposed "irritant" effect upon intestinal and colonic muscle, is the basis of ricinoleic acid's effectiveness as a cathartic.

Absorption of Calcium, Magnesium, and Phosphate

Calcium is ingested in food and drink of an ordinary diet in the amount of 1,000 mg/day. Calcium also enters the gut as calcium secreted by intestinal cells, as calcium contained in digestive secretions and as calcium contained in desquamated cells. During active absorption of glucose, intercellular spaces become swollen, and calcium then moves through the tight junctions into the lumen. Total calcium entering the lumen by these means amounts to approximately 600 mg/day. Consequently, 1,600 mg is available for absorption, and about 700 mg is absorbed, leaving 900 mg to be excreted in the stool. In this example, net absorption is only 100 mg, although 700 mg has actually been absorbed. If the amount ingested is decreased, or if the amount excreted in the stool is increased, negative calcium balance can occur. Calcium is absorbed throughout the small intestine and in the colon as well, but most absorption occurs in the ileum.

In the human small intestine, calcium is actively absorbed when its concentration in the lumen is 1–5 mM. At the highest concentration, the absorbing mechanism is saturated, and at concentrations above 5 mM additional absorption is by passive diffusion.

Calcium is absorbed against a concentration gradient by a two-step active transport mechanism, which uses high energy phosphate. Calcium is bound to a specific calcium-binding protein whose synthesis is controlled by the physiologically active form of vitamin D: 1α,25-dihydroxyvitamin D. Calcium-binding protein is synthesized and secreted into the intestinal lumen by mucus-secreting goblet cells. In the lumen, the protein binds calcium, and the complex is then absorbed through the microvillar membrane into the epithelial cells. Calcium is sequestered within the cells, either by the calcium-binding protein or by cell organelles, because it is important for the function of cells that the concentration of free calcium in the cytosol be maintained at about 10^{-6} M. Calcium is then extruded, by diffusion or by means of a carrier, across the basolateral border of the cells into interstitial fluid. Parathyroid hormone has a small effect upon calcium absorption by increasing the rate of release of calcium across the basolateral membrane.

The net absorption of calcium is a function of dietary calcium, and it is facultatively regulated to meet the body's needs. Calcium absorption in man is zero or negative when the daily calcium intake is 0.1 mM (4 mg) per kg of body weight, and it rises linearly to a maximum of about 3 mM (120 mg) per kg a day when dietary intake is high. Rate of absorption is regulated in part by plasma calcium concentration, and IV infusion of calcium immediately suppresses absorption from the duodenum and jejunum of a normal subject. Reproductive status and age also affect calcium absorption. Female animals absorb calcium more readily when they are pregnant or lactating than when they are not. Persons over 65 years of age have decreased calcium absorption as the result of inadequate conversion of 25-hydroxyvitamin D to 1α,25-dihydroxyvitamin D.

Bile acids enhance calcium absorption in addition to their effect on the absorption of vitamin D. In bile acid deficiency, unabsorbed free fatty acids are excreted as insoluble calcium salts; this results in negative calcium balance and osteomalacia. A patient whose terminal ileum has been removed will often have hyperoxaluria and renal calcium oxalate stones. Such a patient absorbs dietary oxalate 5 times as readily as does a normal subject. The reason is that removal of the terminal ileum reduces bile acid reabsorption, decreases the size of the bile acid pool, and increases excretion of calcium as calcium soaps. Consequently, there is less free calcium in the intestinal lumen to precipitate oxalic acid as an unabsorbable calcium salt. In addition, dihydroxy bile acids increase the permeability of the colon for oxalate.

The average daily diet contains about 10 mM (243 mg) of magnesium. In normal human subjects, net absorption of magnesium is 50% of intake, and it is entirely independent of calcium. When the magnesium contained in a breakfast was tagged with radioactive magnesium (^{28}Mg), absorption was found to begin after an hour, to continue for 8–12 hours, and then to stop, a pattern consistent with absorption through the whole of the small intestine. In perfusion studies, magnesium has been found to be absorbed equally well by the human jejunum and ileum. The rate of absorption rose linearly with increasing concentration of magnesium in the perfusate from zero to 10 mM, but there was a tendency for absorption to level off at concentrations from 10 to 20 mM. Perfusion studies on normal children have given data suggesting that there is a saturable mechanism upon which is superimposed a diffusional component at high luminal magnesium concentrations. That constitutes the evidence that magnesium might be absorbed by an active process in man. It is not known whether magnesium can be absorbed against its electrochemical gradient, nor is it known whether absorption increases in a patient with magnesium deficiency. Absorption is grossly reduced in a patient with chronic renal failure, and primary hypomagnesiumemia is a rare disease of infancy, carried as an autosomal recessive, which is expressed as net malabsorption of magnesium.

About 1 mM, or 0.8% of the body's content of magnesium, appears in the intestinal lumen each day as the result of secretion or desquamation.

Phosphate is absorbed by all segments of the small intestine, and absorption occurs both by passive diffusion and active transport. Concurrent absorption of glucose has no effect on that of phosphate, and phosphorylation of sugar is not a necessary step in the absorption of phosphate.

Absorption of Iron and Copper

A carnivorous man eats 15–20 mg of iron in his daily diet; almost all the iron is contained in hemoglobin and myoglobin. Of this he absorbs 0.5–1 mg. A woman subject to loss of iron by menstrual bleeding absorbs 1–1.5 mg/day. In absence of hemorrhage, loss of iron in deciduous cells, chiefly those of the intestinal mucosa, equals the rate of absorption, and the body's load of iron remains constant at approximately 4 gm. The absolute amount of iron absorbed increases with the dose (Fig 13–4) over the short run, but the fraction of the dose absorbed decreases. At the dose of 100 mg (the largest shown in Fig 13–4), 12.6 mg was absorbed. In a normal person the amount absorbed is, over the long run, related to the amount required, for prolonged overloading depresses absorption. Patients with iron deficiency absorb 2–5 times more of a given dose than do normal subjects. After acute bleeding, the iron requirement temporarily increases, but increased absorption does not begin until 3–4 days after hemorrhage.

In man, most iron is absorbed in the heme of hemoglobin and myoglobin, which have been split from their proteins in the lumen. Iron is released within the epithelial cells by heme oxidase, which catalyzes the conversion of heme to bilirubin, carbon monoxide, and free iron. Of inorganic iron, ferrous is more rapidly absorbed than ferric. Inorganic iron binds to receptors on the brush border and is absorbed by a saturable active process. Cobalt and manganese are absorbed by the same process, and any one of the metals competitively inhibits absorption of the other two. The mixture of foodstuffs in the intestinal lumen has more influence on absorption of inorganic iron than does the form in which iron is fed. Phosphoproteins and insoluble calcium phosphate bind iron, and ferric iron forms an insoluble, unabsorbable ferric phosphate. By reducing iron, ascorbic acid promotes absorption, and at low pH it forms a chelate with ferrous iron, which remains soluble and available for absorption at the pH prevailing in the small intestine. Neutralized gastric juice mixed with iron salts or with heme promotes absorption; the factor responsible is absent from the gastric juice of patients with pernicious anemia. Salicylates (e.g., aspirin) chelate iron, but they do not reduce absorption. Iron chelated with ethylenediaminetetraacetate is rapidly absorbed, but it is lost equally rapidly by renal excretion. Iron in wine is absorbed no more readily than iron in water, alas!

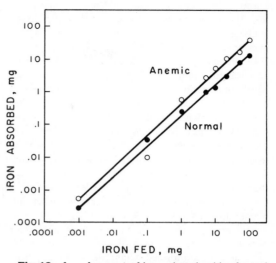

Fig 13–4.—Amount of iron absorbed by fasted normal and anemic human subjects following a single oral dose of iron given with ascorbic acid. The normal subjects had 12.3–16 gm of hemoglobin per 100 ml of blood and 51–190 μg iron per 100 ml of plasma. The anemic subjects had 5–12 gm hemoglobin per 100 ml of blood and 10–25 μg iron per 100 ml of plasma. (Plotted from data of Smith M.D., Pannacciulli I.M.: *Br. J. Haematol.* 4:428, 1958.)

The high iron content of the beer manufactured by the Bantus is said to account for the siderosis prevalent in those tribes.

The duodenum and jejunum are the sites of iron absorption; only trivial amounts are absorbed in the stomach, lower small intestine, and colon. When iron is placed in the duodenum, the initial rate of uptake by mucosal cells is rapid, but the rate falls off to zero in 30–60 min. The ability of mucosal cells to absorb iron is determined by their iron content; and cells that have recently absorbed iron but have not yet delivered it to the plasma temporarily lose their ability to absorb more.

Absorption of heme and inorganic iron into epithelial cell relatively rapid, but exit into the blood is slow. Within the cells, iron split from heme and absorbed inorganic iron mix with the intracellular pool of iron. Consequently, iron accumulates within the cells, and accumulation is a factor inhibiting further absorption. Iron exists within the cells in at least two forms, one diffusely distributed throughout the cytoplasm. This iron is probably chelated with amino acids. The second is as iron hydroxide micelles in ferritin molecules. Ninety percent of freshly absorbed iron enters ferritin. Some persons think that iron is only temporarily stored in ferritin; others believe it is sequestered within ferritin for the life of the cell. Much of the absorbed iron is actively transported from cell to plasma in 2–4 hours. Transfer of iron from cell to plasma is regulated by the body's need for iron. If the need is high, as during pregnancy or recovery from hemorrhage, a large proportion of the iron contained in the mucosal cells moves to plasma, where it is carried by the globulin transferrin; when the need is low, a small portion leaves the cells. Because of the short life span of the mucosal cells, iron remaining in them is soon lost in the feces.

The iron content of mucosal cells is also determined by the plasma level of iron prevailing at the time the cells are formed within the crypts. If plasma iron is high, the content of the newly formed cells is likewise high; and when these cells have migrated up the villi, their ability to absorb iron is relatively low. If, as after hemorrhage, plasma iron is diverted to the bone marrow to participate in erythropoiesis, the iron content of new mucosal cells is low. Iron deficiency also induces additional iron receptors in the brush border of new cells. When they have migrated up the villi, the cells absorb iron avidly. Dependence of iron absorption upon the iron content of cells at the time of their formation accounts for the 3–4 day interval between hemorrhage and increased rate of absorption.

Iron entering cells from the plasma is lost when they desquamate. Therefore, the intestinal mucosa is a pathway for loss of body iron; and because the cells' iron content reflects the plasma iron concentration, intestinal iron excretion is regulated, to some extent, by the body's iron requirements.

Iron is actively secreted into the lumen by the cells of the rat's jejunum. If this occurs in man, it is another means of regulating the body's store of iron. Iron loaded macrophages in the systemic circulation of a patient with hemochromatosis or transfusion siderosis migrate across the epithelium of intestinal villi into the lumen and provide an additional means of reducing the load of iron.

The process of iron absorption is summarized in Figure 13–5.

Approximately 40% of the daily intake of copper is absorbed in the stomach and duodenum, and the absolute amount absorbed increases with increasing dose. In normal man, the body load of copper is constant at 60–100 mg; therefore, excretion must equal intake. Copper is secreted in the bile at an average rate of 25 μg/kg · day (range, 9–53). The rate of biliary secretion is constant over 24 hours, and there is no relation between copper secretion and bile acid output. Copper secreted in bile, together with unabsorbed copper, is eventually excreted in the feces. Consequently, the copper load of a healthy man is regulated by biliary secretion which equals absorption. In persons with a genetically determined failure to excrete copper at a rate equal to absorption, copper

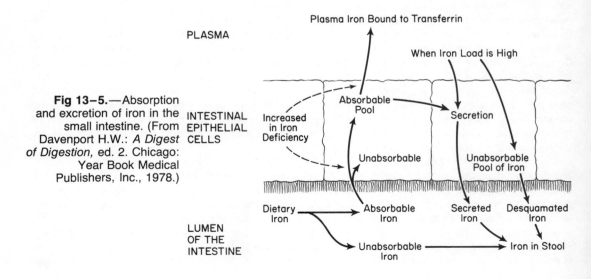

Fig 13–5.—Absorption and excretion of iron in the small intestine. (From Davenport H.W.: *A Digest of Digestion,* ed. 2. Chicago: Year Book Medical Publishers, Inc., 1978.)

accumulates in many tissues, particularly the liver, central nervous system, and the lens. Penicillamine (β-dimethylcysteine), by chelating copper, reduces its absorption and promotes its excretion, and penicillamine is used therapeutically and prophylactically in patients who accumulate copper.

Absorption of Vitamin B₁₂ and Other Water-Soluble Vitamins

Vitamin B_{12}, or cobalamin, a highly charged cobalt-containing compound whose molecular weight is 1,357, is probably the largest water-soluble essential nutrient absorbed intact through the intestinal mucosa. Other B vitamins (nicotinic acid and nicotinamide, pantothenic acid) are small enough to be absorbed in adequate amounts by passive diffusion, but vitamin B_{12} requires a special and still poorly understood system to insure absorption of the 1 μg/day required by an adult man.

In a patient with pernicious anemia who lacks this system, 1–10 mg, or 1,000–10,000 times the daily requirement, must be given by mouth to permit absorption of an adequate amount. Vitamin B_{12} is always present in the diet as one or more of its coenzyme forms bound to protein. All forms are absorbed in the same way.

Two proteins are important in the absorption of cobalamin. The first is R protein, so called on account of its rapid mobility on electrophoresis. R protein is secreted in saliva, gastric juice, and bile. The second is a mucoprotein, the *intrinsic factor*. In man, intrinsic factor, which is ultimately responsible for absorption of cobalamin, is secreted by the oxyntic glandular mucosa, probably by the oxyntic cells. When cobalamin labeled with ^{57}Co is incubated with slices from this area, radioautography shows the isotope to be attached to the oxyntic cells. Upon infusion of histamine, there is an immediate high peak of intrinsic factor output followed by steady secretion at a low level. In the rat, intrinsic factor is secreted by the chief cells; in the hog, by the pyloric glandular mucosa. Some but not all patients with gastric atrophy and achlorhydria fail to secrete intrinsic factor. The defect leads to pernicious anemia, for the patients fail to absorb enough cobalamin to allow normal maturation of erythrocytes. The serum of 90% of such patients has within its γ-globulin fraction antibodies against oxyntic cells. Parenteral administration of cobalamin bypasses the absorptive defect and pre-

vents macrocytic anemia. Before the availability of pure cobalamin, such patients were fed enormous amounts of raw liver. Liver contains so many milligrams of cobalamin that a patient eating it could absorb the microgram quantities of cobalamin needed by some process not depending upon intrinsic factor. Some of the patients would rather have been dead than continue eating liver.

The absorptive process begins with transfer of the vitamin from dietary protein to R protein, whose affinity for cobalamin is far greater than that of intrinsic factor. As the mixture of R protein-bound cobalamin and intrinsic factor reaches the duodenum, the R protein is partially digested by pancreatic enzymes. Cobalamin then transfers to intrinsic factor. The complex attaches to specific receptor sites on the microvilli of epithelial cells in the terminal ileum and is absorbed into the cells. Several hours pass before the vitamin leaves the cells for plasma in which it is carried by transcobalamin, a specific transport protein. After a dose of 1 μg of cobalamin labeled with ^{60}Co is taken by mouth, no isotope enters the blood for 3 hours. Then radioactivity appears, reaching its peak concentration at 8 hours, and disappears after 12 hours. The vitamin is secreted in the bile and partially reabsorbed in the terminal ileum.

Cobalamin is exposed to many risks before it is successfully absorbed. If R protein is not digested by pancreatic enzymes, cobalamin is not transferred to intrinsic factor and is not absorbed. Consequently, pancreatic insufficiency results in cobalamin deficiency. Bacteria residing in diverticula or above intestinal strictures may successfully compete with intrinsic factor for the vitamin and deny it to their host. A tapeworm secretes a compound which, by removing cobalamin from intrinsic factor, makes the vitamin available to the worm but not to the person infested with it.

The stomach is essential for life, not because of its digestive function but because it secretes intrinsic factor required for cobalamin absorption. The normal human liver contains 75,000–225,000 μg of the vitamin, and at the daily rate of utilization of 1 μg, this is enough to last 2–6 years. This estimate is reasonably accurate, because four patients not receiving cobalamin parenterally were found to develop megaloblastic anemia 4–7 years after total gastrectomy.

Dietary folate is a mixture of conjugates of reduced folic acid or methyl folic acid. Before absorption, these are hydrolyzed to monoglutamic folate, and this compound appears to be absorbed by a specific, active process in the proximal small intestine. Monoglutamic folate is converted within intestinal cells to reduce monoglutamic methyl folate, and this is the only form of absorbed folate appearing in the blood.

L-Ascorbic acid is absorbed by a sodium-dependent, active process in the human ileum. Thiamine is absorbed by an active process when its concentration in the lumen is 1.5 μM or less, and above that concentration additional thiamine is absorbed by diffusion. Although thiamine is phosphorylated within the mucosal cells, free thiamine is the only form released into the blood.

Riboflavin and biotin are absorbed by special mechanisms whose details are unknown.

REFERENCES

Allen R.H., Seetharam B., Podell E., et al.: Effect of proteolytic enzymes on the binding of cobalamin to R protein and intrinsic factor, *J. Clin. Invest.* 61:47, 1978.

Armstrong W.McD., Garcia-Diaz J.F., O'Doherty J., et al.: Transmucosal Na$^+$ electrochemical potential difference and solute accumulation in epithelial cells of the small intestine. *Fed. Proc.* 38:2722, 1979.

Barrowman J.A.: *Physiology of the Gastro-Intestinal Lymphatic System.* Cambridge, Cambridge University Press, 1978.

van Berge Henegouwen G.P., Tangedahl T.N., Hofmann A.F., et al.: Biliary secretion of copper in healthy man. *Gastroenterology* 72:1228, 1977.

Castle W.B.: Gastric intrinsic factor and vitamin B$_{12}$ absorption in Code C.F.(ed.): *Handbook of Physiology:* Sec. 6. *Alimentary Canal*, vol. 3. Washington, D.C., American Physiological Society, 1968, pp. 1529-1552.

Crosby W.H.: Iron absorption, in Code C.F. (ed.): *Handbook of Physiology:* Sec. 6. *Alimentary Canal*, vol. 3. Washington, D.C., American Physiological Society, 1968, pp. 1553-1570.

Curran P.F., Schultz S.G.: Transport across membranes: General principles, in Code C.F. (ed.): *Handbook of Physiology:* Sec. 6. *Alimentary Canal*, vol. 3. Washington, D.C., American Physiological Society, 1968, pp. 1217-1244.

Dietschy J.M., Sallee V.L., Wilson F.A.: Unstirred layers and absorption across the intestinal mucosa. *Gastroenterology* 61:932, 1971.

Field M., Fordtran J.S., Schultz S.G. (eds): *Secretory Diarrhea.* Washington, D.C., American Physiological Society, 1980.

Fordtran J.S.: Stimulation of active and passive sodium absorption by sugars in the human jejunum. *J. Clin. Invest.* 55:728, 1975.

Fordtran J.S., Ingelfinger F.J.: Absorption of water, electrolytes, and sugars from the human gut, in Code C.F. (ed.): *Handbook of Physiology:* Sec. 6. *Alimentary Canal*, vol. 3. Washington, D.C., American Physiological Society, 1968, pp. 1457-1490.

Glass G.B.J.: Gastric intrinsic factor and its function in the metabolism of vitamin B$_{12}$. *Physiol. Rev.* 43:529, 1963.

Hallbäck D-A.: Fluid and electrolyte transport in the small intestine as related to the countercurrent exchanger. Göteborg, Sweden, Univeristy of Göteborg, 1979.

Krejs G.J., Fordtran J.S.: Physiology and pathophysiology of ion and water movement in the human intestine, in Sleisenger M.H., Fordtran J.S. (eds.): *Gastrointestinal Disease*, ed. 2. Philadelphia, W. B. Saunders Co., 1978, pp. 297-313.

Linder M.C., Munro H.N.: The mechanism of iron absorption and its regulation. *Fed. Proc.* 36:2017, 1977.

Love A.H.G., Rohde J.E., Abrams M.E., et al.: The measurement of bidirectional fluxes across the intestinal wall in man using whole gut perfusion. *Clin. Sci.* 44:267, 1973.

Rindi G., Ventura U.: Thiamine intestinal transport. *Physiol. Rev.* 52:821, 1972.

Rosenberg I.H., Godwin H.A.: The digestion and absorption of dietary folate. *Gastroenterology* 60:445, 1971.

Schultz S.G., Curran P.F.: Intestinal absorption of sodium chloride and water, in Code C.F. (ed.): *Handbook of Physiology:* Sec. 6. *Alimentary Canal*, vol. 3. Washington, D.C., American Physiological Society, 1968, pp. 1245-1276.

Svanvik J.: Mucosal blood circulation and its influence on passive absorption in the small intestine. *Acta Physiol. Scand.* [suppl.] 385:1, 1973.

Toskes P.P., Deren J.J.: Vitamin B$_{12}$ absorption and malabsorption. *Gastroenterology* 65:662, 1973.

Wilson T.H.: *Intestinal Absorption.* Philadelphia, W. B. Saunders Co., 1962.

Wiseman G.: *Absorption from the Intestine.* New York, Academic Press, 1964.

Zerwekh J.E.: Vitamin D-dependent intestinal absorption of calcium. *Gastroenterology* 76:404, 1979.

14

Intestinal Digestion
and Absorption
of Carbohydrate

ALTHOUGH MOST of the carbohydrates eaten are oligosaccharides or polysaccharides, monosaccharides are delivered by the intestinal mucosa to the portal blood. Dietary carbohydrate is entirely dispensable; nevertheless it is 50%-60% of the American mixed diet, and in many countries is a larger percentage. Carbohydrate intake ranges from 250 to 800 gm/day, and it supplies 1,000–2,500 calories.

The major dietary form of carbohydrate is plant starch composed of straight and branched chains of glucose (Fig 14–1). The straight chains are held together by 1,4'-α-glycosidic linkages, and at the branch points the linkages are 1,6'-α-glycosidic. The polysaccharide cellulose linked in the 1,4'-β-configuration is not attacked by the enzymes of animals. Cellulose and insoluble hemicellulose are digested by microbial flora in the stomach, rumen, and cecum of herbivorous animals, and the products of digestion contribute substantially to the animals' nutrition. In man, 80% of ingested cellulose and 96% of ingested hemicellulose are digested. Lignin is not digested in the human colon, and consequently cellulose contained in cells with lignified walls is not hydrolyzed. Most of the digestion occurs in the colon, and it is not known how much of the products of digestion are absorbed and how much used by colonic flora. Small amounts of other glucose homopolysaccharides, including animal glycogen; some homopolysaccharides of galactose, mannose, arabinose, and xylose; and some heteropolysaccharides are eaten. Of the oligosaccharides, the quantitatively most important are sucrose (glucose-fructose), maltose (glucose-glucose), and lactose of milk (glucose-galactose). Only small amounts of monosaccharides occur in the normal diet. Some hexose alcohols, pentoses, and amine derivatives of carbohydrates are ingested as parts of complex molecules.

Hydrolysis of starch is catalyzed by salivary and pancreatic amylases; disaccharides are split at the brush border of mucosal cells as they are being absorbed. Glucose and galactose are the major sugars actively absorbed; fructose is absorbed by diffusion.

Digestion of Starch

Salivary α-amylase attacks dietary starch, and the extent to which salivary digestion proceeds depends upon the failure of the stomach to mix its contents with acid and on the delay of gastric emptying. About 50% of the starch may be broken down in the stomach before it enters the duodenum. There, starch mixes with pancreatic amylase in a pH environment favorable for action of the enzyme. Most dietary starch is amylopectin, consisting of long straight chains of glucose molecules linked at α-1,4- points and branches attached to the chains by linkage at α-1,6- points. Salivary and pancreatic amylase hydrolyze only α-1,4- linkages within the chain; they do not attack terminal α-1,4-linkages, nor do they break the α-1,6- branching links. Consequently, the products of amylytic digestion of amylopectin within the lumen are maltose (glucose-glucose; α-1,4- linkage), maltotriose (glucose-glucose-glucose; α-1,4- linkages), and a mixture of dextrins containing the α-1,6- branches and averaging six glucose residues per molecule. Hydrolysis of glycogen, which also contains α-1,6- branches is similar to that of amylopectin. Amylose, a minor constituent of starch, contains only

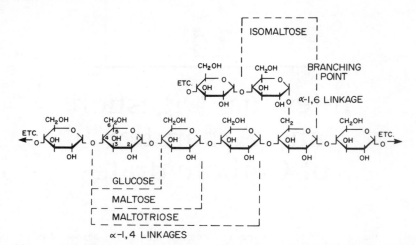

Fig 14–1.—The structure of starch. Hydrolysis catalyzed by pancreatic amylase occurs at the α-1,4- linkage, and the products of hydrolysis are straight-chain oligosaccharides. Since pancreatic amylase does not catalyze hydrolysis of the α-1,6- linkages, isomaltose is also a product of hy-drolysis. Further hydrolysis is catalyzed by the maltases and the isomaltase of the brush border of intestinal epithelial cells. (From Davenport H.W.: *A Digest of Digestion,* ed. 2. Chicago, Year Book Medical Publishers, Inc., 1978.)

α-1,4- linkages, and its digestion products are maltose and maltotriose. Hydrolysis in the duodenum is very rapid. Measurement of enzyme concentration in samples recovered from the human intestine after a meal shows that amylase is maximally concentrated in the duodenum; the amylase contained in only 1 ml of duodenal contents can hydrolyze 5 gm of starch per hour at 37 C. Within 10 min after starch enters the duodenum it is converted to fragments containing, on the average, three hexoses, and complete hydrolysis soon follows. The products of starch digestion, unlike starch itself, are osmotically active; hydrolysis of 10 gm of starch to glucose creates 60 mOsm. Were not glucose absorbed almost as rapidly as starch enters the duodenum 200 ml of water would be required to bring duodenal contents to isotonicity after hydrolysis of this small amount of starch.

Hydrolysis of Oligosaccharides

The activity of enzymes that hydrolyze oligosaccharides is very low in intestinal contents; the enzymes present are not secreted but come from desquamated cells. Consequently, oligosaccharides are not split in the lumen as they are liberated from starch. On the other hand, only monosaccharides appear as absorption products in the portal blood.

Major enzymes catalyzing hydrolysis of oligosaccharides are maltase (maltose and maltotriose hydrolyzed to glucose), lactase (lactose to glucose and galactose), sucrase (also called invertase; sucrose to glucose and fructose and maltose to glucose), and isomaltase (also called oligo-1,6- glucosidase or α-dextrinase; branched dextrins to glucose and maltose). Although samples of intestinal mucosa taken by biopsy from the whole length of the small intestine of normal subjects contain all these enzymes, their concentration per gram wet weight of tissue is highest in midjejunum and upper ileum and lowest in the duodenum and terminal ileum. The enzymes are components of the brush border of epithelial cells; and two of them, sucrase and maltase, have been identified as 60A particles attached to the luminal surface of the microvilli. Brush borders (microvilli and terminal web) can be isolated virtually uncontaminated by other cell components, and such preparations contain all the oligosaccharidases of the cells and phosphomonoesterases as well. Enzymes of the brush border also hydrolyze cellobiose, trehalose, α-glycerophosphate, hexose diphosphate, adenosine triphosphate, and a variety of polyphosphates.

Oligosaccharides are hydrolyzed on contact with the brush border, and their hydrolysis products can be absorbed or can return to the lumen. Glucose and galactose molecules liberated are picked up by an active transport system that resides in cell membranes close to the hydrolytic enzymes, and the two hexoses accumulate within epithelial cells before passing into portal blood. Fructose, not being actively transported, is absorbed more slowly. In man, the rate of hydrolysis is faster than the rate of absorption of glucose, and glucose is absorbed 3–6 times faster than fructose. The result is that some glucose and much more fructose transiently appear in luminal contents during sucrose digestion. Maltose is hydrolyzed as rapidly as sucrose, but lactose is split only half as quickly.

Data on hydrolysis and absorption of sugars by man have been obtained by sampling, at various times and levels, the contents of the small intestine by means of an indwelling tube after a 500-ml liquid meal has been placed in the stomach. When the meal contained 20 gm of lactose and 55 gm of glucose (in addition to 30 gm of fat and 25 gm of protein), up to

40% of the carbohydrate was absorbed in the duodenum (Fig 14–2). All carbohydrate was absorbed by the time the residue of the meal reached the end of the jejunum. When lactose was given as the only carbohydrate, it was absorbed at the same rate in the same place. The lactase activity of the cells of the upper small intestine is adequate for hydrolysis of all lactose of the diet. Maltose and isomaltose are not absorbed in the duodenum; 50%–70% of the maltose fed in a 400-gm meal containing 56.3 gm of maltose, 21.8 gm of protein, 27.3 gm of fat, and 295 gm of water was absorbed in the jejunum and the rest in the upper ileum. Most maltose of the meal was absorbed in 4 hours. When sucrose-fat-protein meals were given to 5 human subjects, 90%–100% absorption of sucrose occurred in the jejunum. Capacity to hydrolyze and absorb sucrose is lower in the ileum than in the jejunum; the human jejunum absorbs sucrose from a solution of 73 or 146 mM sucrose made iso-

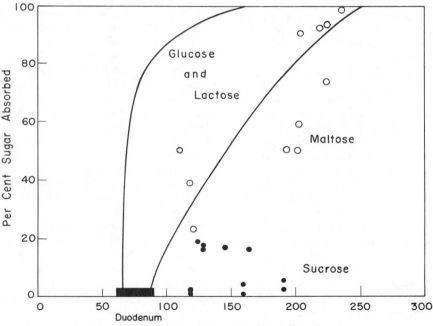

Fig 14–2.—Absorption of mono- and disaccharides by the human intestine. Area between the solid lines encloses 95% of the measurements of absorption of glucose and lactose in 16 separate experiments in which more than 100 samples were obtained. The meal given contained 75 gm of carbohydrate as glucose and lactose in 500 ml. *Open circles* show absorption of maltose from a 400-gm liquid meal containing 56.3 gm of maltose as the only sugar. *Filled circles* show absorption of sucrose from a similar meal. (Adapted from Borgstrom B., et al.: *J. Clin. Invest.* 36:1521, 1957; and Dahlqvist A., Borgstrom B.: *Biochem. J.* 81:411, 1961.)

tonic with NaCl at the rate of 128 $\pm$ 26 mM/hr, whereas the ileum absorbs at the rate of 55 $\pm$ 24 mM/ hr.

Induction and Deficiency of Oligosaccharidases

The concentration of oligosaccharidases depends to some extent upon the diet. In human subjects whose jejunal mucosa was repeatedly sampled by biopsy, a change from a diet containing no sucrose to a high-sucrose diet was followed in 2–5 days by a doubling of sucrase and maltase activity. Sucrase is induced in the crypts, and enzyme activity is expressed as new cells migrate up the villi. Lactose, galactose, and maltose do not induce changes in enzyme activity.

With rare exceptions, all newborn human infants have the enzyme lactase in their intestinal brush border, and they can hydrolyze lactose received in milk and absorb its constituent glucose and galactose. In children of a large number of ethnic groups comprising the majority of mankind, lactase disappears from the brush border at 2–6 years. Thereafter, children become lactose-intolerant, and they remain intolerant the rest of their lives. They have a flat blood sugar curve when given lactose by mouth. If they ingest milk, but not if they ingest aged cheese or fermented milk products from which lactose has disappeared, they have abdominal discomfort, cramps, and loose stools. Most of the lactose they ingest appears in the stool either as lactose or as lactic and fatty acids, products of fermentation of lactose by colonic flora. The flora make gas as well, and a person with lactase deficiency is grossly flatulent. High osmotic pressure of colonic contents and stimulation of peristalsis by acid products cause diarrhea, dehydration, and negative electrolyte balance. Consequently, a person with lactase deficiency voluntarily reduces his consumption of lactose-rich products to amounts he can manage without symptoms.

Northern European whites are among the few who are not lactase deficient in adult life. Adult members of milk-drinking African tribes can hydrolyze lactose, but those who do not habitually use milk products cannot. For example, less than 20% of Batutsi are milk-intolerant, but 100% of adult Zambians are. Because most American blacks are descendants of lactase-deficient ancestors, more than 70% of adult American blacks are likewise lactase deficient.* Other groups with very high frequencies of lactose intolerance include Ashkenazic Jews, Arabs, Greek Cypriots, Japanese, Formosans, and Filipinos. Although only 5%–15% of adult white Americans of North European descent are deficient in lactase, milk intolerance is frequently encountered in the mixed American population. Eighty-six of 166 unselected patients in one Veterans Administration Hospital were found to be lactose-intolerant, and most of them refused to drink the milk offered on their luncheon trays. Consequently, the physician who orders milk for his patients must consider the implications of the fact that a large fraction of his patients may not tolerate it.

A few babies are born with lactase deficiency, and unless the state is quickly recognized, they die of dehydration. Lactase also catalyzes hydrolysis of the rare sugar cellobiose, and children intolerant of lactose cannot digest and absorb cellobiose either. However, they can hydrolyze any of the other disaccharides, and they can survive if their difficulty is correctly diagnosed and if sucrose is substituted for lactose in their diet. Other children who have no invertase in their intestinal mucosa cannot tolerate sucrose; if they are fed sucrose, half of it appears unhydrolyzed in the stool and the other half appears as lactic and fatty acids. Absence of isomaltase causes isomaltose intolerance. Trehalose, which occurs in mushrooms and insects, is not a major form of carbohydrate in the human diet, but deficiency of trehalase has been identified. All these intolerances are the result of the genetically determined absence of a single enzyme. Isomaltose intolerance and sucrose intolerance often occur together, although the activities of the two enzymes, isomaltase and invertase, are independent of each other. Because maltose is hydrolyzed by at least four separate enzymes, intolerance of it can occur only if all four are missing, and such intolerance has not been discovered

In addition to genetic factors, there is a problem of development as well. Some infants with well-documented lactose intolerance lose it about the age of 6 years, and one child who had no lactase in his intestinal mucosa shortly after birth did have the enzyme at 2 years.

---

*This is the incidence repeatedly given by authorities. The American blacks in the medical classes of The University of Michigan say that the frequency of lactose intolerance among them is far lower. Perhaps this is another example of discrimination by the Admissions Committee.

The intestinal mucosa of sea lion pups contains no oligosaccharidases, and they have diarrhea when fed lactose or sucrose; fortunately for them, the milk of their mothers contains no detectable carbohydrate.

Not all untoward consequences of milk drinking are the result of lactose intolerance. In some children and adults, signs and symptoms of milk intolerance, including flat blood sugar test curves after feeding lactose or even glucose, can be induced by feeding milk proteins.

In Vitro Studies of Sugar Absorption

In vitro preparations of rat, guinea pig, and hamster intestine have given knowledge of the means by which sugars are absorbed. One method of preparation is to suspend a length of intestine in a well-oxygenated salt solution whose composition is close to that of extracellular fluid. Through the lumen is perfused, with or without recirculation, another solution containing the sugar to be studied. The blood vessels of such a preparation may be independently perfused. The amount of sugar transported is determined by analyzing the fluid in the lumen and the fluid or fluids perfusing the serosal side of the epithelial cells. Another method is to evert a short piece of intestine so that its mucosal side is outward, fill it with fluid of known composition, tie it to form a sausage-like sac, and incubate it in a solution containing sugar. Again, analysis of the two solutions allows calculation of the rate of sugar movement across the intestine. These methods demonstrate that passage of some sugars through the mucosa is a passive process.

When net water absorption occurs, passively absorbed sugars may be carried by solvent drag. In the absence of water absorption, a passively absorbed sugar is not transported against a concentration gradient; its movement is only down a gradient, and its rate of movement is proportional to the magnitude of the gradient. It moves as well from serosal to mucosal sides as in the opposite direction. Anaerobiosis or metabolic poisons do not decrease these rates.

Actively transported sugars are moved from mucosal to serosal sides against a concentration gradient (Fig 14–3). The transport mechanism is highly specific. Glucose and galactose are the chief naturally occurring hexoses actively absorbed. Glucose derivatives with some modifications about carbon 1 are

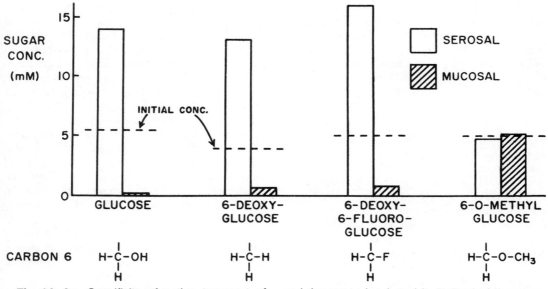

Fig 14–3.—Specificity of active transport of sugar by intestine; transport of sugars with modifications around carbon 6. Sacs of hamster intestine were everted and filled with salt solution containing the sugar at the concentration indicated, and they were incubated in 3–5 ml of the same solution. Concentrations of sugar on each side of the intestine were determined after 60 or 90 min. (From Wilson T.H., Landau B.R.: *Am. J. Physiol.* 198:99, 1960.)

transported; but when the substituted group is too large, the mucosa is incapable of transporting the compound. Gold thioglucose, in which —S-Au is subtituted for —OH on carbon 1, is not transported. Any modification of the hydroxyl group at carbon 2 (for example, in 2-deoxyglucose) impairs transport. Modification of the hydrogen substituent at position 2 does not affect transport unless the substituent group is too large. In a like manner, substitution of large groups, but not of small ones, at other carbon positions prevents active transport. Steric orientation of substituents is important; reversal of orientation of hydroxyl groups at carbon 3 (allose) or carbon 4 (galactose) does not prevent transport, but reversal of both (gulose) does.

Xylose is the chief pentose actively absorbed.

Active Transport of Glucose and Galactose

Three processes occur in active absorption of glucose and galactose: diffusion through the unstrirred layer, transport across the cell membrane, and extrusion from the cell into the interstitial fluid.

A sugar diffuses through the unstirred layer to the cell membrane within the brush border. The rate of diffusion, J, is proportional to the diffusion constant of the sugar, D, divided by the thickness of the unstirred layer, d, and the difference between the sugar's concentration in the bulk phase, C_1, and its concentration at the interface between the unstirred layer and the cell membrane, C_2.

$$J = D/d \, (C_1 - C_2) \qquad (14.1)$$

At the cell membrane the sugar combines with a carrier, which actively transports the sugar across the cell membrane into the cell. Although the nature of the transport machinery is incompletely known, it has well-defined characteristics. Transport is rapid but capable of saturation; one carrier transports glucose and galactose, which consequently are in competition with each other; and the transport machinery uses metabolic energy. As the concentration, C_2, of the sugar at the cell membrane rises, the quantity transported per unit time rises along a hyperbolic curve until a maximum rate of transport is reached. The relation between velocity and concentration is expressed by the equation

$$J = \frac{V'_{max}(C_2)}{K'_a + C_2} \qquad (14.2)$$

V'_{max} is the apparent maximal transport velocity, and K'_a is the apparent Michaelis-Menten constant for the carrier-mediated process. Equation (14.2) is similar in form to the Michaelis-Menten equation, which describes the relation between the velocity of an enzyme-catalyzed reaction and the substrate concentration.

Diffusion and carrier-mediated transport are in series, and therefore

$$J = D/d(C_1 - C_2) = \frac{V'_{max}(C_2)}{K'_a + C_2} \qquad (14.3)$$

Diffusion of glucose and galactose through the unstirred layer limits their rates of absorption, for the subsequent process of active transport into the cell is capable of far higher rates. In man, the unstirred layer is calculated to be 632 μm thick.

Glucose and galactose accumulate in the cell during absorption. They can cross the basolateral border of the cell in both directions: from cell to interstitial fluid and from interstitial fluid to cell. At least part of their flux across the basolateral border is by facilitated diffusion. However, during active absorption their concentration in interstitial fluid can be high, and it is possible that some extrusion of sugars from the cell is by active transport as well as by diffusion.

The mechanism transporting glucose into the cell is immediately adjacent to the enzyme catalyzing the hydrolysis of sucrose, and glucose liberated from sucrose is more rapidly absorbed than is free glucose.

Glucose and galactose compete for the same absorbing mechanism; absorption of one sugar depresses the rate at which the other is absorbed. The magnitude of mutual interference can be predicted from the transport constants. Because glucose has a greater affinity for the transport mechanism, that is, a lower concentration at which its rate of transport is half maximal, its absorption should depress galactose absorption relatively more than galactose absorption depresses transport of glucose. When galactose alone is present on the mucosal side of a 10-cm length of guinea pig intestine, it is absorbed at the rate of 37μM/hr. The addition of an equal concentration of glucose reduces absorption of galactose by 80%; addition of galactose to a preparation absorbing glucose reduces the rate of glucose transport 20%–30%. Other actively transported sugars also inhibit galactose transport, but fructose and other sugars that are not actively absorbed have no effect.

Although the intestine of fetal and newborn hamsters can transport glucose anaerobically, absorption by older animals requires oxygen. Glucose absorption is inhibited by 2,4-dinitrophenol, by phlorhizin, and by the cardiac glycoside ouabain. Ouabain probably affects glucose absorption by virtue of its ability to inhibit Na^+-K^+-activated ATPase and thereby block sodium transport associated with glucose absorption.

Sodium and Glucose Absorption

In a normal person, presence of glucose in the lumen of the intestine is not necessary for sodium absorption. However, sodium is required for at least part of the absorption of glucose.

The concentration of sodium within the lumen is high: 100–140 mEq/kg of water. Its concentration within intestinal epithelial cells is low: 50 mEq or less per kg of cell water. This concentration gradient is maintained by continuous pumping of sodium out of the lateral borders of the cells into interstitial fluid. In addition, there is an electrical gradient across the apical surface of epithelial cells, the cell interior being about 10 mV negative with respect to the luminal fluid. The combination of chemical and electrical gradients comprises an electrochemical gradient tending to drive sodium from lumen to cell interior.

A barrier to free diffusion of sodium exists in the plasma membrane at the apical border of the epithelial cells, immediately beneath the microvilli. Sodium crosses this membranous barrier by attaching to a carrier at the membrane's luminal face. The carrier loaded with sodium moves to the cytoplasmic face of the barrier and there discharges the sodium ion. The carrier then returns to the luminal surface to pick up another sodium ion. Energy consumed in this transport is provided by sodium's electrochemical gradient.

The postulation that the sodium carrier has at least one other binding site to which glucose (and galactose) can attach accounts for the coupling of hexose absorption with sodium. The sugar molecule cannot attach to the carrier unless the carrier is already loaded with sodium. Once the carrier has combined with sodium and then with glucose, sodium's electrochemical gradient drives it from the luminal to the cytoplasmic face of the membrane, where it discharges sodium at a low concentration and glucose at a high one. Accumulation of glucose within cell water occurs at the expense of energy provided by the sodium gradient. Glucose, having a high concentration within the cell, leaks or is extruded into interstitial fluid. This scheme is described in Figure 14–4.

Transport of glucose from luminal to interstitial

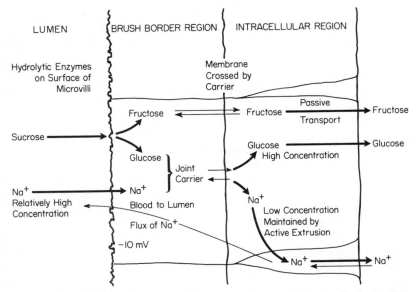

Fig 14–4.—Scheme showing coupling of glucose absorption by the intestinal epithelial cell with sodium transport along sodium's electrochemical gradient. (Adapted from Crane R.K.: Absorption of sugars, in Code C.F. (ed.): *Handbook of Physiology:* Sec. 6. *Alimentary Canal,* vol. 3. Washington, D.C., American Physiological Society, 1968, pp. 1323–1351.)

fluid in the complete absence of sodium can be demonstrated to occur in isolated pieces of rabbit ileum. In the following section, evidence that a similar sodium-independent absorption of glucose may occur in man is summarized.

Absorption of water is a necessary consequence of rapid active absorption of glucose, for the transport of osmotically active particles from lumen to interstitial fluid sets up a gradient upon which water flows. This is useful in the management of cholera. In this disease, a very large volume of fluid pours into the jejunum and ileum. Diarrhea may occur at the rate of 600 ml/hr, and stool composition is about Na^+ 140, K^+ 10, Cl^- 110, and HCO_3^- 40 mEq/L. Intravenous administration of fluid of the composition of that lost in the stool restores and maintains extracellular fluid volume. The same fluid taken orally is not absorbed but is added to the stool. If, however, the patient is allowed to drink a solution of Na^+ 100, K^+ 10, Cl^- 70, and HCO_3^- 40 mEq/L to which 120 mM/L of glucose is added, all fluid drunk is absorbed. Galactose, which is also rapidly and actively absorbed, is as effective as glucose; but fructose, which is more slowly absorbed, is not. Glucose-facilitated water absorption reverses the jejunal secretion of water induced by coliform enterotoxins.

Malabsorption of glucose and galactose, a rare disease in man, is caused by a reduction in the number of functioning sugar carriers in the microvillar membrane of the intestinal epithelial cells.

Glucose Absorption in Man

The concentration of glucose in the intestine during digestion of a starch-containing meal may be 75 mM, and in this circumstance a fraction of the glucose is absorbed by diffusion.

When the rate of glucose absorption by normal subjects is plotted against the luminal concentration, a curve is obtained that can be fitted by equation (14.3) (Fig 14–5). It is difficult to determine how much of the absorption depends upon coupling of glucose and sodium for the reason that the concentration of sodium on both sides of the epithelial cells cannot be manipulated, as it can in experimental animals or in vitro preparations.

Evidence for the role of sodium in absorption of glucose by the human small intestine has been obtained in experiments in which the ileum was perfused with sodium-free solutions. The ileal mucosa, in contrast with that of the duodenum and jejunum, is

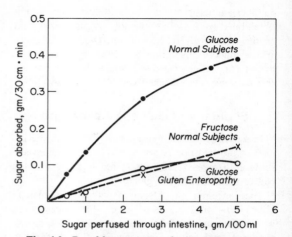

Fig 14–5.—Mean rates of absorption of glucose and fructose in the jejunum of normal subjects and of glucose in the jejunum of patients with gluten enteropathy. The perfusion fluid was delivered to a 30-cm length of jejunum at the rate of 20 ml/min. Data for glucose absorption in the normal subjects can be fitted by equation (14.1), suggesting that absorption occurs by way of a saturable transport system. (From Holdsworth C.D., Dawson A.M.: *Gut* 6:387, 1966.)

sufficiently impermeable to sodium so that, when perfused with a sodium-free solution, the concentration of sodium in the luminal fluid may be less than 3 mN. When the ileum is perfused with a glucose-containing solution in which sodium chloride is replaced by mannitol, the rate of glucose absorption is reduced by only 20%. This does not demonstrate that sodium is unnecessary for glucose absorption, because the concentration of sodium in the unstirred layer may be high. The microvilli of the epithelial cells are covered with a fuzzy coat, or glycocalyx, consisting of copiously sulfated polysaccharides. This fixed layer is negatively charged, and electrical neutrality is attained by the presence of cations in the aqueous phase of the glycocalyx. Under normal circumstances, these cations are chiefly sodium ions. It is probable that even during perfusion of the lumen with a sodium-free solution, there is still a high concentration of sodium within the glycocalyx at the surface of the epithelial cells. If those ions are free to participate in glucose absorption, the fact that glucose is rapidly absorbed from a solution that in the bulk phase is essentially sodium-free cannot be taken as evidence that sodium is unnecessary for glucose absorption.

If the ileum is perfused with a glucose-containing solution in which sodium has been replaced by magnesium and not by mannitol, glucose absorption is reduced by 45%. Magnesium ions, which are only very slowly absorbed in the ileum, can replace sodium ions in the glycocalyx. In this circumstance, there will be fewer sodium ions at the surface of the epithelial cells to participate in glucose absorption, and that part of glucose absorption that depends upon simultaneous absorption of sodium is reduced. If this argument is correct, a substantial part of glucose absorption in the human ileum does not require sodium.

The major part of glucose absorption in man occurs in the duodenum and upper jejunum, and the sodium concentration in the lumen of these segments is always about 140 mN. Consequently, it is impossible to determine how much absorption of glucose in the human intestine does not depend upon simultaneous absorption of sodium.

When a man drinks glucose solutions ranging from isotonic 310 mOsm (5.4%) to 3,100 mOsm (54%), the more concentrated solutions are slightly diluted in the stomach. As the stomach empties concentrated glucose solutions, the contents of the duodenum and jejunum become hypertonic, and gastric emptying is delayed. (In the cat there are slowly adapting receptors whose afferent fibers are in the vagus; they respond to perfusion of the duodenum and upper jejunum with 55–1,100 mOsm glucose solutions. The receptors are almost specific for glucose, and their fibers may be the afferent limb of a reflex slowing gastric emptying.) By the time the solutions reach the middle or lower jejunum or the ileum, they become isotonic. Hypertonic solutions increase intestinal motility, so that they are quickly dispersed through the intestine, exposed to a large mucosal surface, and brought to isotonicity by dilution and absorption. Adjustment of tonicity in the duodenum is chiefly by dilution and in the jejunum by absorption. When a drink of 100 ml of 50% glucose is taken, the rate of absorption in the duodenum is 6–20 gm/hr. Rarely as much as half of a drink of 25–100 gm of glucose taken as 5%–50% solution reaches the jejunum, and in most instances the amount passing into the jejunum is less than 5% of that ingested.

Hydrolase-Related Transport of Hexoses

Hexose components of disaccharides are also accumulated by intestinal epithelial cells by a process that does not depend on sodium transport. The sugar to be accumulated must arrive at the brush border as a glycoside capable of being hydrolyzed by an enzyme present in the membrane of the microvilli. The glycoside binds to the enzyme and is split into its component hexoses. These then are transported into the cells; almost none of the products of hydrolysis escapes into the lumen. Thus, sucrose is hydrolyzed to glucose and fructose, both of which are immediately transported into the cells. Likewise, maltose is hydrolyzed, and its component glucose molecules are accumulated. In fact, the glucose contained in maltose is absorbed faster than free glucose. It is likely that the hydrolytic enzymes are themselves the carriers that move sugars. This process of hydrolase-related transport is additive with sodium-dependent and sodium-independent transport.

Glucose Metabolism and Transport

In the course of being absorbed, glucose enters the cells of the mucosa, but there is no evidence that intestinal glucose uptake is facilitated by insulin. On the other hand, glucose in the course of being absorbed stimulates the release of insulin from the pancreatic islets. The mediator is probably gastric inhibitory polypeptide (GIP) released during absorption of glucose. Once within the cells, some glucose enters into cellular metabolism; 10% or more of the glucose absorbed may be completely oxidized, chiefly by the phosphogluconate pathway. Mucosal cells have a high rate of aerobic glycolysis, and the lactate produced is transported to the serosal side of the intestine. An isolated piece of intestine can establish a lactate concentration gradient, the concentration on the serosal side being 4–20 times that on the mucosal side. However, conversion of glucose to lactate is not the chief means of glucose absorption. When glucose that is uniformly labeled with radioactive carbon (^{14}C) is placed in a loop of dog intestine and the total venous blood draining the loop is collected, only 7%–17% of the glucose absorbed is recovered as lactate; the remainder appears as glucose. Lactate itself is absorbed in the jejunum, more rapidly at pH 2.8 than at physiological pH, because at low pH lactic acid is unionized and is relatively fat-soluble.

In man, IV infusion of somatostatin reduces glucose, and amino acid reduces absorption by the jejunum without affecting water or electrolyte absorption.

Fructose Absorption

In the range of concentrations in which fructose is present in the intestine during absorption of sucrose, fructose is absorbed at a rate directly proportional to its luminal concentration (see Fig 14–5). This rate is greater than that of passively absorbed mannose and less than that of actively absorbed glucose. When, in perfusion studies, the luminal concentration of fructose is raised to 100 mM or more, the curve relating rate of absorption to luminal concentration becomes concave downward, demonstrating that the absorption process is saturable.

In intestinal epithelial cells, fructose is largely converted to glucose or to lactic acid.

Children placed on a diet in which the carbohydrate is 70% fructose exhibit a slight fructosuria and retain more nitrogen than when the dietary fructose is less than 5%; clearly, a considerable fraction of the fructose must have been absorbed as such. Some adults with diabetes mellitus have the same response to oral as to IV fructose, and in them, too, fructose must be absorbed as fructose.

REFERENCES

Alpers D.H., Seetharam B.: Pathophysiology of diseases involving brush-border proteins. *N. Engl. J. Med.* 296:1047, 1977

Armstrong W.McD., Garcia-Diaz J.F., O'Doherty J., et al.: Transmucosal Na⁺ electrochemical potential difference and solute accumulation in epithelial cells of the small intestine. *Fed. Proc.* 38:2722, 1979.

Bayless T.M., Rothfeld B., Massa C., et al.: Lactose and milk intolerance: Clinical implications. *N. Engl. J. Med.* 292:1156, 1975.

Bieberdorf F.A., Morawaski S., Fordtran J.S.: Effect of sodium, mannitol, and magnesium on glucose, galactose, 3-O-methylglucose, and fructose absorption in the human ileum. *Gastroenterology* 68:58, 1975.

Crane R.K.: Absorption of sugars, in Code C.F. (ed.): *Handbook of Physiology:* Sec. 6. *Alimentary Canal,* vol. 3. Washington, D.C., American Physiological Society, 1968, pp. 1323-1351.

Crane R.K.: A concept of digestive-absorptive surface of the small intestine, in Code C.F. (ed.): *Handbook of Physiology:* Sec. 6. *Alimentary Canal,* vol. 5. Washington, D.C., American Physiological Society, 1968, pp. 2535-2542.

Fisher R.B., Gardner M.L.G.: Dependence of intestinal glucose absorption on sodium, studied with a new arterial infusion technique. *J. Physiol.* 241:235, 1974.

Gardner M.L.G.: Metabolic energy dependence of glucose, water and sodium absorption in the presence and absence of "downhill" sodium gradients across the isolated rat small intestine. *J. Physiol.* 265:231, 1977.

Kimmich G.A., Carter-Su C., Randles J.: Energetics of Na⁺-dependent transport by intestinal cells: Evidence for a major role for membrane potentials. *Am. J. Physiol.* 233:E357, 1977.

Sahi T.: Dietary lactose and the aetiology of human small-intestinal hypolactasia. *Gut* 19:1074, 1978.

Simoons F.L.: The geographic hypothesis and lactose malabsorption, a weighing of the evidence. *Am. J. Dig. Dis.* 23:963, 1978.

Welsh J.D., Poley J.R., Bhatia M., et al.: Intestinal disaccharidase activity in relation to age, race, and mucosal damage. *Gastroenterology* 75:847, 1978.

Wilson T.H.: *Intestinal Absorption.* Philadelphia, W.B. Saunders Co., 1962.

Wiseman G.: *Absorption from the Intestine.* New York, Academic Press, 1964.

15

Intestinal Digestion
and Absorption of Protein

ADULTS REQUIRE approximately 0.5–0.7 gm of protein per kg of body weight per day to remain in nitrogen balance; growing children require 4 gm per kg of body weight per day at the age of 1–3 years. All but a negligible fraction of ingested protein is broken down completely and absorbed as amino acids and small polypeptides. Ten to 30 gm of protein contained in digestive juices and about 25 gm of protein derived from desquamated cells are added to ingested protein, and these, too, are digested and absorbed. Only about 10% of the daily intake escapes into the stool in the form of bacteria, desquamated cells, and mucoproteins.

Gastric Digestion of Protein

Gastric digestion of protein is dispensable, for persons with achlorhydria (atrophy of the gastric mucosa resulting in failure to secrete acid and pepsin) can remain in nitrogen balance while eating protein in its usual forms. When the stomach is functioning normally, the extent of gastric digestion varies widely, depending, as it does, on factors affecting secretion, the size and division of protein foods swallowed, mixing in the body and antrum, and the rate of gastric emptying. At the worst, only a small amount of protein may be attacked in the stomach; and at best, 10%–15% may be broken down to amino acids. Therefore the protein of the diet delivered to the intestine is a mixture of completely undigested bundles of muscle fibers, native protein in solution, and products of peptic digestion, ranging from large polypeptides to a few free amino acids. Only the last are ready for absorption; consequently, the major part of protein digestion must occur in the intestine.

Absorption of Native Protein

Mammalian fetuses do not synthesize their own antibodies, and in some species, including man, passive immunity at birth results from placental transfer of maternal antibodies. This does not occur in ruminants or rodents, and the plasma of their newborn is devoid of γ-globulin.

Whey of colostrum of ruminants contains IgG antibodies, and these are ingested by the suckling young. The globulins are protected against digestion by trypsin inhibitor in colostrum and by neonatal achlorhydria. During the first 36 hours of life, the IgG antibodies are absorbed through intestinal epithelial cells by pinocytosis.

In the rat, the process of absorption is selective and prolonged. Ingested IgG molecules bind to receptors on the microvillar membrane of jejunal epithelial cells, and binding is apparently responsible for specificity of absorption. Rat intestine absorbs rat globulins 50 times faster than albumin; it absorbs rat globulins twice as fast as globulins from monkey or rabbit; and it does not absorb cow or fowl globulins at all. Invagination of the surface of the cells produces membrane-bound vesicles, called phagosomes, within the epithelial cells, and these migrate toward the base of the cells. During migration, there is some digestion of antibody within the phagosomes, but most of the antibody is delivered to the extracellular space by exocytosis. Absorbed globulins reach the general circulation by way of the lymph. Ability to absorb large quantities of antibodies lasts 18–20 days after birth. In this period, administration of cortisone alters the morphology of intestinal cells so that they are indistinguishable from adult ones, and it partially suppresses their ability to absorb native protein. Ability to absorb large molecules is retained by the adult rat. If one is fed tritiated bovine serum albumin, 2% of the amount fed can be identified as intact albumin in the lymph draining the intestine.

Human infants do not absorb antibodies from maternal colostrum. Nevertheless, γ-globulins and other macromolecules can enter an infant's intestinal epithelial cells by nonselective pinocytosis. Only a very small fraction of these reaches the circulation, for most are digested within the cells. Native proteins absorbed in minute amounts can be immunologically important. Babies who had never received egg white were sensitized to it by intradermal injection of egg albumin. When the babies were subsequently fed the same protein, a wheal, developing at the site of sensitization, showed that absorption of native egg albumin had occurred.

Most adults can absorb immunologically detectable amounts of whole protein. Peyer's patches, the clusters of lymphoid follicles in the proximal small intestine, are penetrated by proteins which are subsequently phagocytized by subendothelial macrophages. Asbestos fibers and insoluble particles as large as 2–5 μm are also absorbed through Peyer's patches. Although the amounts of protein absorbed are nutritionally negligible, they may at times be responsible for local intestinal or systemic disorders.

The absorption of trypsin has been described in chapter 10. There is evidence that chymotrypsin is likewise absorbed and resecreted.

Intestinal Digestion of Protein

The rate at which protein is delivered to the duodenum, and therefore its rate of digestion and absorption, varies widely. Fifty percent of a finely divided lean-meat meal may leave the stomach in the first hour, and 83% by the end of the third hour. Once in the duodenum, protein is rapidly digested. Fluid aspirated from the human duodenum shortly after feeding contains 200–800 μg of trypsin and chymotrypsin per ml. This is enough to convert 50% of the protein in duodenal contents to trichloroacetic acid-soluble material in 10 min. The trypsin concentration in contents of the terminal ileum is the same as that in the proximal jejunum; but because the volume of intestinal contents falls as chyme moves down the intestine, the amount of trypsin in the terminal ileum is

only 8% of that in the proximal jejunum.

The time course and extent of protein digestion in man have been followed by aspirating samples through an indwelling tube from various levels of the intestine. When the protein content of a mixed meal is 5%, the concentration of fed protein at the end of the duodenum ranges for 1% to 4%, with the most frequent value being about 2%. By the time the residue reaches the end of the duodenum, 50%–60% of the fed protein has been digested and absorbed. Fed protein appears in the jejunum 30 min after feeding and is detectable there for 4 hours. It does not appear in the ileum until 2 hours have passed, and its concentration in the ileum reaches a peak at 4 hours. When a meal containing 50 gm of protein is eaten, about 0.5 gm of the protein escapes digestion in the small intestine and enters the colon, where its digestion is completed by microorganisms. In addition, about 1.2 gm of free amino acids and oligopeptides derived from the ingested protein passes into the colon 4 hours after the meal is eaten. The protein contained in the stool, equalling about 10% of the protein eaten, is not food protein but comes from bacteria and cellular debris.

Endogenous Protein

Between 10 and 30 gm of protein enters the intestinal lumen in secretions each day, and 80–90 gm of desquamated cells contributes another 10 gm. In addition, plasma proteins enter the digestive tract. Capillaries of the gut are more permeable to proteins than are capillaries of muscle, and the interstitial fluid of the gut contains a high concentration of plasma proteins. The concentration of plasma albumin in gastric interstitial fluid is about 80% of that in plasma, and the concentration of fibrinogen is about 30%. Some plasma proteins escape into the lumen and are digested. In normal persons, an average of 1.9 gm of albumin is shed into the stomach; this represents 11% of the daily degradation of albumin. In protein-losing gastroenteropathies, the rate of plasma protein loss may be enormous, and hypoalbuminemia results.

Enzymatic Hydrolysis of Protein

Proteins in the intestinal lumen are digested by pancreatic enzymes, by enzymes in the brush border of intact epithelial cells, by brush border enzymes shed into the lumen, and by peptidases of desquamated cells. Pancreatic enzymes do not hydrolyze dipeptides in the intestine; those are split by the other enzymes. Jejunal enzymes completing hydrolysis come from the cytoplasm of intestinal cells, but enzymes in the lumen of the ileum come chiefly from the brush border. In addition, polypeptides liberated by peptic and tryptic digestion are hydrolyzed to free amino acids and oligopeptides by the peptidases in the brush border of intact epithelial cells. The process of hydrolysis at the brush border, called "membrane hydrolysis" by some, yields 30% basic and neutral amino acids and 70% small peptides.

Pancreatic proteolytic enzymes are responsible for the major part of protein digestion within the intestinal lumen. The flow of pancreatic juice begins 10–20 min after a meal is eaten, and the concentration of pancreatic enzymes remains high throughout the jejunum and ileum over the whole period of digestion and absorption. Pancreatic enzymes are present in the stool of normal men, dogs and rats; but the fraction of the total quantity secreted by the pancreas that is lost in the stool is very small. Pancreatic enzymes in the stool contribute to pruritus and if the anus is not adequately cleaned after defecation.

In the absence of pancreatic juice, protein digestion is impaired. When the pancreatic ducts of dogs are ligated or the pancrease is completely removed, the percentage absorption of fed protein ranges between 22 and 85. The quantity absorbed varies from day to day in the same animal for unknown reasons. Although much fed protein and unchanged muscle fibers may appear in their stools, dogs completely devoid of pancreatic juice can be made to absorb enough protein to support normal growth simply by increasing total protein intake. Children with cystic fibrosis, in whom pancreatic enzymes are almost completely absent, have low absorption of protein, but absorption can often be improved by feeding pancreatic enzymes. In two tests on a human subject whose pancreas had been completely resected for carcinoma of its head, 38% and 54% of 75 gm of fed protein were excreted in the stool. When the subject was fed pancreatic enzymes, protein loss fell to 16% and 23%. Cimetidine, given along with pancreatic enzymes, inhibits acid secretion, and higher duodenal concentrations of trypsin and lipase are reached than when pancreatic enzymes are given alone or with sodium bicarbonate.

Absorption of Amino Acids

Absorption is an active process supported by metabolic energy. Three transport systems have been identified by in vitro methods similar to those used in the study of sugar absorptions:

1. Neutral amino acids are carried by a single transport system, and they compete with each other for absorption. There is some, but not absolute, optical specificity; D-methionine is transported at about one third the rate of L-methionine, and high concentrations of the D-forms inhibit absorption of the L-forms. The acid transported must have a free carboxyl group; esterified acids or those reduced to an alcohol are not transported. There must be a free α-hydrogen; no substitution is allowed. The side chain must be neutral. In the aliphatic series, almost anything from H (glycine) to NH_2-CO-CH_2-CH_2-CH_2 (citrulline) is allowed, and in the aromatic series, permitted groups range from phenyl (phenylalanine) to imidazole (tryptophan). The affinity of the amino acid (K_t) for its carrier decreases with increasing polarity of the side chain.

2. Basic amino acids are carried by a separate system at 5%–10% of the rate of transport of neutral acids. Amino acids transported include L-arginine, L-lysine, DL-ornithine, and L-cystine.

3. A third system transports L-proline, hydroxyproline, sarcosine, dimethylglycine, and betaine. Proline and hydroxyproline have a much stronger affinity for this carrier system than for the neutral one; therefore, they are always transported by it rather than by the neutral system, for they would be blocked from the neutral carrier by other amino acids always present in the intestine.

Absorption of free amino acids is a carrier-mediated, saturable process. An amino acid combines with a carrier in the membrane and enters the cell. Sodium can combine with the same carrier, or both an amino acid and a sodium ion can do so together. There is no preferred order of combination.

Rabbit ileal mucosa contains a sodium-independent transport system for neutral amino acids, and substantial absorption of lysine by diffusion through the paracellular path has been demonstrated in rat intestine. It is not known whether these routes of absorption are present in the human small intestine.

During absorption, amino acids accumulate in the mucosal cells; they enter the mucosal side faster than they leave the basal side to enter the blood. Entry into the cells is the most rapid step in absorption; when absorption of one amino acid is competitively inhibited by another, its concentration within the mucosal cells is lower than that prevailing during uninhibited absorption. Most amino acids, once inside the mucosal cells, are not extensively metabolized, the exceptions being glutamic and aspartic acids. A fraction of these two undergoes transamination with pyruvic acid so that alanine is formed and released into portal blood during their absorption. Glutamine is rapidly absorbed, and it is metabolized within the cells. More than half is oxidized to carbon dioxide, and the rest is converted into tissue proteins, citrulline, proline, and amino acids. A third of the nitrogen in the absorbed glutamine goes to citrulline, and the rest goes to alanine, proline, and ammonia. Half of the ammonia released from the gut into blood is derived from glutamine in the jejunum and ileum.

Absorption of Oligopeptides

Enzymatic hydrolysis of proteins is slow, and in the small intestine proteins are not reduced to their constituent free amino acids in the 4 hours or so in which protein digestion and absorption occur. Nevertheless, absorption of protein products is essentially complete because di- and tripeptides are absorbed faster than are free amino acids. Di- and tripeptides are absorbed across the brush border into the cytosol of intestinal epithelial cells where they are hydrolyzed to free amino acids by four intracellular peptidases, each with substrate specificity. Tetrapeptides are not absorbed as such; they are shortened by brush-border enzymes. For example, when tetraglycine is the substrate, two amino acids are first removed from the N-terminus by a brush-border hydrolase, and then the free amino acids and the dipeptide are absorbed. Brush-border hydrolysis is the limiting step in tetraglycine absorption. Trileucine is hydrolyzed to leucine and a dipeptide before absorption.

With few exceptions, free amino acids derived from absorbed peptides, not peptides themselves, are delivered to portal blood. Glycylglycine, which is difficult to hydrolyze, is one of the peptides appearing in portal blood.

The individual amino acids of glycylmethionine compete with each other for absorption by a sodium-dependent process. When they are absorbed as the dipeptide by a sodium-independent mechanism, they cannot compete with each other. Peptides have little effect upon absorption of free amino acids, but they compete with each other for absorption.

Peptides are absorbed most rapidly in the jejunum, where they are abundantly produced in the initial stages of protein hydrolysis. Free amino acids, the ultimate products of hydrolysis, are absorbed most rapidly in the ileum.

The two absorption systems, that for peptides and that for free amino acids, assimilate the products of protein digestion with equal efficiency. When the absorption by human subjects of 50 gm of a partial hydrolysate of protein was compared with that of an equivalent amino acid mixture, 74% of the first and 72% of the second were found to be absorbed 160 cm from the mouth. Although absorption of peptides was more complete at 30 and 60 min, peptides and amino acids were absorbed to the same extent by 3 hours.

Hydrolysis and Absorption of Nucleoproteins

The cell surface and the intracellular milieu are the loci of hydrolysis of other compounds. Phosphatases occur on the surface of mucosal cells. Pyrimidine nucleotides in contact with the mucosa are hydrolyzed to nucleosides and inorganic phosphate, although the hydrolytic enzymes reponsible cannot be found in the intestinal contents. Other compounds hydrolyzed on the surface of the intestinal cells include glucose-1-phosphate, phenyl phosphate, β-glycerophosphate, adenylic acid, hexose diphosphate, adenosine triphosphate, fructose-6-phosphate, and many polyphosphates. The course of digestion and absorption of nucleoproteins is summarized in Figure 15–1.

Hypoxanthine and xanthine are both oxidized to urate within the mucosal cells, a process that can be inhibited by allopurinol. In the hamster, urate is actually secreted into the lumen; whether similar secretion in man provides for extrarenal excretion of uric acid is unknown. The methylated xanthine caffeine, because it is fat-soluble, is rapidly absorbed by diffusion.

Defects in Protein Absorption

In pancreatic and biliary deficiency resulting in steatorrhea, loss of dietary protein nitrogen in the stool strictly parallels loss of dietary fat.

There are functional similarities between the small

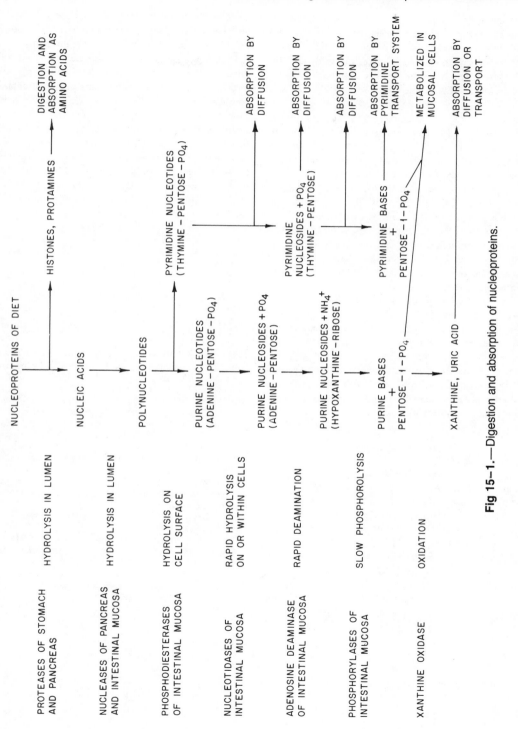

Fig 15–1.—Digestion and absorption of nucleoproteins.

intestine and the renal tubules with respect to amino acid absorption. In each, transport of amino acids from luminal fluid to blood is inhibited by probenecid. Patients with cystinuria have impaired renal tubular absorption of cystine, lysine, ornithine, and arginine; their ability to absorb the same amino acids in their intestines is also reduced. Some of the unabsorbed amino acids are converted in the colon to cadavarine and putrescine, which are absorbed and appear in the urine. The aminoaciduria of Hartnup disease involves, among many other amino acids, threonine, phenylalanine, tyrosine, tryptophan, and histidine. In a patient with Hartnup disease, intestinal absorption of free glycine, histidine, tyrosine, and serine was found to be severely impaired, and absorption of valine, leucine, isoleucine, methionine, phenylalanine, glutamic acid, and alanine was subnormal. However, the intestinal defect in Hartnup disease and other forms of renal tubular malabsorption of amino acids appears to be confined to the absorption of free amino acids. The amino acids in peptides can be absorbed, and, for example, a patient with Hartnup disease can absorb phenylalanine and tryptophan in dipeptides but not as free amino acids. Consequently, the defects are nutritionally insignificant. However, unabsorbed amino acids are degraded by bacterial metabolism to potentially toxic products, and absorption of these in the colon may account for neurological abnormalities.

Gluten Enteropathy

In gluten enteropathy (also called celiac disease, idiopathic steatorrhea, and nontropical sprue), the intestinal mucosal surface is flat, and the luxuriant array of long, thin microvilli on the surface of the epithelial cells is usually replaced by a few clubbed microvilli. There is malabsorption of all nutrients. The underlying defect appears to be the absence from the brush border of peptidases required for the complete hydrolysis of gluten of wheat, rye, and oats. If gluten is completely eliminated from the diet, there is prompt and lasting remission. If gluten is reintroduced into the diet, most patients relapse, but some may show no signs or symptoms despite continued morphological deterioration of their intestinal epithelium.

Glutamine-containing polypeptides obtained from gliadin after exhaustive digestion with pepsin and trypsin evoke signs and symptoms of gluten enteropathy when fed in small amounts to a patient in remission. Damage to the intestinal epithelial cells occurs within 8 hours, and it progresses to the basement membrane and capillaries. Disaccharidases of the brush border are depressed within 24 hours. Many patients challenged with gliadin respond by secreting an IgM gliadin antibody, and antibodies to gliadin, and to cow's milk as well, are found in their serum.

REFERENCES

Chung Y.C., Kim Y.S., Shadchehr A., et al.: Protein digestion and absorption in the human small intestine. *Gastroenterology* 76:1415, 1979.

LeFevre M.E., Hammer R., Joel D.D.: Macrophages of the mammalian small intestine: A Review. *J. Reticuloendothel. Soc.* 26:553, 1979.

Matthews D.M.: Protein absorption—then and now. *Gastroenterology* 73:1267, 1977.

Matthews D.M., Adibi S.A.: Peptide absorption. *Gastroenterology* 71:151, 1976.

Milne M.D.: Genetic disorders of intestinal amino acid transport, in Code C.F. (ed.): *Handbook of Physiology: Sec. 6. Alimentary Canal*, vol. 3. Washington, D.C., American Physiological Society, 1968, pp. 1309-1322.

Morris C.C.: Gamma globulin absorption in the newborn, in Code C.F. (ed.): *Handbook of Physiology: Sec. 6. Alimentary Canal*, vol. 3. Washington, D.C., American Physiological Society, 1968, pp. 1491-1512.

Silk D.B.A., Nicholson J.A., Kim Y.S.: Hydrolysis of peptides within the lumen of the small intestine. *Am. J. Physiol.* 231:1322, 1976.

Walker W.A., Isselbacher K.J.: Uptake and transport of macromolecules by the intestine. Possible role in clinical disorders. *Gastroenterology* 67:531, 1974.

Wilson T.H.: *Intestinal Absorption*. Philadelphia, W. B. Saunders Co., 1962.

Wiseman G.: Absorption of amino acids, in Code C.F. (ed.): *Handbook of Physiology: Sec. 6. Alimentary Canal*, vol. 3. Washington, D.C., American Physiological Society, 1968, pp. 1277-1308.

Wiseman G.: *Absorption from the Intestine*. New York, Academic Press, 1964.

16

Intestinal Digestion and Absorption of Fat

IN ADDITION to being an important source of calories, fat is the major structural component of the body. Because fat is insoluble in water, fat forms the membranes of cells and cell organelles; without insoluble fat the body would thaw and resolve itself into a dew. In order for insoluble fat of the diet to become avail-able for uses of the body, fat must undergo a complex physical and chemical process of digestion and absorption (Fig 16–1), and the process can go wrong in many ways, resulting in maldigestion and malabsorption of fat.

Most ingested fat is delivered unchanged to the

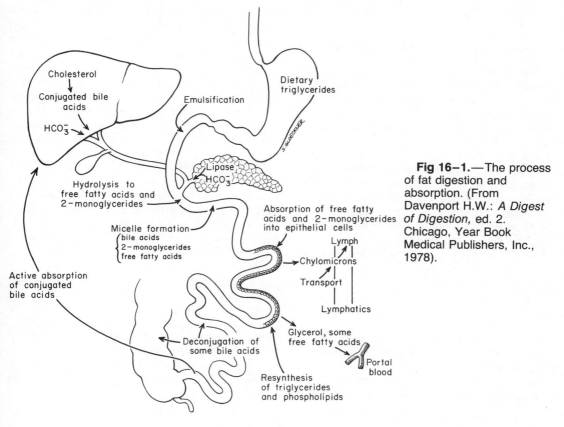

Cholesterol

Conjugated bile acids

HCO_3^-

Emulsification

Dietary triglycerides

Lipase
HCO_3^-

Hydrolysis to free fatty acids and 2–monoglycerides

Micelle formation
bile acids
2–monoglycerides
free fatty acids

Active absorption of conjugated bile acids

Absorption of free fatty acids and 2–monoglycerides into epithelial cells

Lymph

Chylomicrons

Transport

Lymphatics

Deconjugation of some bile acids

Glycerol, some free fatty acids

Portal blood

Resynthesis of triglycerides and phospholipids

Fig 16–1.—The process of fat digestion and absorption. (From Davenport H.W.: *A Digest of Digestion,* ed. 2. Chicago, Year Book Medical Publishers, Inc., 1978).

duodenum because little fat digestion occurs in the stomach. The extremes of human daily fat intake range from below 25 gm (or less than 12% of the caloric intake) among rice-eating peoples such as Japanese coal miners to 140–160 gm (or 42% of caloric intake) among Los Angeles funeral directors. The difference in the intake of fats is due to the use, by the latter group, of animal fats containing saturated fatty acids, for populations increase their fat intake by raising their use of meat and dairy fats. The intake of unsaturated fatty acids is nearly the same in the two extreme groups cited.

Because gastric emptying is slowed by fat in the duodenum, the rate at which fat enters the intestine is self-regulated. The emptying of a very fatty meal containing 50 gm of fat is spread out over 4–6 hours, so that the upper limit of the burden placed on the intestine's ability to deal with fat is between 8 and 12 gm/hr. A normal intestine absorbs all ingested fat; the fat that is always present in the feces is not unabsorbed residue but is derived from the intestinal and colonic mucosa or is synthesized by bacteria.

Emulsification and Hydrolysis of Fat

Most dietary fat of either animal or vegetable origin consists of triglycerides: glycerol combined in low-energy ester linkages with three fatty acids. The acids contain an even number of carbon atoms, the saturated acids being almost entirely palmitic (C_{16}) and stearic (C_{18}) and the unsaturated ones being oleic (C_{18}, 1 double bond) and linoleic (C_{18}, 2 double bonds). Only milk fat, which contains 3%–9% of C_4 to C_{14} acids, contributes any significant quantity of shorter chain lengths. In general, the melting points of fats containing longer chain, saturated fatty acids are higher than those containing shorter chain, unsaturated fatty acids. Phospholipids in which 1 glycerol alcohol is linked to a phosphoric ester of an organic base, most frequently inositol or choline, also occur in the diet. Daily intake is 1–3 gm. About 11 gm/day of the lecithin of human bile, α-palmityl-β-oleyllecithin, mixes with intestinal contents and is digested and absorbed along with exogenous fat.

Fats are insoluble in water and immiscible with chyme, and they are prepared for hydrolysis and absorption by emulsification in the duodenum. Emulsifying agents are fatty acids, monoglycerides, lecithin and lysolecithin derived from it, proteins, and bile salts.

If emulsions in food are not broken by acid in the stomach, triglycerides are delivered to the duodenum emulsified by proteins. Milk fat, for example, is emulsified by β-lactoglobulin. In the stomach, about 10% of dietary triglycerides are hydrolyzed to free fatty acids and monoglycerides by gastric lipase and by lingual lipase secreted by lingual and pharyngeal glands. These products of hydrolysis assist in emulsifying fat. In the duodenum, lecithin and lysolecithin are additional effective emulsifying agents. Bile salts themselves are poor emulsifying agents, but a mixture of bile salts and polar lipids—lecithin, lysolecithin, and monoglycerides—has a much greater emulsifying power. The result is that fats are dispersed in the duodenum as droplets 0.5–1.0 μ in diameter, completely covered by emulsifying agents and negatively charged. The emulsion is stable over the intestinal pH range of 6.0–8.5. Because the droplets are larger than a wavelength of light, the emulsion is cloudy. Emulsions are most readily formed with liquid fat; and a fat like tristearin, which is solid at body temperature, is by itself poorly emulsified and incompletely digested and absorbed. The mixing of liquid with solid fats reduces the mixture's melting point and brings solid fats within range of intestinal digestion.

In order to catalyze the hydrolysis of triglycerides, pancreatic lipase must be spread over the surface of a fat droplet. Because the droplet is covered with emulsifying agents, pancreatic lipase itself cannot reach its substrate. Pancreatic lipase is inactive against an emulsion stabilized with bile salts; this is the reason that bile salts are said to "inhibit" pancreatic lipase. Along with lipase, the pancreas secretes colipase, a protein of about 11,000 MW. Colipase binds to the surface of a fat droplet, displacing the emulsifying agents. One molecule of colipase acts as an anchor for one molecule of lipase to be absorbed to the surface of a droplet.

Pancreatic lipase is a group-specific esterase which hydrolyzes triglycerides to fatty acids and 2-monoglycerides (Fig 16–2). Because the energy of ester bonds is low, there is a tendency for liberated fatty acids to recombine with glycerol alcohol groups. Only protonated fatty acids can recombine. Those free fatty acids within the oil drop are protonated, but most of the free fatty acids, as soon as they are liberated, move into micelles where they are partially ionized and unavailable for resynthesis. There is slow, spontaneous migration of fatty acids from one

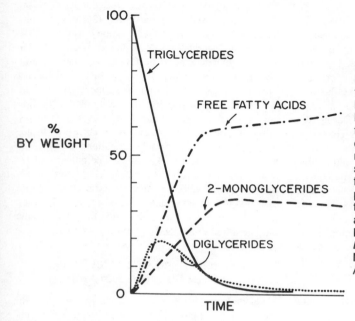

Fig 16-2.—The course of hydrolysis of a triglyceride catalyzed by pancreatic lipase. There is a transient appearance of diglycerides, but the final products of enzymatic hydrolysis are 2-monoglycerides and free fatty acids. Then there is a slow decline in the amount of 2-monoglycerides and a corresponding slow appearance of additional free fatty acids as the ester bond on the 2-position of glycerol is hydrolyzed. At the same time there is a slow appearance of free glycerol. (From Davenport H.W.: *A Digest of Digestion,* ed. 2. Chicago, Year Book Medical Publishers, Inc., 1978. Adapted from F. H. Mattson.)

alcohol group to another. Consequently, a 2-monoglyceride may be converted into a 1-monoglyceride, which is rapidly hydrolyzed by pancreatic lipase. This migration is probably responsible for the fact that pancreatic lipase appears to hydrolyze 2-monoglycerides at the rate of 1%–2% of that at which it breaks primary ester bonds.

In the course of digestion and absorption, triglycerides are partially hydrolyzed and resynthesized. If a triglyceride containing a fatty acid labeled with ^{14}C in position 1 is ingested, the ^{14}C fatty acid is found distributed among positions 1, 2, and 3 in the triglycerides recovered from lymph. Most of the ester bonds in the 1 and 1' position are hydrolyzed. A triglyceride can be doubly labeled, with one isotope marking the glycerol moiety and another isotope identifying the fatty acid in the 2 position. When such a triglyceride is fed and the triglycerides of the lymph are analyzed, the two labels are found to be contained in many of the same molecules. This demonstrates that a large fraction, 50%–75%, of the ester bonds in the 2 position of triglycerides escapes hydrolysis throughout digestion and absorption.

Pancreatic lipase also catalyzes the hydrolysis of the primary ester bond of lecithin of food and bile to form 2-lysolecithin. Pancreatic juice contains a zymogen which, when activated by trypsin, becomes phospholipase A$_2$. This breaks the bond in the second position of lecithin, the product being 1-lysolecithin. The result is that duodenal contents have a high concentration of lysolecithins during digestion of fat.

Fat Digestion in Man

Samples of adult human intestinal contents have been obtained by transintestinal intubation. A tube having an internal diameter of 2 mm is passed the whole length of the intestine; and through a hole in its wall, samples are withdrawn for analysis. By means of marks on the tube, the distance between the hole and the nose is determined, and from fluoroscopic observations of the relation of the hole to intestinal landmarks, the site from which samples are drawn is deduced. A fat meal of known composition, emulsified and stabilized with protein, is placed in the stomach. If the appropriate tags are added to the meal, its change in volume and composition resulting from addition by secretion, subtraction by absorption, or mutation by digestion can be calculated. When a 500-ml liquid meal containing 30 gm of corn oil is placed in the stomach, emptying occurs over 4 hours.

The concentration of fat and fatty acids, which was 6% in the original meal, is between 1% and 3% in the upper duodenum as a result of dilution; and although the total volume of intestinal contents falls sharply in the duodenum, fat concentration in the upper jejunum is below 1%. By the time the ileum is reached, 4 hours after feeding, the fat in the small residual volume is 0.3%–0.7%. Pancreatic secretion begins 10–20 min after feeding, and lipase concentration in intestinal contents is highest in the first hour. The gallbladder empties within 30 min, and bile concentration is also high (Fig 16–3). During the course of digestion, 4–5 gm of bile salts enters the intestine; since this quantity is larger than the body's pool of bile salts, the salts recirculate during the digestion of a single meal. Phospholipids and cholesterol contained in bile are added to intestinal contents. Most of the meal is absorbed from a milieu containing a high concentration of lipase, bile salts, and lysolecithin derived from action of pancreatic lecithinase-A on the phospholipids. A mixture of digestive products, the exact composition depending on the nature of fat fed, the locus of sampling, and the elapsed time, accumulates in the lumen.

The concentration of fat digestion products falls sharply as chyme moves through the duodenum into the jejunum, and by midjejunum almost all fat of the diet is absorbed.

In one experiment, normal men were fed pure triolein as the only glyceride, and in it was dissolved a trace of free fatty acid labeled with isotopic carbon

(^{13}C). Samples collected near the duodenal-jejunal junction contained, on the average, 42% of their total fat as free fatty acids, 8% as 1-monoglycerides, 10% as 2-monoglycerides, 16% as diglycerides, and 24% as triglycerides. The concentration of labeled free fatty acid in the mixture had diminished, showing that free fatty acids had been absorbed more rapidly than the other components; but the accumulation of free fatty acids demonstrated that hydrolysis was occurring faster than the free fatty acids were absorbed. In addition, ^{13}C-labeled fatty acid was found esterified in the mono-, di-, and triglycerides. The quantity of labeled fatty acid exchanged with unlabeled fatty acids in ester bonds was 30% of the maximal value to be expected if complete randomization of labeled among unlabeled acids had taken place. On the average, one fatty acid had exchanged in each triglyceride.

Physical and Chemical State of Fat During Absorption

Fat undergoing digestion and absorption is distributed among the emulsified fat droplets, micelles that are small, hydrated polymolecular aggregates, and molecular solution.

If intestinal content is heated to destroy lipase and centrifuged at high speed (100,000 × g for 12 hours at 37 C), an oily top phase and a completely clear

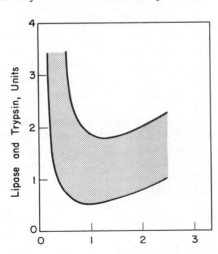 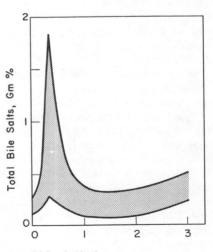

Time, Hours after 500 ml Meal

Fig 16–3.—Left, concentration of lipase and trypsin in contents of human duodenum and proximal jejunum following a 500-ml liquid meal containing fat, protein, and carbohydrate. **Right,** concentration of bile salts in the same samples. (Adapted from Borgstrom B., et al.: *J. Clin. Invest.* 36:1521, 1957.)

bottom phase are obtained. In the duodenum, the oil phase is derived from the large, relatively unstable fat droplets. It contains all the triglycerides of intestinal contents, most of the diglycerides, some free fatty acids, and very little monoglyceride. In samples obtained from the upper jejunum, the oil phase consists almost entirely of free fatty acids, for by the time fat reaches the jejunum, pancreatic lipolysis is almost complete. The fact that there are free fatty acids in the oil phase means that lipolysis is faster than dispersion of free fatty acids in the micellar phase. The oil phase disappears in midjejunum.

Although the lower aqueous phase shows no opalescence or light scattering, it contains as much fat as does the oil: most of the monoglycerides and fatty acids, a large fraction of the cholesterol, only a small amount of diglycerides, and almost no triglycerides. The fat is dispersed in highly stable micelles, which, being 4–6 mμ in diameter, have one millionth the volume of fat droplets of the oily phase. Their diameter is a little less than twice the length of the paraffin chains of the fatty acids; each probably consists of about 20 fat molecules whose hydrocarbon chains interdigitate within a fluid interior and whose polar groups form a negatively charged spherical shell surrounded by cations in aqueous solution.

When the concentration of a bile salt is above a certain value, its *critical micellar concentration,* the bile salt spontaneously aggregates with monoglycerides to form micelles. Figure 16–4 shows the relation between the amount of monoglyceride carried in solution and the concentration of three conjugated bile salts. The critical micellar concentration of the two dihydroxy-conjugates is about 0.25 mM; above that concentration they associate with a large amount of monoglyceride. The critical micellar concentration of the trihydroxy-conjugate is 4 mM. Under normal circumstances during the digestion of a meal, the concentration of conjugated bile salts in intestinal contents is always above the critical micellar concentration. Consequently, monoglycerides, as they are liberated from triglycerides, form micelles with bile salts. Micelles obtained from human intestinal contents during digestion of a meal were found to contain 1.4 M of fatty acids, 0.15 M of lysolecithin, and 0.06 M of cholesterol for each mole of bile salt. On a molar basis, their content of free fatty acids was greater than their content of monoglycerides.

Unconjugated bile salts have essentially the same critical micellar concentration as their corresponding conjugates. However, they are less soluble in intes-

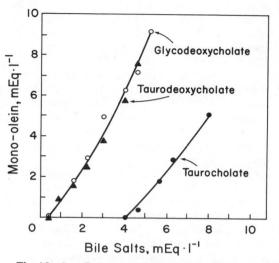

Fig 16–4.—Relation between bile salt concentration and quantity of monoglyceride brought into micellar solution at pH 6.3, 37 C and [Na$^+$] of 150 mN. The critical concentration for the two dihydroxy-conjugates is 0.25 mM and for the trihydroxy-conjugate is 4 mM. (Adapted from Hofmann A.F., Borgstrom B.: *Fed. Proc.* 21:43, 1962.)

tinal contents. When extensive deconjugation occurs in the intestine, the deconjugated bile salts tend to precipitate and therefore fall below their critical micellar concentration in solution.

Each bile salt has a temperature, its Krafft point, below which it cannot form micelles. The Krafft point of all naturally occurring bile salts, except lithocholic acid, is well below body temperature.

The concentration of free fatty acids and 2-monoglycerides in molecular solution in intestinal contents is not precisely known; it is probably of the order of 0.01 mM. There is very rapid exchange of fatty acids and monoglycerides between micelles and molecular solution; the mean residence time of a particular molecule in a micelle is about 10 msec. Exchange between micelles and solution keeps the solution saturated.

Absorption of Fat Digestion Products

Fat digestion products accumulate in the bulk phase of intestinal contents, and the bulk phase is well mixed by segmentation and peristalsis. The contents of the lumen are progressively less and less well mixed in the radial direction from the center of the lumen to the surface of the intestinal epithelial cells. A thin layer, 0.05–2 mm thick, on the surface of the epithelial cells is poorly mixed or mixed not at all. This is the unstirred layer.

Lipid molecules can diffuse through the lipoprotein membranes of the epithelial cells, because they are fat soluble. Once inside the cells, free fatty acids and monoglycerides are picked up by the metabolic machinery in the endoplasmic reticulum and are rapidly resynthesized to triglycerides and phospholipids. Consequently, the concentration of fat digestion products at the membranous surface of the epithelial cells is nearly zero. There is a diffusion gradient from the well-stirred bulk phase through the unstirred layer to the surface of the epithelial cells. Free fatty acids and 2-monoglycerides diffuse down the gradient through the unstirred layer to be absorbed by the epithelial cells. The flux of molecules through the layer is the product of the diffusion gradient and the diffusion constant of the molecules in solution. The concentration of free fatty acids and monoglycerides in solution is so low that their flux through the unstirred layer is not great enough to achieve the rate of absorption that actually occurs.

Most of the free fatty acids and monoglycerides diffuse through the unstirred layer aggregated in micelles. Diffusion coefficients decrease as the molecular weight increases. The weight of micelles is about 300–400 times as great as that of an individual fatty acid, and the diffusion coefficient of a micelle is approximately one seventh that of a fatty acid. However, the concentration of fatty acids and 2-monoglycerides in micelles is 1,000 times as great as their concentration in solution. Therefore, the flux of fatty acids in micelles through the unstirred layer is more than 100 times as great as the flux of fatty acids in solution (Table 16–1; Fig 16–5). Formation of micelles from the products of triglyceride hydrolysis prevents segregation of hydrolysis products into poorly soluble oils, keeps the luminal solution satu-

rated, and provides the vehicle that carries most of the free fatty acids and 2-monoglycerides through the unstirred layer to the absorbing surface of the intestinal epithelial cells.

The constituents of micelles are absorbed at different rates at different sites; micelles themselves are not absorbed. Monoglycerides and free fatty acids are absorbed chiefly in the duodenum and upper jejunum, and most of the conjugated bile acids are absorbed in the terminal ileum. Because cholesterol is absorbed more slowly than are free fatty acids or monoglycerides, the concentration of cholesterol rises in micelles as chyme moves down the intestine.

Lipids that are relatively soluble in water need not be carried in micelles. Short-chain and medium-chain fatty acids and the triglycerides of those acids are sufficiently soluble in molecular form so that diffusion through the unstirred layer is rapid enough to allow their absorption without micelle formation.

Lecithin entering the intestine from the diet or bile is hydrolyzed within the lumen to lysolecithin, which is absorbed as such into the cells. There its free alcohol group is acylated, and the resulting phospholipid appears in chylomicrons. In addition, about 20%–30% of the phospholipid in chylomicrons is synthesized from α-glycerophosphate and free fatty acids within the epithelial cells.

Sphyngomyelin, the phospholipid composed of a fatty acid, choline, phosphoric acid, and either sphyngosine or dihydrosphyngosine, appears to be absorbed as such into the epithelial cells where it is degraded. None occurs in chylomicrons.

Intracellular Fat Transport and Metabolism

As fat is being absorbed, the apical cells at the tip of an intestinal villus contain more fat than do the cells at the sides, and the fat is restricted to the supranuclear parts of the cell. Fat digestion products enter the epithelial cell as individual molecules, and they

TABLE 16–1.—Diffusion of Fatty Acids as Molecules in Solution, as Free Fatty Acids, and as Constituents of 2-Monoglycerides Through an Unstirred Layer

FORM	DIFFUSION GRADIENT (μM CM^{-3})	DIFFUSION COEFFICIENT (CM^2 SEC^{-1} 10^{-6})	FLUX OF FATTY ACIDS (μM CM^{-1} SEC^{-1} 10^{-6})	RELATIVE FLUX
Molecules	0.01	7	0.07	1
Micelles	10.00	1	10.00	142

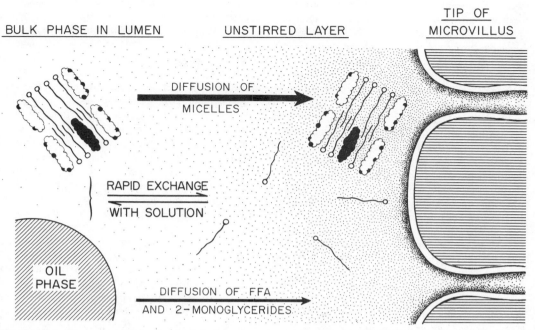

BULK PHASE IN LUMEN   **UNSTIRRED LAYER**   **TIP OF MICROVILLUS**

DIFFUSION OF MICELLES

RAPID EXCHANGE WITH SOLUTION

OIL PHASE

DIFFUSION OF FFA AND 2-MONOGLYCERIDES

Fig 16–5.—The role of micelles in fat absorption: holding the products of lipolysis in micellar solution, supplying free fatty acids (FFA) and 2-monoglycerides for solution in the aqueous phase and carrying fat digestion products through the unstirred layer. So long as there is an oil phase, free fatty acids and 2-monoglycerides exchange between the oil phase and molecular solution. Because there is exchange between micelles and solution and between oil phase and solution, there is exchange between micelles and oil phase.

pass through the terminal web before aggregating. During fat digestion, the smooth endoplasmic reticulum, beginning just below the terminal web, becomes filled with triglyceride droplets. Microsomes derived from the endoplasmic reticulum contain the enzymes that resynthesize triglycerides, and the fat droplets within the reticulum are newly synthesized by the adjacent membrane. The droplets are contained within a labyrinthine system of small cavities connected by slender trabecular bridges. In addition, irregular masses of fat not included in a membranous envelope lie in the cytoplasm. There is, however, no evidence that the cytoplasmic matrix contains any increased concentration of fatty acids or dissolved fat; resynthesis apparently keeps pace with absorption. Triglyceride droplets are transported through the reticulum to the Golgi apparatus in the supranuclear part of the cell. There chylomicron assembly is completed, and 20 or more nascent chylomicrons are enclosed within the membrane of a secretory vesicle. The secretory

vesicle migrates to the lateral cell membrane, where the membrane of the secretory vesicle fuses with the plasma membrane of the cell. Chylomicrons are then discharged into the intercellular space.

Most monoglycerides absorbed into the mucosal cells are re-esterified to triglycerides or phospholipids without further hydrolysis. The lipase contained in mucosal cells hydrolyzes ester bonds of short-chain acids, e.g., trioctanoin, but not those of long-chain acids, such as tripalmitin (C_{16}); therefore, it plays little role in the intracellular metabolism of most dietary fat. Human subjects with cannulated thoracic lymph ducts have been fed 2-monoglycerides labeled with ^{3}H in the glycerol and with ^{14}C in the fatty acid; total recoveries of the label in lymph ranged from 35% to 55%. In the first 6 hours after feeding, triglycerides in lymph had the same ratio of labels as that of the monoglyceride fed, and the labeled fatty acid was in the 2-position. After 6 hours, the proportion of labeled glycerol to labeled fatty acid decreased, and there was some change in position of the fatty acid in the triglyceride; some hydrolysis and isomerization had occurred. The small quantity of 1-monoglycerides produced during digestion is absorbed and re-esterified to triglycerides without intermediate hydrolysis.

A few fatty acids entering the cell are oxidized to provide energy; the fate of the rest depends upon their chain length. About 10%–15% of the longest fatty acids escape into portal blood, where, bound to albumin, they are carried to the liver. The proportion of free fatty acids going to the liver in portal blood increases as chain length decreases; most medium-chain and all short-chain fatty acids become blood-borne. The remaining fatty acids are resynthesized into triglycerides and phospholipids before they leave the mucosal cells to enter the lymph. In one experiment in which free palmitic acid (C_{16}) labeled with ^{14}C was fed to rats, either alone or mixed with triglycerides, 92% of the labeled acid was recovered in the lymph. Of this, 80%–90% was incorporated into triglycerides and the remainder into phospholipids. None of the labeled acid occurred as free fatty acid. The same results have been obtained when labeled long-chain acids esterified as triglycerides were fed; despite very extensive intraluminal hydrolysis, all labeled acids recovered in lymph were in ester form. Experiments using human subjects from whom intestinal lymph can be recovered give qualitatively similar results.

Synthesis of triglyceride ester bonds is not simple reversal of lipase hydrolysis. The ester bonds of 2,2-dimethylstearin are not split by pancreatic lipase, and when 2,2-dimethylstearic acid is mixed with triglycerides and pancreatic lipase, either in vitro or in the lumen of the intestine, the acid is not incorporated into triglycerides. Nevertheless 2,2-dimethylstearic acid is absorbed, and it appears in lymph in triglycerides and phospholipids. Catalyzed by a kinase requiring Mg^{++}, fatty acids react with ATP and coenzyme A to form fatty acyl~CoA; this product in turn reacts with monoglycerides to give diglycerides or with diglycerides to give triglycerides. One fatty acyl~CoA can also react with L-α-glycerophosphate to form lysophosphatidic acid. This product can react with a cytidine intermediate to become a phospholipid, or it can lose its phosphate and become a diglyceride. Mucosal cells can also synthesize long-chain fatty acids, chiefly stearic, (C_{18}), from acetic acid, and they can lengthen palmitic acid by two carbons to form stearic acid.

Chylomicrons are assembled and packaged in se-cretory vesicles within the Golgi apparatus. Chylomicrons are 0.1–3.5 mμ in diameter, and they are covered with a mosaic membrane of protein, free cholesterol, and saturated triglycerides in a monolayer of phospholipid. The phospholipid, which constitutes only 5%–9% of the total lipid of a chylomicron, covers 40%–64% of the surface; none is in the core of the chylomicron. An adequate supply of phospholipid is essential for chylomicron formation. Seventy to eighty percent of the phospholipid of chylomicrons is derived from absorbed lysolecithin, which in turn is derived from lecithin. Consequently, an adequate luminal supply of lecithin is important for the translocation of absorbed fat into the lymph. The protein coating chylomicrons is β-lipoprotein whose apoprotein is synthesized with the mucosal cells. (Other apoproteins destined to become part of high-density lipoproteins are also secreted by the mucosa during fat absorption.) There is only enough protein to cover 10% of the particle's surface. The droplets contain 89%–93% triglyceride by weight, 5%–9% phospholipids, 0.7%–1.5% cholesterol, and 1%–7% free fatty acids. Fat in chylomicrons only remotely reflects the composition of dietary fat, for most short- and medium-chain fatty acids and some glycerol are shunted to the portal blood, and saturated, long-chain fatty acids are added by mucosal synthesis.

Part of the glycerol liberated by hydrolysis of triglycerides in the intestinal lumen is oxidized by the mucosal cells to carbon dioxide, and some finds its way to the liver, where it is transformed to glycogen. The remainder is used for resynthesis of triglycerides. When 1.67 gm of glycerol-1-^{14}C was fed along with free fatty acids to a patient with chyluria, over 93% of the ^{14}C was recovered in the glycerol of lipids of the urine in the next 24 hours. In two experiments, 14% and 19% of all the triglycerides recovered in 24 hours were labeled, and as much as 60% of the lipids of a single sample contained glycerol-1-^{14}C. Glycerol used for resynthesis is also derived from glucose, the immediate precursor being L-α-glycerophosphate.

During fat absorption, the phospholipids of the intestinal mucosa turn over rapidly, although their concentration remains constant. When phosphate labeled with ^{32}P is fed along with triglyceride, the specific activity of the phospholipids of the mucosa rises rapidly. Peak concentration of ^{32}P in phospholipids is reached 4 hours after feeding, but the various fatty acids are incorporated into phospholipids at different rates. Stearic acid enters more rapidly than do palmitic and myristic acids. Intracellular phospholipid metabolism may be part of the process by which fat droplets are provided with an envelope. Phospholipid turnover is greatly accelerated during transport of particles through the endoplasmic reticulum, and phospholipids may be a major component of the membranes surrounding the particles.

Fat hydrolysis and resynthesis are summarized in Figure 16–6.

Absorption of Fat Into Lymph and Portal Blood

Upon being extruded from the intestinal epithelial cells, chylomicrons accumulate in the intercellular spaces and eventually reach lymph channels. Chylomicrons are too large to diffuse the distance, and they are probably carried by the bulk flow of water moving in the same direction during absorption. Chylomicrons first cross the basement membrane which lines the basal surface of the epithelial cells; whether they pass through the membrane itself or through holes in the membrane is unknown. Some electron micrographs show chylomicrons densely packed on the cellular side of the membrane and absent from the other side, implying that the membrane is a physical barrier for them. Other electron micrographs show chylomicrons on both sides of an apparently intact membrane, but holes in the membrane could be out of the plane of the section. Capillaries of the villi are fenestrated, but the fenestrae are stopped by an uninterrupted basement membrane. These fenestrea may be the ports of entry of water-soluble compounds into capillary blood, but the basement membrane prevents the entry of chylomicrons. On the other hand, chylomicrons enter lacteals through open channels. In the terminal lymph sacs of fasting animals, adjacent endothelial cells interdigitate, forming a continuous cellular lining of the sac. There is no basement membrane surrounding the sacs. During absorption, the sacs are expanded; the epithelial cells are stretched out so that there are distinct gaps between cells. Chylomicrons enter lymph channels through these gaps. They may also travel through lymphatic endothelial cells in large vesicles. Once within the terminal lacteals, the fat droplets are carried centrally by the tidal flow of lymph caused by contraction of smooth muscle fibers of the villi or by more gross movements of the mucosa. Once delivered into the blood, chylomicrons do not recirculate in the lymph.

Lymph flow increases during fat absorption, and absorbed fat may appear in the lymph as long as 13 hours after fat has entered the intestine. The fat content of the lymph may be as high as 6%, and only a few hundred milliters of lymph is required to carry the absorbed fat. When the thoracic duct is ligated, absorption of fat is temporarily reduced; after a few days, absorption returns to normal, probably because other lymph-to-vein channels open. However, fat absorption is grossly impaired when intestinal lipodystrophy (Whipple's disease), in which condition the lymphatics are filled by accumulation of a mucopolysaccharide.

Medium-Chain Triglycerides

Short-chain and medium-chain fatty acids are not abundant in the diet. Depot fat contains few fatty acids shorter than 10 carbons, and less than 10% of the fatty acids in milk are shorter than 14 carbons. Synthetic triglycerides containing medium-chain (C_6 to C_{12}) fatty acids are fed as a source of energy to patients who are unable to digest and absorb triglycerides made of long-chain fatty acids.

Intraluminal hydrolysis of medium-chain triglycerides is faster than that of long-chain ones; and, because rapid isomerization between the 2- and 1-positions of glycerol occurs, hydrolysis is more complete. However, prior hydrolysis and micelle formation are not required for absorption. Medium-chain triglycerides, absorbed intact, are hydrolyzed within the mucosal cells by intestinal lipase, and most of their constituent glycerol and fatty acids pass into the portal blood. Very few medium-chain fatty acids are re-esterified. When the rare patient with chyluria whose abdominal lymph vessels drain into his renal pelvis is fed triglycerides of mixed long-chain and medium-chain fatty acids, no fatty acids of 8 or 10 carbons appear in the triglycerides of his urine, and the proportion of lauric (C_{12}) and myristic (C_{14}) fatty acids to the longer fatty acids is less than that in the ingested triglycerides. Only an occasional medium-chain fatty acid may be incorporated into a triglyceride containing two long-chain fatty acids and find its way into the chylomicrons of lymph.

Malabsorption: Pancreatic Deficiency

A classification of the physiological disturbances leading of malabsorption of fat is given in Table 16–2.

Steatorrhea is defined as the excretion of more than 6 gm of fat per day in the stool. Excess fat may be

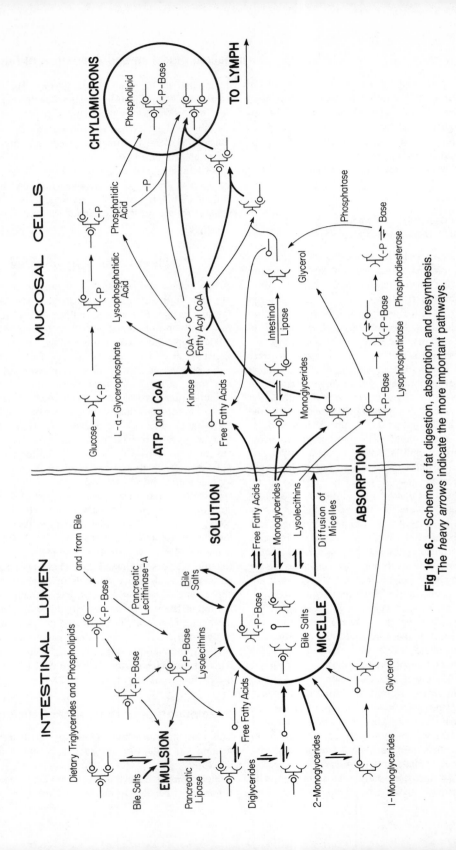

Fig 16–6.—Scheme of fat digestion, absorption, and resynthesis. The *heavy arrows* indicate the more important pathways.

TABLE 16–2.—Errors in Fat Digestion and Absorption Leading to Steatorrhea*

LOCATION	STEP	PHYSIOLOGICAL DISTURBANCE	DISEASE STATE
Intraluminal	Emulsification of triglycerides	Impaired emulsification	Deficiency of conjugated bile salts: biliary fistula or obstruction, iliectomy, bacterial deconjugation
	Hydrolysis to fatty acids and monoglycerides	Pancreatic lipase deficiency	Pancreatic disease
		Absolute or relative bicarbonate deficiency	Pancreatic disease or gastric hypersecretion
	Formation of micellar phase	Conjugated bile salt deficiency	Biliary fistula or obstruction; bacterial deconjugation
		Altered partition of fatty acids between oil and micellar phases from absolute or relative deficiency of bicarbonate	Pancreatic disease or gastric hypersecretion
Intraluminal to cellular	Absorption of fatty acids and monoglycerides by mucosal cells	Decreased cell uptake; reduction in number, activity, or surface area of cells	Intestinal resection of bypass; tropical sprue; gluten enteropathy
		Cells already saturated with fatty acids and monoglycerides	Alteration in cellular steps or triglyceride synthesis, chylomicron formation, or transport from cell
		Decreased time of contact of micellar phase with mucosal cells	Increased transit time
	Resynthesis of fatty acids and monoglycerides to triglycerides	Deficiency of resynthesizing enzymes	Not described
Cellular	Chylomicron formation	Deficiency in synthesis of chylomicron protein	A-β-lipoproteinemia
Cellular to extracellular	Transport of chylomicrons from cell via lymph channels to blood	Lymphatic obstruction or lymphangiectasia	Lymphosarcoma; intestinal lipodystrophy; protein-losing enteropathy

*Adapted from Hofmann A.F.: *Gastroenterology* 50:56, 1966.

present in oil droplets as unhydrolyzed triglycerides or as a mixture of triglycerides and free fatty acids. Azotorrhea, excessive excretion of undigested and unabsorbed protein in the stool, closely parallels steatorrhea.

Normal men are capable of absorbing up to 150–200 gm of fat a day. No more than 2–4 gm of the 3–6 of fat excreted in the stool each day is of dietary origin; the remainder comes from intestinal secretions and desquamation and from the colonic flora. About 40% of this fecal fat is in the bodies of bacteria; the remainder is in nonbacterial solids. In the complete absence of bile or pancreatic juice, fat escapes into the stool roughly in proportion to the amount in the diet (Fig 16–7). In biliary deficiency, half the dietary fat is not absorbed, and in pancreatic deficiency two thirds is not. Although fat in the stool, or steatorrhea, demonstrates absorptive defects, the really striking fact is that fat can actually be absorbed at all when there is no pancreatic lipase or bile acids.

There is a large pancreatic reserve, and for pancreatic deficiency to affect fat digestion and absorption adversely, the amount of lipase in the intestinal lumen must be less than 10% of the normal value. Four fifths to nine tenths of the gland may be removed without interfering with fat digestion or absorption. Only 15% of patients with disturbances of pancreatic secretion show malabsorption of fat, and those are the most advanced cases. In steatorrhea, the composition of fecal fat differs from that of dietary fat. Short-chain fatty acids are absent, and abnormal fats produced by bacterial modification of unabsorbed

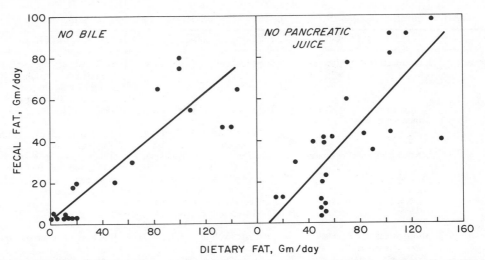

Fig 16–7.—Relation between dietary fat and total fecal fat in man. **Left,** in complete absence of bile. Slope of the regression line is 0.46 ± 0.05. **Right,** in complete absence of pancreatic juice. Slope of the regression line is 0.47 ± 0.09. (Adapted from Annegers J.H.: *Q. Bull. Northwestern Univ. Med. School* 23:198, 1949.)

fats are present. One such product is a hydroxy fatty acid structurally similar to ricinoleic acid, the fatty acid of castor oil; it may cause diarrhea.

In pancreatic deficiency, a large fraction of fat is hydrolyzed and absorbed. The lipase concentration of jejunal contents of one patient with 75% pancreatectomy was found to be 70% of normal value; this patient did not have steatorrhea on a high-fat diet. Seven patients, each having greater that 95% complete pancreatectomy, had no demonstrable functioning pancreatic tissue. They were fed 83–130 gm of triglycerides a day, and they absorbed 61–91% of the fat. In them, the lipase concentration of jejunal contents was one tenth that of control subjects. In two of these patients, one with moderate and the other with minimal steatorrhea, jejunal contents were aspirated after a fatty meal. The aspirates contained, on a molar basis, 8% of the fatty acids in triglycerides, 14% and 25% in diglycerides, and 22% and 39% in monoglycerides; 39% and 45% of the fatty acids were free. The extent of hydrolysis was similar to that of normal men.

Feeding a preparation of porcine pancreatic enzymes with meals usually decreases fecal fat in patients with pancreatic deficiency. Unfortunately, fed enzymes are inactivated or digested in an acid medium. Exogenous pancreatic enzymes are most effec-

tive if fed with meals and when acid secretion is inhibited by concurrent treatment with cimetidine.

Steatorrhea as the result of pancreatic enzyme deficiency occurs when enzymes secreted by a normal pancreas are inactived by acid flooding the duodenum in hypersecretory states such as the Zollinger-Ellison syndrome.

When bile salts are below their critical micellar concentration in the intestinal lumen, fat absorption is reduced, and voluminous fatty stools are passed that contain approximately half the dietary fat. The defect is not in fat hydrolysis but in absorption. Hydrolysis is adequate, for between 80% and 100% of the fat in the stool is in the form of free fatty acids, although only triglycerides have been ingested.

All the steps down to micelle formation can occur in the absence of bile. Fat is emulsified by protein, lecithin, and lysolecithin, and colipase and lipase attach to emulsified fat. Once hydrolysis begins, monoglycerides assist emulsification. Although some micelles can be formed by lysolecithin and monoglycerides, not enough are formed to carry all fat that digestion produces rapidly through the unstirred layer. Some hydrolyzed fat can be absorbed from micelles and from molecular solution, for the defect is relative rather than absolute. The larger part of hydrolyzed fat remains in the oil phase.

When ileal disease or ileal resection interrupts the enterohepatic circulation, deficiency of bile acids is periodic. After the first meal of the day, duodenal bile acid concentration is about half normal. After subsequent meals, the concentration is very low, for the first meal has flushed out the pool of acids accumulating overnight. During bacterial overgrowth, the concentration of conjugated bile acids is reduced. In both states, bile acid concentration is below the critical micellar concentration; the percentage of luminal fat in the micellar phase may be only 2%, compared with the normal value of 50%–60%. Consequently, fat absorption is seriously impaired.

Other Absorptive Defects

Malabsorption of fat occurs widely in the tropics, and the disease tropical sprue has been attributed to deficiency of pteroylglutamic acid, vitamin B_{12}, and adequate protein. In prolonged deficiency, irreversible atrophy of the small intestinal mucosa occurs. Nontropical sprue, or idiopathic steatorrhea, occurs in temperate climates and is unrelated to dietary deficiency. There are thinning and irregular dilatation of the small intestinal mucosa, with atrophy of both mucosa and muscularis. Microvilli on the epithelial cells become short and clubbed. Degeneration of myenteric and submucous plexuses may also occur. The total fat content of the stool amounts to 40%–50% on a dry-weight basis, or about 50 gm/day. This equals 450 calories lost. Both the absorptive and the morphologic defects can be reversed by elimination of gluten from the diet. In persons who have been relieved of steatorrhea in this way, another attack can be precipitated by feeding purified gluten or fragments of it produced by peptic digestion. For this reason, the name idiopathic steatorrhea is being dropped in favor of gluten enteropathy.

Patients with steatorrhea may have osteomalacia as the result of reduced calcium absorption and a consequent negative calcium balance. Loss of calcium in fatty stools has been attributed to vitamin D deficiency and to the carriage of calcium into the stool as calcium soaps. However, the patients often do not respond to IM-administered vitamin D, and there is no consistent relation between the degree of steatorrhea and loss of calcium.

Patients with severe malabsorption of fat often have calcium oxalate kidney stones. They absorb dietary oxalate much more readily than do normal persons with the result that they have hyperoxaluria, whose intensity correlates well with the severity of their fat absorption defect. There are two reasons for increased absorption of dietary oxalate by the colon: (1) Fatty acids bind calcium, and when substantial amounts of free fatty acids reach the colon, less calcium is available to precipitate oxalate in unabsorbable form. (2) Dihydroxy bile acids increase oxalate permeability of the colon, but not of the jejunum, about ten times, and this effect is independent of calcium. Therefore, in those forms of steatorrhea in which a large amount of dihydroxy bile acids reaches the colon, passive absorption of oxalate is increased.

Cholesterol Absorption

Cholesterol, which is present in all diets, chiefly as the free alcohol, is absorbed by adult human beings at the rate of about 10 mg/kg day. When 10–20 eggs are fed, cholesterol absorption is doubled. The upper limit of absorption when a large amount is fed is about 2 gm/day. Peak delivery of cholesterol into the circulation by way of the lymph occurs, in man, 9 hours after feeding, compared with the peak of fat absorption 3 hours earlier. When cholesterol absorption is studied by means of trace amounts of cholesterol-^{14}C, the compound is found to be rapidly absorbed into the intestinal mucosa, reaching its highest concentration in 3 hours. Then it is slowly released into the lymph.

Cholesterol in the intestinal lumen is derived from two sources in addition to the diet: from the bile, which in the adult man carries 1–2 gm/day, and from the secretions or desquamated intestinal mucosal cells. The latter source has been found, in three adults with complete obstruction of biliary and pancreatic ducts, to be 0.25–0.4 gm/day. When a patient with chyluria was fed a low-cholesterol diet containing 0.04 gm/day, she excreted 0.34 gm into her urine. Her shunt from intestinal lymphatic channels to renal pelvis was estimated to divert 40% of her intestinal lymph; therefore, she absorbed and synthesized in her mucosa about 0.85 gm/day, or 21 times as much as she ate. Cholesterol not absorbed enters the colon, where it is reduced by colonic flora to coprostanol. No reduction occurs in the small intestine, for coprostanol is not present in fluid collected from an ileostomy. The fecal excretion of total sterols by a normal 70-kg man on a mixed diet is 0.5 gm/day, with a range of 0.12–0.90 gm.

Most or all esterified cholesterol is hydrolyzed by pancreatic cholesterolesterase before it is absorbed. When a cholesterol ester of a labeled fatty acid is fed, less than 2% of the label is recovered in the cholesterol esters of lymph. The more rapidly a fed cholesterol ester is hydrolyzed by the esterase, the more rapidly it is absorbed.

Cholesterol must be incorporated into micelles in order to be absorbed. When bile is absent and micelle formation greatly reduced, no cholesterol is absorbed. Less than half the cholesterol in the lumen is micellar. In one sample of intestinal contents recovered from the duodenum and upper jejunum of a human subject and centrifuged at $100,000 \times g$ for 24 hours, the supernatant oil contained 0.19 mg of cholesterol per ml, and the subnatant micellar solution contained 0.15 mg/ml, of which only 0.04 mg was esterified. Differences in the capacity of sterols to form micelles probably account for selectivity in their absorption. Ergosterol, which is poorly absorbed, does not form micelles as readily as cholesterol; the critical concentration of bile acids required to carry ergosterol into micelles is more than 3 times that needed for cholesterol, and ergosterol's concentration in micelles is about half that of cholesterol. Cholesterol is not absorbed at the same rate as other constituents of micelles. When a 50-cm length of upper jejunum of a human subject was perfused with a micellar solution made of bile acids, 1-monoglycerides, and cholesterol, none of the bile acids, all of the monoglycerides, and 73% of the cholesterol were absorbed. A quarter of the cholesterol in the solution was recovered at the lower end of the perfused segment sedimented during high-speed centrifugation, showing that this fraction had been liberated from micelles and was no longer soluble. Concurrent absorption of exogenous fat is not required for cholesterol absorption. Cholesterol fed in a fat-free diet can be absorbed, probably because endogenous fat sufficient for micelle formation is always present in intestinal contents. Simultaneous absorption of unsaturated fats stimulates cholesterol absorption to a greater degree than does absorption of an equivalent quantity of saturated fats.

Once absorbed into the mucosal cells of the duodenum and upper jejunum, cholesterol mixes with the large pool of cholesterol within them. All epithelial cells from stomach to colon synthesize cholesterol from acetate, those of the ileum and distal colon being most active. A tracer dose of cholesterol-^{14}C remains detectable within the mucosal pool for 48–72 hours. In the fasting human subject, intramucosal turnover results from continuous transfer of intracellular cholesterol to lymph and its continuous replacement by new synthesis and by absorption of cholesterol secreted in bile. A total of 1–3 gm of cholesterol from all sources is delivered from the intestinal tract to the lymph each day. Synthesis of cholesterol in the liver adds to the circulating pool; when absorption of cholesterol is low, hepatic synthesis increases. Fed cholesterol is therefor diluted by cholesterol already circulating and synthesized by intestine and liver.

Although by far the largest fraction of cholesterol within the mucosa is the free alcohol, cholesterol is esterified in the cells, and two thirds or more of it in lymph is esterified with fatty acids drawn both from fed fat and from the intracellular fatty acid pool. In cholesterol esterification, a fatty acid is transferred from carnitine to CoA and then to cholesterol. Oleic acid is the acid most frequently used. Cholesterol leaves the cells in chylomicrons or in very low density lipoproteins; about 70% is in chylomicrons and the rest in the aqueous phase. Chylomicrons of human lymph contain 0.7%–3.5% cholesterol ester and about 1% free cholesterol.

Absorption of Other Sterols and Vitamins

Other ingested sterols are absorbed to varying extents. The rate of absorption is roughly related to the sterol's structure: the more lipoidal the greater the rate; the more polar and water soluble, the lower. Phytosterols and ergosterol with additional double bonds in the B ring are very poorly absorbed. When phytosterols uniformly labeled with ^{14}C are fed to a rat, they enter the intestinal mucosa, where they occur as free alcohols. Only 2% of the label on these phytosterols appears in the lymph in 24 hours, and about one third of the phytosterols is metabolically degraded in the intestinal mucosa until it is no longer identifiable as sterol. When the human small intestine is perfused with a micellar solution of bile acids, monoglycerides, and cholesterol, phytosterols appear in the perfusate; they are apparently excreted by the mucosal cells. Administration of these and other related sterols blocks absorption of cholesterol, so that fecal excretion of cholesterol rises and its lymph concentration falls.

The several forms of vitamins A, D, E, and K and their precursors are fat-soluble and only slightly if at all water-soluble. Consequently, their incorporation into micelles is essential for their transport through the unstirred layer and for their adequate absorption. In the absence of bile they are poorly absorbed. This is particularly important in the case of vitamin K which is not stored in appreciable amount. Being fat-soluble, the vitamins are absorbed into the intestinal epithelial cells by passive diffusion throughout the length of the small intestine and colon. β-carotene derived from plants is absorbed into the intestinal epithelial cells where it is converted to retinol, first by cleavage to retinaldehyde and then by reduction to retinol. The long-chain esters of retinol derived from animal tissues are hydrolyzed in the intestinal lumen, and free retinol is absorbed. In the epithelial cells, retinol from both sources is re-esterified and incorporated in chylomicrons to be delivered to the liver. Both vitamins D_2 and D_3 are absorbed and incorporated into chylomicrons for transport to the liver where the first step in conversion to the physiologically active derivative occurs.

The sterol derivatives—testosterone, cortisol, cortisone acetate and corticosterone—are not transported in lymph after absorption; they enter the circulation by way of the portal blood.

REFERENCES

Borgström, B.: Importance of phospholipids, pancreatic phospholipase A_2, and fatty acid for the digestion of dietary fat. *Gastroenterology* 78:954, 1980.

Dietschy J.M.: The role of the intestine in the control of cholesterol metabolism. *Gastroenterology* 57:461, 1969.

Dietschy J.M.: Mechanisms of bile acid and fatty acid absorption across the unstirred layer and brush border of the intestine. *Helv. Med. Acta* 37:89, 1972.

Dietschy J.M., Gotto A.M. Jr., Ontko J.A. (eds.): *Disturbances in Lipid and Lipoprotein Metabolism*. Bethesda, American Physiological Society, 1978.

Hofmann A.F.: Functions of bile in the alimentary canal, in Code C.F. (ed.): *Handbook of Physiology:* Sec. 6. *Alimentary Canal,* vol. 5. Washington, D.C., American Physiological Society, 1968, pp. 2507–2533.

Johnson J.M.: Mechanism of fat absorption, in Code C.F. (ed.): *Handbook of Physiology:* Sec. 6. *Alimentary Canal,* vol. 3. Washington, D.C., American Physiological Society, 1968, pp. 1353–1376.

Kayden H.J., Senior J.R. Mattson F.H.: The monoglyceride pathway of fat absorption in man. *J. Clin. Invest.* 46:1695, 1967.

Nervi F.O., Dietschy J.M.: The mechanisms of and the interrelationship between bile acid and chylomicron-mediated regulation of hepatic cholesterol synthesis in the liver of the rat. *J. Clin. Invest.* 61:895, 1978.

Regan P.T., Malagelada J.R., DiMagno E.P., et al.: Comparative effects of antacids, cimetidine and enteric coating on the therapeutic response to oral enzymes in severe pancreatic deficiency. *N. Engl. J. Med.* 297:854, 1977.

Rommel K., Goebell H. (eds.): *Lipid Absorption: Biochemical and Clinical Aspects*. Baltimore, University Park Press, 1976.

Sabesin S.M., Frase S.: Electronmicroscopic studies of the assembly, intracellular transport, and secretion of chylomicrons by rat intestine. *J. Lipid Res.* 18:496, 1977.

Treadwell C.R., Vahouny, G.V.: Cholesterol absorption, in Code C.F. (ed.): *Handbook of Physiology,* Sec. 6. *Alimentary Canal,* vol 3. Washington, D.C., American Physiological Society, 1968, pp. 1407–1438.

Weiner I.M., Lack L.: Bile salt absorption; enterohepatic circulation, in Code C.F. (ed.): *Handbook of Physiology:* Sec. 6. *Alimentary Canal,* vol. 3. Washington, D.C., American Physiological Society, 1968, pp. 1439–1456.

Wilson F.A., Dietschy J.M.: Characterization of bile acid absorption across the unstirred water layer and brush border of the rat jejunum. *J. Clin. Invest.* 51:3015, 1972.

Wilson T.H.: *Intestinal Absorption*. Philadelphia, W.B. Saunders Co., 1962.

Wiseman G.: *Absorption from the Intestine*. New York, Academic Press, 1964.

17

Absorption and Excretion
by the Colon

THE COLON receives more than 500 ml of chyme a day from the ileum; from the chyme the colon absorbs sodium and water, and to the residuum the colon adds potassium and bicarbonate. About 100 ml of water, containing sodium and potassium in amounts less than the dietary intake, is lost in the stool each day; but when diarrhea occurs, losses of water, sodium, and potassium may be catastrophic. The abundant microflora inhabiting the colon of omnivorous animals, such as man, partially digests the residues of plant cells and contributes trace nutrients. The microbial mass makes up 60% of the wet weight of contents of the colon. Bacterial fermentation adds hydrogen and, in 3 of 10 persons, methane to the gas arriving from the small intestine, and about 1,000 cc/day of gas is passed as flatus. Frequency of bowel movements is extremely variable, but at some low frequency the symptoms of constipation appear. These are largely the result of chronic distention of the rectum. Under normal circumstances, any potentially toxic compound absorbed from the colon is prevented by the liver from entering the general circulation.

Ability of the Human Colon to Absorb

The rate at which water and electrolytes are delivered to the human colon has been estimated in two ways: by transintestinal intubation and by measurement of ileostomy fluid. For the first, a double lumen tube is passed by mouth into the ileum. A solution containing the marker polyethylene glycol is infused into the lower ileum at a known rate, and ileal contents are sampled just above the ileocecal valve. Analysis of the fluid recovered gives its electrolyte composition, and dilution of the marker gives the rate of flow of fluid at the site of sampling. Ileal flow estimated in this way may be somewhat higher than that occurring in unintubated subjects, because the polyethylene glycol used as a marker increases the osmotic pressure of the luminal contents and therefore retains some water in the lumen. At the same time, stools are collected and analyzed. The difference between ileal flow and excretion in the stool is the amount absorbed. The data from such observations given in Table 17–1 show the importance of the colon in conserving water, sodium, and chloride.

When ileal flow was measured over 24 hours, it was found to be least overnight, averaging 0.1–1.7 ml/min. It rose to a peak of 6–8 ml/min after a meal. In order to mimic these changes in flow, the experiments were performed, the results of which are given in Table 17–2. By means of transintestinal intubation, fluid whose composition was similar to that of ileal fluid was infused into the cecum at the rates given in the table. Stools and urine were collected and analyzed. Again, the capacity of the colon to absorb, even at high rates of inflow, was demonstrated.

A patient with an ileostomy is denied the conservative function of the colon. Immediately after an ileostomy is created, fluid loss is high. Eventually, fluid loss falls; the average in 21 patients with well-established ileostomies was found to be 690 ml/day. Some clinicians believe that patients learn to reduce their fluid intake and therefore their fluid loss; others have found no relation between fluid intake and ileostomy loss. The mass of sodium lost in ileostomy fluid averages 85 mEq/day, with a range of 44–197. Because this loss is equal to or greater than the usual sodium intake, a patient with an ileostomy is in danger of dehydration. The mass of potassium lost is far

TABLE 17–1.—Fluid and Electrolyte Load and Absorption
in the Colon of 5 Normal Human Subjects*

	CONCENTRATION IN TERMINAL ILEAL FLUID (mEQ/L)	QUANTITY (PER 24 HR)		
		TERMINAL ILEUM	STOOL	ABSORBED
Water Mean	—	1,524 ml	39 ml	1,485 ml
Range	—	1,255–1,751	22–43	1,192–1,718
Na$^+$ Mean	127	196 mEq	1 mEq	195 mEq
Range	101–139	139–243	0–2	137–242
K$^+$ Mean	6	9 mEq	5 mEq	5 mEq
Range	5–7	7–11	1–8	1–7
Cl$^-$ Mean	67	103 mEq	1 mEq	103 mEq
Range	50–89	63–123	0–1	62–123

*Adapted from Phillips S.F., Geller J.: *J. Lab. Clin. Med.* 81:733, 1973.

TABLE 17–2.—Ability of the Human Colon to Absorb
Fluid Infused into the Cecum*†

	RATES OF INFUSION			
	Low (2,000 ml/day)		High (4,000 ml/day)	
	50% in 4 2-hr periods at 2 ml/min; rest at 1 ml/min		50% in 4 2-hr periods at 4.5 ml/min; rest at 2 ml/min	
	Load	*Output*	*Load*	*Output*
Na$^+$	278 mEg/day	4	570 mEq/day	28
K$^+$	20	7	40	11
Infused water appearing in urine (%)	88–92		81–84	

*Adapted from Debongnie J.C., Phillips S.F.:*Gastroenterology* 74:698, 1978.
†Composition of fluid infused in mEq/L: Na$^+$, 145; K$^+$, 10; Cl$^-$, 110; HCO$_3^-$, 45.

below the normal intake, and there is little possibility that negative potassium balance will occur as the result of an ileostomy.

Most absorption occurs in the cecum and transverse colon. Nevertheless, the sigmoid colon and the rectum have substantial ability to absorb, for feces retained there are dried to hard balls.

Secretion

Some animals—e.g., the racoon, black bear, and skunk—have no ileocecal valve; their intestine is continuous, without interruption, from duodenum to rectum. This emphasizes the fact that the secretory and absorptive behavior of the colonic mucosa is only quantitatively different from that of the rest of the intestine. In the intestine of man, villi do not occur be-

low the ileocecal valve, and the colonic mucosa consists of crypts, with the free surface between crypts covered by columnar epithelial cells. Crypts and epithelium have a high density of mucus-containing goblet cells, and colonic secretion is rich in mucus. The aqueous secretion is scanty. Reabsorption predominates in the colon; when colonic contents remain stagnant, the aqueous phase is absorbed, leaving only an inspissated mass. However, in pathological conditions, the colonic mucosa can be stimulated to secrete up to 500 ml/day, and that volume contributes to diarrheal fluid. The major stimuli for such secretion are the dihydroxy bile acids.

The small amount of fluid normally secreted by the colon is alkaline, for bicarbonate is secreted as chloride is reabsorbed. Bicarbonate neutralizes the products of fecal fermentation, for the surface of a fecal mass is neutral while the pH of its center may be 4.8. The secretion contains no enterokinase, invertase, lipase, or trypsin; lysozyme is present.

Slimy, viscid mucus composes only 0.4% of fresh colonic secretion. Its electrolyte composition is K^+ 146–200, Na^+ 3–10, and HCO_3^- 87–155 mEq/L. When the goblet cell is stimulated, concentrated mucus leaves through the cell's stoma. As the mucus becomes free, the cell's nucleus becomes round and its goblet part less bellied out. More of the cytoplasm is revealed; and, as mucus is lost, the cell comes to resemble a simple columnar epithelial cell. At this stage, a few granules of mucus may remain in the cytoplasm just above the nucleus. With continued stimulation, the cell loses its cytoplasm and becomes first cuboidal and finally squamous.

The cells do not readily desquamate even under strenuous stimulation by surface irritants. Surface irritants (mustard oil and the like) stimulate secretion, and so does the mechanical irritation caused by the colon's rubbing against itself or over the fecal mass. This secretion occurs independently of extrinsic innervation. Secretion is evoked by stimulation of peripheral ends of the nervi erigentes and by central stimulation of one cut nervus erigens with the other remaining intact. Acetylcholine is the mediator; and parasympathomimetic drugs, as well as pilocarpine, cause secretion, while atropine blocks it. Parasympathetic stimulation increases motility of the colon, but concomitant increase in secretion is not merely secondary to movement. Blood flow through the mucosa increases; and, in general, movement, secretion, and blood flow are parallel one with another. Stimulation of the sympathetic supply, which relaxes colonic movement and causes vasoconstriction, also reduces any ongoing secretion.

Mechanisms of Electrolyte Exchange in the Colon

The most important absorptive function of the colon is the absorption of sodium ions. Absorption of sodium conserves the major electrolyte of the extracellular fluid, and absorption of sodium is responsible for the excretion of potassium. Absorption of water depends largely upon the absorption of sodium, because water follows the major osmotically active particle from the lumen into the blood.

There are electrogenic sodium pumps on the baso-lateral borders of the colonic epithelial cells which actively pump sodium ions from within the cells into the interstitial fluid. Removal of sodium from the cells creates a sodium gradient across the mucosal border of the cells, and sodium diffuses into the cells when its concentration in the lumen is 25 mN or greater. This passage across the luminal membrane is not glucose-dependent. Sodium pumping creates an electrical gradient of about 10 mV across the colonic mucosa, the mucosal surface being negative with respect to the serosal surface. This establishes an electrochemical gradient for potassium which diffuses passively, probably by the paracellular pathway, from interstitial fluid to lumen. Potassium secretion is, therefore, entirely dependent upon sodium absorption.

Aldosterone has only a trivial effect, if any, upon electrolyte transport in the small intestine, but in the colon aldosterone induces an increase in sodium permeability of the mucosal border of the epithelial cells. Consequently, sodium ions diffuse more readily into the cells when the aldosterone concentration increases; they become more readily available to the sodium pumps, which then extrude them across the basolateral borders of the cells. In this way, the concentration of aldosterone influences sodium absorption without acting directly upon the sodium pumps. Because sodium is more rapidly pumped when the concentration of aldosterone increases, the electrochemical gradient for diffusion of potassium into the lumen is enhanced. Thus, aldosterone indirectly influences the secretion of potassium.

Secretion of potassium is increased in hyperkalemia and reduced in hypokalemia, but these effects are entirely secondary to those of hyperkalemia and hypokalemia upon adrenal cortical secretion of aldosterone.

Net absorption of sodium, of course, requires net absorption of anions, chiefly chloride. The electrochemical gradient established by sodium absorption favors passive absorption of chloride through the paracellular pathway. Nevertheless, there is active absorption of chloride in the colon, and active absorption of chloride is coupled 1:1 with secretion of bicarbonate. Because colonic contents are acid as the result of fermentation by colonic flora, secreted bicarbonate is converted to carbon dioxide, which escapes in the flatus. As the result of neutralization of bicarbonate in the lumen, two osmotically active particles, a bicarbonate ion and a hydrogen ion, disappear from the luminal fluid, and consequently a corresponding amount of water can diffuse from lumen to interstitium. The chloride-bicarbonate exchange, in effect, results in net absorption of an anion and water.

The colon does not actively absorb glucose or amino acids, and the value of nutritional enemas containing those compounds is doubtful. Short-chain fatty acids, acetic, propionic, and butyric, are produced by fermentation in the colon. Their anions are in high concentration in the colonic contents of man. In the un-ionized form, they are rapidly absorbed by passive diffusion. Oxygen uptake by the colon increases when the fatty acids are absorbed, and absorption of acetic acid increases blood flow through the colon.

Absorption of water and electrolytes, including ammonia, is reduced when portal blood pressure rises above 10 mm Hg.

Diarrhea

Diarrhea is defined, for Western man at least, as stool volume greater than 500 ml/day. It occurs when the absorptive capacity of the colon is overwhelmed by fluid delivered from the small intestine or when the absorptive capacity of the colon is reduced. This latter state may be secondary to some malfunction in the small intestine.

If ileal fluid is delivered to the colon at a steady rate, the colon of a normal man can absorb more than 2 L/day. One patient with pancreatic cholera passed 2,390 ml of water in his feces, but the colon had already absorbed 6,394 ml of water delivered to it by the ileum. If the rate of delivery is greater or is grossly irregular, the colon cannot absorb the fluid it receives.

Infusion of 250 ml of isotonic fluid into the cecum in 30 min did not increase stool size, but infusion of 500 ml in the same time usually produced a liquid stool. When a man forces himself to drink an isotonic solution (Na$^+$ 140, K$^+$ 10, Cl$^-$ 105, HCO$_3^-$ 35 mEq/L) at the rate of 2–6 L/hr, his fluid output per rectum ranges from 1 to 5 L/hr. The toxin of *V. cholerae* causes the mucosa of the jejunum and ileum to pour an enormous volume of an isotonic solution of sodium, potassium, chloride, and bicarbonate into the lumen of the small intestine. A cholera victim may lose as much as 60 L of water containing more than 8,000 mEq of sodium and 1,200 mEq of potassium in

a 5-day period. Diarrhea of intestinal origin also occurs when ileal contents delivered to the colon, in volume within the normal range, contain something that alters the function of the colon. A simple example is the watery stool that follows ingesting of magnesium sulfate. This osmotically active compound is poorly absorbed in the colon as well as in the small intestine. Those persons who, because of intestinal deficiency of oligosaccharidases, fail to digest and absorb one or another carbohydrate, have diarrheal stools containing acid fermentation products.

Because bile acids are well absorbed both actively and passively in the small intestine, less than 5% of those undergoing an enterohepatic cycle enters the colon. However, there are usually so many cycles that the usual loss into the colon is about 500 mg/day or approximately one quarter of the bile acid pool. Hepatic synthesis keeps the pool size constant. If as much as 20% of circulating bile acids fails to be absorbed in the small intestine, hepatic synthesis is still able to keep the bile acid pool at nearly normal size. Fat digestion and absorption are only minimally impaired, and fat occurs in the stool at less than 16 gm/day. However, a 20% loss of bile acids means that between 800 and 5,000 mg of bile acids may escape to the colon, the amount depending upon the number of cycles occurring in a day. In the colon, bile acids are deconjugated and dehydroxylated. Dihydroxy bile acids, with the apparent exception of ursodeoxycholic acid, inhibit sodium absorption in the colon and therefore also inhibit water absorption. They also stimulate secretion. When the amount of bile acids entering the colon is substantially in excess of 500 mg/day, diarrhea results. This kind of diarrhea can be alleviated by feeding cholestyramine, a synthetic resin that adsorbs bile acids.

If hepatic synthesis is not capable of maintaining the bile acid pool, maldigestion and malabsorption of long-chain triglycerides result in steatorrhea of more than 20 gm of fat a day. This causes diarrhea. Bacterial action on unabsorbed long-chain fatty acids produces hydroxy acids; these may stimulate intestinal motility. The diarrhea of gross steatorrhea cannot be controlled by adsorption of bile acids alone; but it can be controlled by a combination of adsorption of bile salts on cholestyramine and substitution of medium-chain triglycerides (which do not require bile acids for digestion and absorption) for long-chain ones in the diet.

Bile acid-induced diarrhea is occasionally encountered in persons with no ileal enteropathy or allied disorder, and once diagnosed is effectively treated with cholestyramine. Cholestyramine also reduces the diarrhea encountered after truncal vagotomy.

A patient with a tumor secreting vasoactive intestinal peptide has secretory diarrhea. VIP stimulates secretion by the colon as well as by the small intestine. The load on the colon is increased, and the colon, in which secretion has been substituted for absorption, is unable to deal with the load.

Surreptitious laxative abuse is a common cause of chronic diarrhea of unknown origin; in one study of 27 patients, 9 were found to be taking drugs which caused their diarrhea.

Stool Size and Composition

Some representative values for stool size passed by British men and women are given in Table 17–3. For the women, there were no variations associated with the menstrual cycle. Observations on residents of the United States have given substantially similar results.

Stool size is strongly influenced by the amount of crude fiber in the diet, and the volume of stool and the frequency of defecation can be regulated by adding bran and vegetables to the diet. Whereas British naval ratings and their wives have an average daily stool weight of 104 gm (range, 39–223 gm), British vegetarians have stool weights and bowel patterns similar to the vegetarian Ugandans. Ugandans as a group hold the Olympic record for stool weight, an average of 470 gm/day, with range of 178–980 gm. In them, ascariasis as well as a high-fiber diet probably contributes to the large stool size. Ten thousand years ago, our ancestors had even larger stools. The dried intestinal residues recovered from bog men have been found to contain 30%–56% plant residues, which helped to ease the evacuation of coarse, bone-studded fecal pellets, sometimes 3 in. in diameter.

Vegetables contain cellulose, hemicellulose, and lignin. Lignin is not digested in the human intestine, but the other two are. In apples, carrots, and cabbage the fiber is chiefly celluose, and it is readily digested. Wheat fiber consists of small cells with lignified cell walls, and consequently wheat fiber is not broken down in the human colon. A patient with an ileostomy excretes about 85% of ingested cellulose and about 28% of ingested hemicellulose in his ileostomy fluid, but a person with an intact colon excretes only 22% and 4% of cellulose and hemicellulose, respectively. Because cellulases are not present in human digestive juices, these substantial degradations of cellulose and hemicellulose in the small intestine and colon must be accomplished by microorganisms.

Some physicians believe that absence of crude fiber from the diet and consequent alteration of colonic function are the causes of appendicitis, diverticulosis, and benign and malignant tumors of the colon.

Many foods, including milk, potatoes, and eggs, which contain little or no indigestible residue, are alleged to increase fecal bulk and frequency of defecation; however, no evidence outside the realm of folklore supports their reputation. Prune juice, which is completely devoid of indigestible residue, produces catharsis, probably on account of its content of magnesium salts.*

---

*Contrary to occasionally expressed opinion, the active principle of prune juice is probably not an isatin derivative. See Hubacher M.H., Doernberg S.: *J. Pharm. Sci.* 53:1067, 1964.

TABLE 17–3.—MEAN STOOL SIZE AND RANGE OF INDIVIDUAL SAMPLES OF STOOLS COLLECTED FROM 10 BRITISH MALES AND 10 BRITISH FEMALES BY REPEATED 5-DAY COLLECTIONS IN 4–6 CONSECUTIVE WEEKS*

MEASUREMENT	MALES	FEMALES
Transit time, hr	66 (34–117)	82 (47–128)
Wet wt, gm/24 hr	131 (58–283)	126 (65–233)
Dry wt, gm/24 hr	30 (15–44)	32 (11–45)
Water, %	75 (66–84)	74 (69–88)
Frequency, motions/24 hr	0.96 (0.3–2.0)	1.16 (0.7–1.8)
Interval between motions, hr	28 (11–70)	23 (13–36)
Size individual stools, gm	142 (33–452)	111 (10–331)

*Adapted from Wyman J.B., et al.: *Gut* 19:146, 1978.

The colon secretes a few mEq of calcium into the stool and absorbs a few mEq of magnesium. Although the net movement of bicarbonate is into the lumen, its concentration in stool water is low because it reacts with organic acids formed by bacterial fermentation. The concentration of organic anions is 133–238 mEq/L. Stool water also contains ammonium (mean 14, range, 2–34 mEq/L), calcium (mean 38, range, 6–72 mEq/L), and magnesium (mean 49, range, 13–98 mEq/L). The stool is hyperosmotic, with a mean osmolality of 376 (range, 336–423) mOsm, because formation of osmotically active organic compounds outstrips osmotic equilibration.

Organic materials entering the colon are mucus, desquamated cells, and enzyme secretions of the upper digestive tract together with undigested food residues. Most of the digestible carbohydrate, fat, and protein has already been attacked and absorbed. When fed protein is tagged with radioactive iodine, little or none of the radioactive isotope reaches the stool. The small fraction of dietary protein that enters the colon is attacked by bacteria; and the major part of stool protein is contributed by bacteria themselves, which compose about 60% of the dry weight of the stool. There are few bacteria within the stomach or small intestine. Most ingested bacteria are killed by acid in the stomach. The bacterial population increases in the ileum and reaches its maximum in the colon. When intestinal stasis occurs, the bacterial population of the static fluid in the lumen rises quickly. It is likely that the relatively slow movement of colonic contents, rather than any particular environmental factor, accounts for its abundant flora.

Trace nutrients synthesized by colonic flora are absorbed. These include riboflavin, nicotinic acid, biotin, and folic acid. In both man and animals, the administration of sparingly soluble bacteriostatic agents such as sulfaguanidine reduces the colonic flora and, when vitamin B intake is low, may precipitate signs of deficiency. The prothrombin time of human infants frequently lengthens in the first few days after birth, when normal intestinal flora, which will later contribute some vitamin K, have not yet become established.

The nature of the diet influences the size and composition of intestinal flora; but, on the other hand, intestinal flora alter the structure and function of the digestive tract. Germ-free animals have shallow crypt glands, taller and more delicate villi, and a thin-walled cecum.

Intestinal Gas

The normal human digestive tract contains about 150 cc (ATPS) of gas. Of this, 50 cc is in the stomach, little in the small intestine, and 100 cc in the colon. The sources of intestinal gas are swallowing, neutralization, fermentation, and diffusion.

Gas in the stomach accumulates from that accompanying food or drink in each swallow, from frothy saliva, and from gas trapped in food. The amount swallowed varies greatly with habit, and it may be as much as 500 cc with a meal. Several liters may be quickly swallowed during a period of anxiety. Most of the gas is either eructed or passed on, but on rare occasions enough may accumulate so that the gas bubble in the stomach fills the entire upper left abdominal quadrant. Sensations of bloating and discomfort are not related to the total volume of gas, for a person with a huge bubble may have no untoward symptoms, whereas one who complains of gas on the stomach may have less than the usual volume. A person who complains that everything he eats turns to gas has no more gas in the bowels than does a normal person; he has instead disordered motility, which does not move the gas rapidly, and abnormal sensitivity to distention.

When the gas bubble in the stomach is large enough to extend below the entrance of the esophagus, gas may escape into the esophagus when the lower esophageal sphincter relaxes. It usually does not pass the hypopharyngeal sphincter. Distention of the esophagus stimulates secondary peristalsis, and the gas is swept back into the stomach. This sequence of filling and emptying may be repeated many times before eructation occurs. Then, when the esophagus is filled with gas, the jaw is thrust forward, intrathoracic pressure rises, and gas is expelled through the partially opened hypopharyngeal sphincter.

Gas not eructed passes into the small bowel, most easily if the subject is supine, and its to-and-fro movement in the antrum or its passage through the pylorus causes gushing and explosive sounds. These and other intestinal gas noises are called borborygmi. Very little gas is normally present in the small intestine, probably because it is rapidly absorbed or passed on to the colon. That present causes soft crepitations or slow rumbles, which occur at the frequency of intestinal movements, 7–12 times a minute. Their absence in the "silent abdomen" is a sign of general cessation of intestinal motility. The colon, in addition to the stomach, is the major source of loud borborygmi.

When 1 mEq of acid reacts with 1 mEq of bicarbonate at body temperature, 25 cc of carbon dioxide is liberated. During emptying of a meal from the stomach, 50 mEq of acid may react with an equivalent amount of bicarbonate, and 1,250 cc of carbon dioxide is formed. The partial pressure of carbon dioxide may rise as high as 700 mm Hg, and most of the carbon dioxide diffuses back into the blood. Carbon dioxide resulting from the reaction of organic acids in the colon with bicarbonate secreted by the colonic mucosa is more slowly absorbed.

Under normal circumstances, very little gas is formed by fermentation in the small intestine. When obstruction occurs, fermentation may produce as much as 3,500 cc of gas in the small intestine in 24 hours. Normally, the chief site of fermentation is the colon. Bacteria producing hydrogen gas are entirely confined to the colon. They require fermentable substances, and therefore hydrogen gas is produced only when carbohydrates are incompletely digested and absorbed. Some hydrogen is absorbed and excreted through the lungs. Much of it is actually consumed by other bacteria so that the amount of hydrogen gas passed in the flatus is only a small fraction of that produced in the colon. Colonic flora of all persons produce a net amount of hydrogen gas at the rate of about 0.6 cc/min. Fourteen percent is absorbed and excreted through the lungs; the rest escapes in the flatus. About one third of the population produces a large amount of methane in the colon; the other two thirds does not. Some methane is absorbed, and its measurement in the breath is a means by which its production can be followed. Only a very small amount of hydrogen sulfide is produced, and it is far less than 0.01% of the flatus. Because the smell of hydrogen sulfide is easily recognized, eggs are blamed for flatulence. Other trace gases giving the characteristic odor to the flatus have not been identified.

The amount of hydrogen and methane produced depends upon the nature of unabsorbed residues. The hulls of beans contain undigestible oligosaccharides, chiefly raffinose and stachyose, and when beans are eaten to the extent of 25% of the caloric intake, the rate of passage of flatus may be as high as 200 cc/hr. Deficiency of lactase leaves lactose to be fermented, and when milk is drunk by a person with lactase deficiency, he makes as much as 4 cc of gas per min.

All varieties of fermentation produce organic acids, which, when neutralized by bicarbonate, liberate carbon dioxide.

Addition of other gases to swallowed nitrogen reduces the partial pressure of nitrogen in the gut, and consequently nitrogen diffuses from blood into the gas phase at the rate of 1–2 cc/min. Any oxygen diffusing from blood is used by colonic flora.

The human colon usually contains about 100 cc (range, 30–200 cc) of gas. The volume increases when ambient pressure is reduced by quick ascent in an unpressurized airplane or by failure of pressurization at high altitude. At the altitude of 40,000 ft, where the barometric pressure is 140 mm Hg, 100 cc of gas saturated with water vapor at sea level becomes 767 cc, so when the stewardess says something about ''the unlikely event of a sudden reduction in cabin pressure,'' the informed passenger has something to think about besides the technique of adjusting an oxygen mask.

There have been no mass surveys upon which a reliable estimate of the volume and composition of the average American flatus can be based. A low figure for volume was provided by 5 male medical students untroubled by flatulence who passed from 380 to 650 cc in 24 hours, and a high figure has been given by a physician who collected 1,500–1,600 cc/day from himself for a week. Individual samples have been analyzed with great precision by gas chromatography or mass spectrometry, but the variation from sample to sample is enormous. Oxygen is usually nearly absent; nitrogen ranges from 12% to 60%; and carbon dioxide may be as high as 40%. Because hydrogen and methane may each compose as much as 20%, flatus containing them will burn with a hard, gem-like flame. The mixture of hydrogen and methane not only burns, it may explode, and patients have been killed by violent detonation of colonic gas during snare removal of a polyp using high-frequency current. Such events tend to be underreported in the medical literature for obvious reasons.

It is not the fat content but the entrained gas bubbles that cause feces to float. Feces may be made to sink or rise like a Cartesian diver by increasing or decreasing the ambient pressure.

Controlled expulsion of gas when there are feces in the rectum is accomplished by contraction of abdominal muscles; this raises pressure in the rectum. At the same time, the external anal sphincter is voluntarily contracted. When pressure in the rectum exceeds pressure in the sphincter, gas escapes through a slit that is too narrow for solid feces to pass. The relation between pressure and velocity of gas in the anal canal obeys Bernoulli's principle, and consequently the lips of the anus vibrate like the double reed of a bassoon, sounding a low-pitched note.

Constipation

"The abnormal action of the bowels in constipation may manifest itself in three different ways: defecation may occur with insufficient frequency, the stools may be insufficient in quantity, or they may be abnormally hard and dry."† This definition of constipation implies that there is some normal standard of frequency, size, and quality of bowel movements, but it is difficult to specify the exact point at which normal variation shades into pathologic function. Some healthy persons open their bowels after every meal; others usually do so once a day; but many who defecate only once a week experience no difficulty attributable to costiveness. The popular notion that one must defecate once a day or be overwhelmed by the disaster of irregularity is false. Nevertheless, at some degree of infrequency the symptoms of constipation appear; these are mental depression, restlessness, dull headache, loss of appetite that is sometimes accompanied by nausea, a foul breath and coated tongue, and abdominal discomfort with heaviness and swelling.

Constipation may be *colic* when transit through the colon is abnormally prolonged, or it may be *dyschezic* when the rectum cannot be emptied. Colic constipation occurs in a person in whom food residues take as long as 72 hours to reach the descending colon. It occurs in the aged in whom the colon may be dilated and atrophic. The gastrocolic reflex is not aroused during anorexia or bed rest, and an entirely bland diet with no residue leaves little to stimulate colonic motility. Movement of the colon is reduced by pain, particularly that arising from intestinal disease. In some constipated persons resistance to flow of feces results from abnormally intense colonic contractions (see Fig 6–3); in them the colon may be spastic with or without pain. Narrowing of the lumen of the colon by strictures, malignant growths, cicatrization of ulcera-

tions, or obstruction by masses of hard, dry feces delays passage of contents to the rectum. Inefficient defecation often occurs when, for one reason or another (haste, lack of privacy, or unpleasant or uncomfortable surroundings), the urge to defecate is ignored. Fear of painful defecation, spasm of the anus induced by anal ulcer, or inflamed hemorrhoids may also depress the defecation reflex and lead to constipation.

In dyschezia, or the inability to defecate completely, the rectum is always filled with feces, even immediatly after defecation; and in the extreme form, defecation without mechanical assistance is impossible. The condition results from insufficient power of the defecation mechanism or from an obstacle to defecation. The latter may be a coprolith or a foreign object. Foreign objects recovered from the colon include a 40-W frosted electric light bulb, a shot glass, a tool case complete with tools (that in a prisoner), and an umbrella (Fig 17–1).

Straining to defecate in the constipated state may

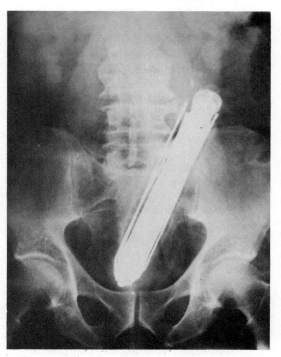

Fig 17–1.—An umbrella in the sigmoid colon. The physician's chief concern when removing it was not to spring the catch.

†Hurst A.F.: *Constipation and Allied Intestinal Disorders* ed. 2. New York, Oxford University Press, 1919). The definition and terminology of *colic constipation* and *dyschezia* (difficult easing) follow Hurst.

cause hernia, hemorrhoids, prolapse of the rectum, or anal ulceration, and elderly persons may faint or have a fatal cardiovascular accident.

Autointoxication

The miserable state of constipation is exacerbated by fear of its consequences. These fears are assiduously cultivated by vendors of laxatives and by quacks who profit from many bizarre forms of treatment, with the result that many persons believe that serious and permanent injury to health will follow the missing of a bowel movement. To most adults, feces are repulsive. (They are not so to infants, who have not yet been acculturated.) It is easy to believe that retention of feces is deleterious, and on this belief is based the theory of autointoxication, which is that toxins absorbed through the colon from retained feces are responsible for all disease, ranging from alopecia to zoanthropy.

In a person with a normal liver, autointoxication is an imaginary condition. Some products of digestion in the small intestine and colon are potential toxins. Among these are histamine, tryptamine, and cadaverine formed by decarboxylation of their parent amino acids. In a normal person they are removed from portal blood by the liver and do not reach toxic concentrations in systemic blood.

Ammonia is produced in most tissues in addition to the digestive tract, and it is converted to urea by the liver. In normal circumstances, hepatic blood flow is a sufficiently large fraction of cardiac output that systemic blood can be cleared of ammonia by the liver (Fig 17–2). Ammonia is also removed from arterial blood by skeletal muscle, which synthesizes glutamate from 50% of the ammonia reaching it. However, ammonia is released by skeletal muscle during exercise.

Ammonia is produced in the lumen of the gut by deamidation of glutamine and by hydrolysis of urea. Each day, about a quarter of the urea pool diffuses into the lumen of the small intestine. About half of it is hydrolyzed to ammonia and carbon dioxide by the urease confined to microorganisms which inhabit the digestive tract from nasopharynx to rectum. The urea escaping hydrolysis in the small intestine is hydrolyzed in the colon whose mucosa is impermeable to urea. Ammonia produced in the digestive tract is absorbed as NH_3 by non-ionic diffusion, and it is carried in portal blood to the liver. In normal persons most

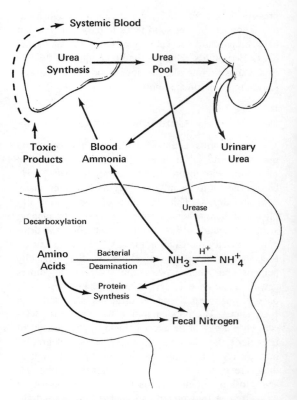

Fig 17–2.—Ammonia metabolism, showing the rationale for treatment of hyperammoniumemia by a low-protein diet, antibiotic therapy to reduce the population of urease-containing bacteria, and by feeding lactulose. About half the gut production of ammonia occurs in the small intestine and the rest of the colon. Ammonia is produced in many tissues in addition to the kidneys.

of the ammonia reaching the liver is synthesized into urea, and the rest is secreted into the bile.

In a person with cirrhosis or portocaval shunt, ammonia escapes into systemic blood, causing encephalopathy, which may be fatal. In such a person, reduction of ammonia production in the digestive tract, and particularly in the colon, can be achieved by (1) a low-protein diet, which curtails the source of ammonia nitrogen; (2) thorough antibiotic therapy with neomycin, which suppresses urease-containing bacteria; and (3) by feeding lactulose.

Lactulose is 1,4-galactosidofructose, an artificial sugar which, because it is not hydrolyzed by digestive enzymes, is fermented in the digestive tract, making the contents of the colon more than normally acidic. It was formerly thought that lactulose is effective in reducing blood ammonia for the reason that, in the acidic environment it produces, ammonia (NH_3) is converted to ammonium (NH_4^+), which is not absorbed by passive diffusion. However, during lactulose treatment, ammonium content in the stool falls rather than rises. Lactulose stimulates protein synthesis by colonic microorganisms, and ammonia and ammonium become innocuous amino nitrogen.

Many experimental observations show that it is not the nature of the colonic contents but their presence in the descending colon and rectum that is responsible for the symptoms of constipation. When defecation is voluntarily restrained for several days, all the familiar effects occur. Immediately following defecation these effects disappear too quickly to be accounted for by the elimination of toxins. The same symptoms of constipation can be produced by stuffing the rectum with cotton or distending it with a balloon. When the mass is removed, the symptoms promptly disappear. Further evidence is provided by the numerous men who have retained enormous masses of feces for years without suffering any bad effects other than the burden of carrying a colon containing from 60–100 lb of fecaliths.‡

Fear of constipation encourages the use of laxative drugs, and anthropological surveys have shown that among certain groups, such as the inhabitants of Nebraska farm communities, it may be the custom for every person to take a dose of laxative every day. When self-medication is omitted, the colon that had been violently and artificially emptied subsides; no bowel movements occur for a day or two. This confirms the subject's suspicion that he is the victim of constipation, and he resorts to still more heroic measures. The result is that the abused colon is not given a chance to function at its normal pace. Patients may complain of "constipation," but because they take laxatives regularly they never have formed stools.

REFERENCES

Berk J.E. (ed.): Gastrointestinal gas. *Ann. N.Y. Acad. Sc.* 150:1, 1968.
Bigard M-A., Gaucher P., Lasalle C.: Fatal colonic explosion during colonoscopic polypectomy. *Gastroenterology* 77:1307, 1979.
Burkitt D.P., Walker A.R.P., Painter N.S.: Effect of dietary fibre on stools and transit-time, and its role in the causation of disease. *Lancet* 2:1408, 1972.
Calloway D.H.: Gas in the alimentary canal, in Code C.F. (ed.): *Handbook of Physiology:* Sec. 6. *Alimentary Canal,* vol. 5. Washington, D.C., American Physiological Society, 1968, pp. 2839–2860.
Cummings J.H.: Laxative abuse. *Gut* 15:758, 1974.
Daniel E.E., Bennett A., Misiewicz J.J., et al. Symposium on colon function. *Gut* 16:298, 1975.
Debongnie J.C., Phillips S.F.: Capacity of the human colon to absorb fluid. *Gastroenterology* 74:698, 1978.
Frizzell R.A., Schultz S.G.: Effect of aldosterone on ion transport by rabbit colon in vitro. *J. Membr. Biol.* 39:1, 1978.
Hofmann A.F.: Bile acids, diarrhea, and antibiotics: Data, speculation, and a unifying hypothesis. *J. Infect. Dis.* 135:S126, 1977.
Kaufman N.M., Schuster M.M.: Colonic motility studies discriminate 3 types of constipation. *Gastroenterology* 76:1166, 1979.
Lasser R.B., Bond J.H., Levitt M.D.: The role of intestinal gas in functional abdominal pain. *N. Engl. J. Med.* 293:524, 1975.
Levitt M.D.: Production and excretion of hydrogen gas in man. *N. Engl. J. Med.* 281:122, 1969.
Levitt M.D.: Intestinal gas production—recent advances in flatology, *N. Engl. J. Med.* 302:1474, 1980.
Levitt M.D., Bond J.H. Jr.: Volume, composition, and source of intestinal gas. *Gastroenterology* 59:921, 1970.
Levitt M.D., Duane W.C.: Floating stools—flatus versus fat. *N. Engl. J. Med.* 286:973, 1972.
Lockwood A.H., McDonald J.M., Reiman R.E., et al.: The dynamics of ammonia metabolism in man. *J. Clin. Invest.* 63:449, 1979.
McCaffery T.D. Jr., Lilly J.O.: The management of foreign affairs of the GI tract. *Am. J. Dig. Dis.* 20:121, 1975.
Morris A.I., Turnberg L.A.: Surreptitious laxative abuse. *Gastroenterology* 77:780, 1979.
Onstad G.R., Zieve L.: What determines blood ammonia? *Gastroenterology* 77:803, 1979.
Phillips S.F., Geller J.: The contribution of the colon to electrolyte and water conservation in man. *J. Lab. Clin Med.* 81:733, 1973.
Politzer J-P., Devroede G., Vasseur C., et al.: The genesis of bowel sounds: Influence of viscus and gastrointestinal content. *Gastroenterology* 71:282, 1976.

---

‡A notable example of failure to defecate for more than a year was described by Geib D., Jones J.D.: *J.A.M.A.* 38:1304, 1902.

Index